Agricultural Meteorology

Agricultural Meteorology

G.S.L.H.V. Prasada Rao

Former Professor and Head
Department of Agricultural Meteorology
College of Horticulture
Kerala Agricultural University
Thrissur

PHI Learning Private Limited

Delhi-110092
2025

In fond memory of ***Shri Asoke K. Ghosh*** *(October 1942 – February 2024), Founder Chairman and Managing Director of PHI Learning, whose vision endlessly inspires.*

The Legacy Continues....

Published by Pushpita Ghosh, PHI Learning Private Limited, Rimjhim House, 111, Patparganj Industrial Estate, Delhi-110092 and Printed by Glorious Printer, Delhi-110092.

₹625.00

AGRICULTURAL METEOROLOGY
G.S.L.H.V. Prasada Rao

ISBN-978-81-203-3338-3 (Print Book)
ISBN-978-93-90464-02-9 (e-Book)

CONTENTS

PREFACE

The syllabus in agricultural meteorology at the undergraduate level covers the primitive aspects of meteorology and its applications in the field of agriculture. It is framed by the Indian Council for Agricultural Research (ICAR), keeping the field's multidisciplinary character in mind, and the same curriculum is followed in all the State Agricultural Universities. Since the literature is scant and very much limited in applied aspects, and no single textbook is available in this field, it is widely felt that there is a need for this type of publication. Hence, this textbook in agricultural meteorology, equipped with the latest content in tune with the syllabus framed at the undergraduate level, serves the purpose for which it is intended.

The textbook contains seventeen chapters, dealing with various basic and applied aspects of the field. It is profusely illustrated with graphs, schematic presentations, and photographs. Information on meteorological instruments is also incorporated at the end of each chapter, wherever relevant, with a brief write-up on each instrument. In addition, each chapter contains its own set of questions, references and suggested readings. The book contains many indigenous examples, especially related to the humid tropics.

The text is designed to cover three areas of study, namely (i) basic aspects of weather elements, (ii) weather elements and their effects, and (iii) weather service. The first area of study consisting of Chapters 1–8 deals with the basic aspects of meteorology and agricultural meteorology. The atmospheric constituents and their role, the difference between meteorology and agricultural meteorology, and the scope and importance of agricultural meteorology and its status in the country, are highlighted in the introductory chapter. The principal meteorological variables that govern the atmosphere, namely radiation, temperature, general circulation (atmospheric pressure and wind), atmospheric moisture, clouds, precipitation, and monsoon are dealt with in the next seven chapters. Strong emphasis is given on radiation and temperature, which influence the general circulation of atmosphere in terms of atmospheric pressure and wind distribution.

The causal factors leading to the global and local weather and climate variations are brought out in detail. The phenomena of stability and instability in the atmosphere, leading to the formation of clouds and precipitation, are explained in Chapters 6 and 7, while the monsoons are covered in Chapter 8. The pro-active monsoon management model, in the form of a flowchart created on the basis of a concept paper written by Dr. M.S. Swaminathan, is provided in Chapter 8 under "Monsoon". Several examples on the influence of various weather elements on crops are given in each chapter, while a separate chapter deals with weather and crops. It helps the reader to understand easily the interaction between the effects of various weather elements and the growth of crops.

The second area of study consisting of Chapters 9–12 deals with the effects and application of weather elements, namely weather and crops, evapotranspiration, drought, and agroclimatic classification. The effects of weather on various crops, in particular the plantation crops, are summarised in Chapter 9.

Evapotranspiration is a very complex system in the soil–plant-atmosphere continuum. The definitions of potential, reference, and actual evapotranspiration, crop and pan coefficients, and irrigation scheduling based on evapotranspiration are dealt with in brief in Chapter 10. The relevance of different kinds of droughts and the drought management strategy are discussed in Chapter 11. The importance of climatic and agroclimatic classification, the agroecological zones, and the relevance of agroclimatological zones with specific reference to crops and regions, are highlighted in Chapter 12.

The third and final area of study, comprising Chapters 13–17, deals with weather forecasting, crop simulation modelling, seasonal outbreaks of crop diseases, climate change, and remote sensing. Various types of weather forecasting techniques, and their pros and cons in terms of reliability and technology, including those for weather service to farmers through different modes, are discussed in Chapter 13. The chapters on crop growth simulation modelling, climate change, and remote sensing are aimed at the beginners in agricultural studies and research. The book provides comprehensive coverage of topics like the origin of the sciences of meteorology and agricultural meteorology, the applications of bioclimatic law, the concept of pan coefficient and its use in irrigation scheduling, monsoon and drought management, the effect of biotic and abiotic stresses, UV-B radiation, the concept of global cooling by aerosols, the analogous technique in rainfall distribution, the concept of growing degree days, the spectral response of coconut, and the influence of weather on crop pests and diseases. The weather abnormalities and the weather modification techniques are dealt with under "climate change". The economic impact of agro-advisory services, and the role of ICT in dissemination of agro-advisory services, are also highlighted under "weather forecasting".

The examples quoted in this textbook are taken mostly from the work carried out by the author as a researcher for the last three decades.

The author's earnest hope is that the textbook would prove valuable for students, teachers, and researchers in the field of agricultural meteorology. Suggestions and comments from students, academicians, and scientists towards the improvement of the text would be most welcome and may be communicated to the author at his email id: kauagmet@yahoo.co.in

G.S.L.H.V. Prasada Rao

1 INTRODUCTION

The word *meteoro* owes its origin from a Greek word that means *above the earth's surface*, which is nothing but atmosphere. The dictionary meaning of *logy* is *a subject of study or interest*. Therefore, the branch of science dealing with the atmosphere is known as *meteorology*. In the strict sense, the lower atmosphere extending up to a height of 20 km from the earth's surface, where frequent physical processes take place, is the concern of meteorology. Of course, the height of lower atmosphere varies with latitude due to geographical conditions, shape of the earth and its position with reference to the sun. It is high in the tropics and low over the poles. Hence, the science of meteorology is defined as *the science that deals with the physical processes of the lower atmosphere, which produce weather*. The weather science concerns the study of the structure, state and behaviour of the atmosphere.

Meteorology is essentially an observational science dealing with different atmospheric variables such as pressure, temperature, wind, humidity, density, cloud amount, radiation and related variables at a given moment. It attempts to predict the behaviour of the atmosphere over a period of a few hours to a few days ahead. The main task of meteorology is to provide weather forecast and communicate the same in time to the users. Another important duty of the science of meteorology is to increase our knowledge of the atmosphere, with constant endeavour through research and development aiming at improving the accuracy of weather forecast, since the present weather forecast skills are inadequate. A flowchart giving the details of the origin of the sciences of meteorology and agricultural meteorology is depicted in Fig. 1.1.

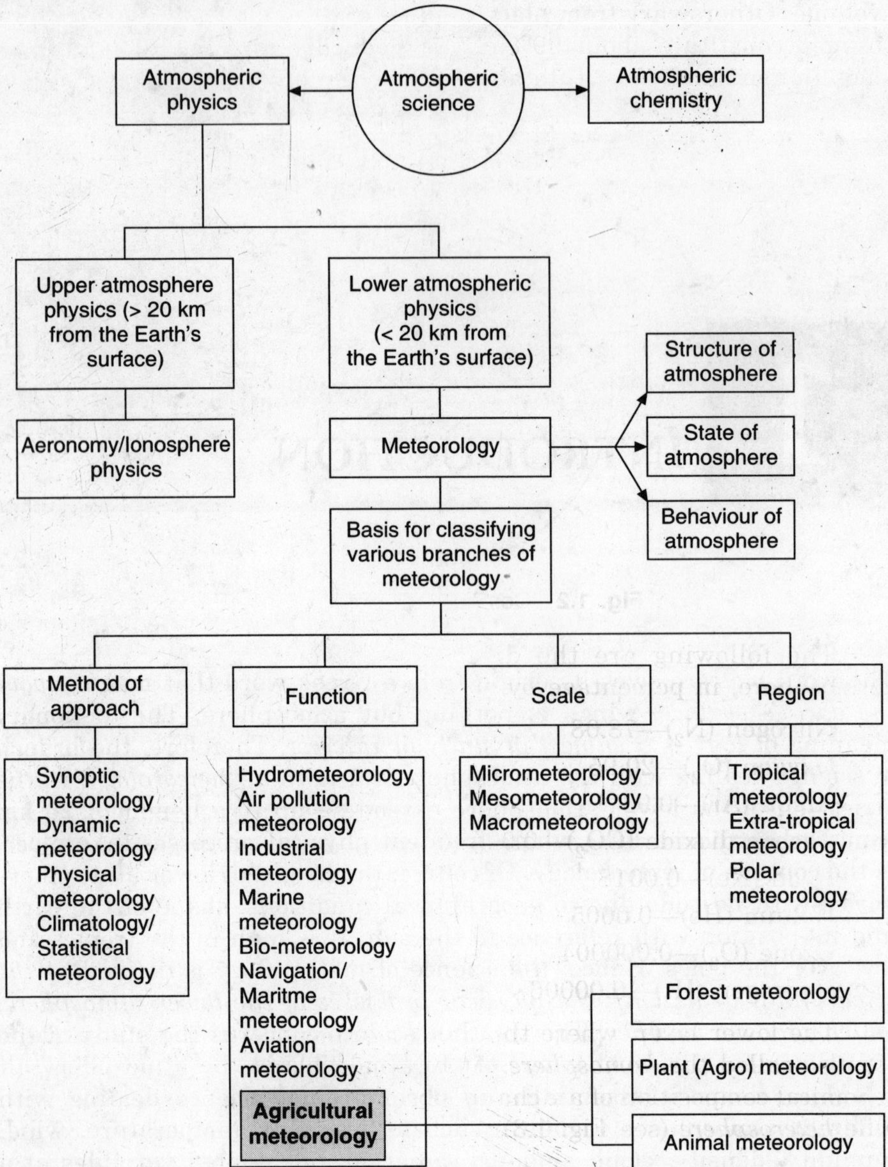

Fig. 1.1 Origin of the sciences of meteorology and agricultural meteorology.

1.1 COMPOSITION OF ATMOSPHERE

The atmosphere is a mechanical mixture of gases. It contains a huge number of solid and liquid particles called *aerosols*. Some gases are permanent atmospheric constituents in fixed proportions to the total gas volume. Others vary from place to place and time to time. Nitrogen and oxygen constitute about 99% of the clean dry air. The remaining are mostly inert and constitute about 1% of the atmosphere (see Fig. 1.2).

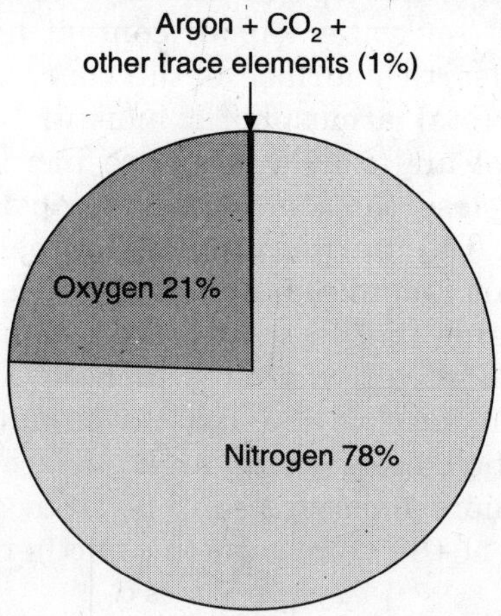

Fig. 1.2 Composition of atmosphere.

The following are the different gases that are present in the atmosphere, in percentage by volume approximately:

Nitrogen (N_2)—78.08	Krypton (Kr)—Trace
Oxygen (O_2)—20.95	Xenon (Xe)—Trace
Argon (Ar)—0.93	Methane (CH_4)—0.00017
Carbon dioxide (CO_2)—0.03	Carbon monoxide (CO)—Trace
Neon (Ne)—0.0018	Ammonia (NH_3)—Trace
Helium (He)—0.0005	Sulphur dioxide (SO_2)—Trace
Ozone (O_3)—0.000004	Nitrous oxide (NO_2)—0.00003
Hydrogen (H_2)—0.00006	Iodine (I)—Trace

The lower layer, where the chemical composition of air is uniform, is also called the *homosphere*. At higher levels in the atmosphere, the chemical composition of air changes considerably. This layer is known as the *heterosphere* (see Fig. 1.3).

1.1.1 Atmospheric Moisture or Water Vapour

Atmospheric moisture is largely derived from oceans through the evaporation of water. It is also derived through *evapotranspiration* from soils, plants and forests. Rivers, canals, lakes, ponds, surface and open wells, and variuos other water bodies are the other sources of water vapour content in the atmosphere. Water vapour plays a major role in the hydrological cycle and it is one of the most variable meteorological elements in the atmosphere. It is the source of all clouds and precipitation.

It is an efficient absorber of both long and shortwave radiations. The outgoing longwave radiation (4 to 80μ) suffers significant absorption by CO_2 and water vapour. Water vapour is a selective absorber in the far red and infrared. A major absorption peak occurs between 5 and 8μ and beyond 25μ. About 90% of water vapour content lies below 6 km of the atmosphere since the earth's surface is the source. It is estimated that less than 1% of the total atmospheric moisture is found only above 10 km. The capacity of air to hold moisture increases with increasing temperature. Nevertheless, the atmospheric temperature is regulated by water vapour content. The diurnal range of temperature (the difference between daily maximum and minimum temperatures) is low in the humid and coastal climates and high in the dry and inland climates. It is due to large quantities of water vapour which is present in the lower atmosphere across the humid and coastal climates. Similarly, water vapour is more over the tropics due to high surface air temperature and its capacity to hold more moisture content. It is relatively less if one moves towards north of the tropics in the northern hemisphere.

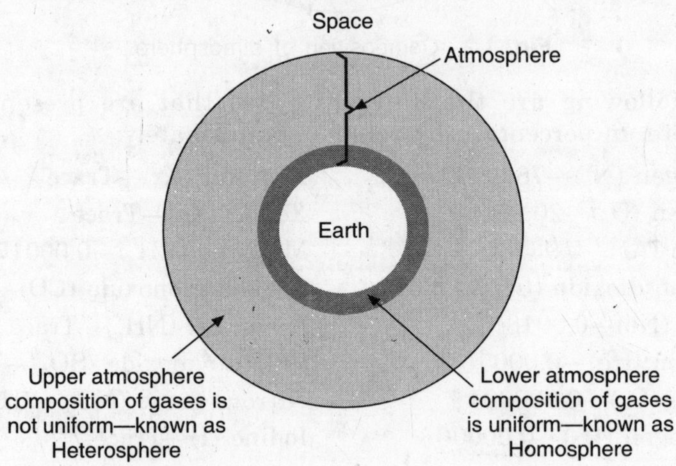

Fig. 1.3 Homosphere and Heterosphere.

1.1.2 Carbon Dioxide (CO_2)

Carbon dioxide constitutes about 0.03% of dry air. All green plants use carbon dioxide for the process of photosynthesis. Being an efficient absorber of longwave radiation from the earth, it is considered to be of great climatic significance. Carbon dioxide has important absorption peaks at about 2, 3 and 4μ and a major absorption peak at about 15μ. This gas emits about half of the absorbed heat back to the earth, influencing the flow of energy between the earth and the atmosphere. CO_2 is a very important atmospheric constituent in the heat budget of the earth and atmosphere. There is a sharp increase in carbon dioxide in recent times due to human-made interventions in terms of burning fossil fuels such as coal, oil and natural gases. The projected concentration of CO_2 in the atmosphere by 2050 is as shown in Table 1.1.

Table 1.1 Projected concentration of CO_2 in the atmosphere

Period	CO_2 *concentration*
Before pre-industrial revolution	250–290 ppm
1958	315 ppm
1984	354 ppm
2000	370 ppm
Projection by 2050	>450 ppm

The rate of increase in CO_2 concentration will lead to increase in the earth's temperature by between 1°C and 3.5°C by the turn of this century. Higher surface temperature will lead to more melting of snow and ice cover on land and sea, thereby causing sea level rise, a great threat to coastal areas. If sea levels rise about 1 metre, about 1% of Egypt, 6% of the Netherlands, 17.5% of Bangladesh and 80% of Majuro Atoll in the Marshall Islands and the coastal areas in India would be under water by the year 2100. Non-availability of fresh water is likely due to saline intrusion day the control belt.

1.1.3 Aerosols

Aerosols include all the solid and dust particles present in the atmosphere. There is great variation in the amount of aerosols over the earth surface. Many aerosols are invisible to the naked eye and are microscopic. They include sea salts from breaking sea waves, pollen, fine seeds of plants, spores, bacteria and various organisms lifted by the wind, smoke and black carbon (soot) from fires, tiny sand particles, volcanic ashes and meteoric dust. Aerosols in the process of scattering contribute to the varied colours of red and orange at sunset and sunrise. The blue of the sky is also due to the selective scattering by microscopic dust particles.

From the meteorological point of view, these tiny solid particles are very important. They absorb a part of incoming shortwave solar energy. Aerosols are tiny particles of about 1 micron in size, which scatter sunlight back to space and thereby cause a regional cooling effect. Such aerosols are sulphates, soot, organic carbon and mineral dust. Some of the dust particles are hygroscopic and act as nuclei of condensation, resulting in cloud formation and thereby precipitation. Clouds may increase reflectivity or *albedo*, which may lead to cooling effect. Black carbon absorbs solar radiation in visible spectrum and reduces the amount of radiation reaching the earth surface. Increase in black carbon may alter large-scale atmospheric circulation and the hydrological cycle. Black carbon is a product of incomplete combustion of coal, diesel fuels, biofuels and outdoor biomass generated from industrial pollution, traffic, outdoor fires and household burning of coal and biomass fuels.

1.1.4 Ozone (O_3)

Another important constituent in the atmosphere is ozone (O_3). It is found only in a very small quantity in the upper atmosphere. The greatest concentration of ozone is found at an average height of 25 km and above. The ozone molecule is made up of 3 atoms of oxygen. It is the most efficient absorber of ultraviolet radiation from the sun and thus protects all life forms on earth. Ozone is an almost perfect absorber of UV radiation at about 0.3μ. Ozone and oxygen have other absorption peaks with the major one due to ozone at 9.6μ. Ozone depletion due to industrialization in recent times is a matter of serious concern for humankind and the larger biological world. Among UV radiations, UV-B is more sensitive to ozone fluctuations and reaches the earth surface if ozone depletes, causing biological damage. A global network has been created to monitor UV-B filtered radiation in terms of Minimum Erythema Dose (MED). More on UV-B radiation is discussed in Chapters 2, 3 and 16. Increase in chlorine levels through chlorofluorocarbons leads to depletion in ozone.

1.1.5 Greenhouse Effect

Carbon dioxide, water vapour, methane, carbon monoxide, sulphur and nitrous oxides, chlorofluorocarbons and chlorofluoromethanes are the atmospheric constituents of major importance. Incoming shortwave radiation is not absorbed by the above atmospheric constituents, and the outgoing longwave radiation from the earth is absorbed by them and reradiate back to the earth surface. In the process, earth warms up and its temperature rises. It is known as the *greenhouse effect*. It is also known as the *glasshouse effect* as an important property of glass is that it allows solar radiation through it and disallows longwave radiation from the earth; thereby warming takes place. The atmospheric constituents which have the property of absorbing longwave radiation and transparent to shortwave radiation are known as the *greenhouse gases* (GHG). Human activities like fossil fuel combustion, production of synthetic chemicals, biomass burning, deforestation, excess use of chemical fertilizers, and use of pesticides change the chemical composition of the atmosphere, thereby enhancing the greenhouse effect. The present growth rate of GHG is 1% in CH_4, 0.4–0.5% in CO_2 and 0.2–0.3% in N_2O as per the Intergovernmental Panel on Climatic Change (IPCC). In addition to the above, chlorofluorocarbons and chlorofluoromethanes contribute significantly to the greenhouse effect. Is the greenhouse effect mitigated by the regional cooling effect from aerosols like black carbon? New research has shown that the aerosol effect could dominate the greenhouse effect over the continents and by and large the effect of greenhouse will be slowed down. In contrast, the ozone-friendly substitutes for chlorofluorocarbons (CFCs) like hydrofluorocarbons (HFCs), hydrochlorofluorocarbons (HCFCs) and perfluorocarbons (PFCs) are also powerful greenhouse gases that contribute to global warming.

1.1.6 Atmospheric Window

The atmosphere is transparent to the spectral region between 8.5 and 11μ, which falls under the longwave radiation. It is understood that the atmosphere absorbs the earth's radiation. In the spectral region of 8.5–11μ, the atmosphere acts as a transparent medium instead of absorbing. Hence, the above spectral region is known as the atmospheric window. The above spectral region is used in microwave remote sensing to monitor the features of the earth surface in the case of overcast sky.

1.2 WEATHER AND CLIMATE

Weather decides crop production and thereby crop productivity, and climate decides crop selection. The tropical and temperate crops are classified on the basis of the thermal regime across the globe.

Weather	*Climate*
Refers to physical state of atmosphere at a given time in terms of temperature, rainfall, relative humidity and sunshine hours, and so on.	Refers to average state of weather (At least 30 years of data is required to find out the climatic normals).
Decides crop production and thereby crop productivity.	Decides geographical distribution of crops and farming systems (E.g. Tropical crops, sub-tropical crops and temperate crops).
Day to day farm operations are dependent on weather.	Farm planning is dependent on climate.
Crop pest and disease incidence is dependent on weather.	Hotspot areas of pest and disease incidence and their geographical distribution can be delineated on the basis of climate.
Extremes like floods and droughts, and cold and heat waves, depend on weather.	Extreme weather prone areas can be delineated on the basis of climate.

1.2.1 Weather

The physical state of atmosphere at a given point in time at a given location is referred to as 'weather'. Weather is described in terms of the instantaneous or short period mean value of the various atmospheric variables such as pressure, temperature, humidity, cloudiness and sunshine, and evaporation and rainfall. Weather determines the period of farming operations, modes and ease of transportation, reservoir storage, housing, clothing, atmospheric pollution and comfort.

1.2.2 Climate

Climate is the long-term regime of the atmospheric variables, or the composite of the day-to-day values of the weather elements over a long period of a given place or area. The period of averaging weather may be several days, weeks, months, years or even centuries. The climate normals are generally worked out for a period of 30 years, which is considered as the standard to express climatic conditions for a given place. The India Meteorological Department (IMD) has prepared climate normals for every 30-year period since 1901.

1.3 SCALES OF CLIMATE

1.3.1 Microclimate

Microclimate deals with the climatic features peculiar to small areas and with the physical processes that take place in the layer of air very near the ground. Soil-ground conditions, nature of vegetation cover, aspect of slopes, state of the soil surface, relief forms—all these may create special local conditions of temperature, humidity, wind and radiation in the layer of air near the ground which differ sharply from general climatic conditions. One of the most important tasks of agricultural meteorology is to study the properties of air near the ground and surface layer of soil, which falls under the microclimate.

1.3.2 Mesoclimate

Mesoclimate falls between micro and macroclimates. It is concerned with the study of climate over relatively smaller areas between 10 and 100 km across.

1.3.3 Macroclimate

Macroclimate deals with the study of atmosphere over large areas of the earth and with the large-scale atmospheric motions that cause weather. It helps in forecasting weather. The scales of air motion in different climates are given in Table 1.2.

If any weather system develops under different types of climate, it persists for longer periods under macroclimates and for smaller periods under microclimates. The duration of weather system is given under the time scale in the table below.

Table 1.2 Scales of meteorological motion systems in different climates

Type of climate	*Horizontal scale* (km)	*Vertical scale* (km)	*Time scale* (hr)
Macroclimate			
Planetary scale	2000–5000 and more	10	200–400 (8.3–16.6 days)
Synoptic scale	500–2000	10	100 (4.16 days)
Mesoclimate	1–100	1–10	1–10
Microclimate	< 0.1	0.2	0.1–0.2 (6–12 min.)

1.4 BRANCHES OF METEOROLOGY

The science of meteorology can be broadly grouped into many branches based on method of approach, function, scale and region. Based on the method of approach, it can be divided into four branches. They are discussed here in brief (also see Fig. 1.1).

1.4.1 Synoptic Meteorology

It deals with the conditions of the atmosphere at a given moment and attempts to predict changes from initial condition to a period of between a few hours and a few days ahead, by means of preparation and analysis of weather maps including upper air charts, aerological diagrams and satellite imagery on cloud movement. By definition, the field of synoptic meteorology studies the relationship between the atmospheric circulations and the surface environment of a region. The basic principle is to understand the large-scale wind systems over a region and give weather forecast region- or station-wise. It concerns the relationship between atmospheric circulation and local climate.

1.4.2 Dynamic Meteorology

It is the backbone of synoptic meteorology. It explains the motions and energy transformations, which occur in the atmosphere, using the tools of mathematics and physics. Dynamic meteorology can be subdivided into the thermodynamics and the dynamics of the atmosphere. The former concerns the effect of thermal influences caused by solar and terrestrial radiations, while the latter describes the resulting motions. The science of dynamic meteorology is meant to develop models based on the hydrodynamics and thermodynamic equations. Once the initial state of the atmosphere is known, the future state can be predicted using differential equations. It is an objective method of weather forecasting, also known as Numerical Weather Prediction.

1.4.3 Physical Meteorology

The approach here is that of practical or even experimental physics, in contrast to the mathematical or theoretical physics with which the dynamic meteorologist feels more comfortable. It is a fringe study, investigating those meteorological phenomena not directly linked with the circulation of atmosphere. Admittedly wide in scope, it connects meteorology with other branches of science. In this respect, it occupies a position analogous to that of physical chemistry, which is a part of both chemistry and physics. If the general principles that govern the transmission, absorption, emission and dispersion of electromagnetic energy through the atmosphere are considered, what happens to the radiation during this time is a concern of physical meteorology.

1.4.4 Statistical Meteorology

This branch deals with the mean state of the physical properties of air. This branch is popularly known as *climatology*. From the operational point of view, climatology and meteorology can be considered as one and the same.

1.5 AGRICULTURAL METEOROLOGY

Agricultural meteorology is the science which deals with the physical characteristics of the environment of growing plant or animal organisms. It is concerned with the physical processes of the above environment, which can be exploited in the interest of agriculture to increase crop production in terms of quantity as well as quality. It is also concerned with the interactions between meteorological and hydrological factors on the one hand and agriculture in the widest sense, including horticulture, animal husbandry and forestry, on the other. In other words, agricultural meteorology is defined as the science which deals with the influence of weather on crop husbandry as well as animal husbandry. Crop husbandry includes tillage to crop harvest.

1.5.1 Meteorology and Agricultural Meteorology

The differences between the two disciplines have been shown as follows.

Meteorology	*Agricultural meteorology*
It is a branch of atmospheric physics.	It can be considered either a branch of applied meteorology, or a branch of agriculture as it deals with agriculture.
It is the weather science.	It is a product of agriculture and weather science.
It is a physical science.	It is a biophysical science.
It aims at weather forecasting.	It aims at improving quantity and quality of crop production through meteorological skills.
Weather service is the concern.	Agroadvisory service to the farmers is the concern based on weather forecast.
It is a linking science to society.	It is a linking science to the farming community.

1.5.2 Importance of Agricultural Meteorology

The success or failure of agricultural production depends upon the chain of five factors viz. seed or variety, soil, weather, technology and farmer. Any weak link in the chain finally determines the crop output. The Indian foodgrains production during the *kharif* season largely depends on the onset of monsoon and its behaviour under the rainfed conditions. The troughs and peaks in foodgrains production continue to exist from year to year in consonance with the weather behaviour despite the technological advances that took place in the field of agriculture. The recent all-India drought in *kharif* 1987 and 2002 due to failure in southwest monsoon adversely affected foodgrains production to a considerable extent (see Fig. 1.4). Even the decline in foodgrains production during the last two years was due to the occurrence of moderate droughts. The target during 2007–08 of attaining the projected

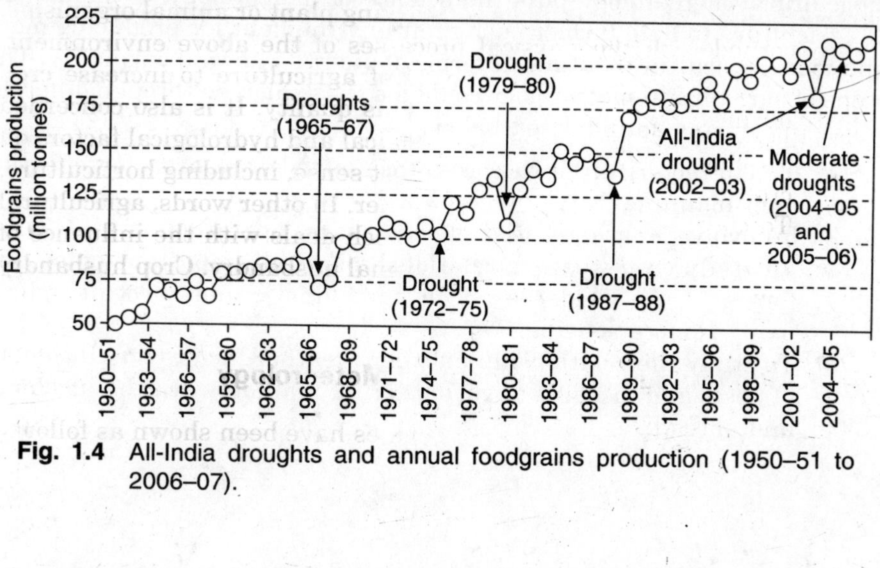

Fig. 1.4 All-India droughts and annual foodgrains production (1950–51 to 2006–07).

foodgrains production level of 220 million tonnes could also not be fulfilled, as several regions across the country experienced droughts and floods.

Similar was the case in the regions in the humid tropics like Kerala too, where the plantation crop production was hit very badly due to unprecedented drought during summer when rains failed from mid-November 1982 to the first week of June 1983. Such trends was also noticed during summer 2003–04 in some districts across Kerala. The water levels in majority of the surface water reservoirs were very poor and led to hydrological drought in summer 2004 due to deficit rainfall since 1999. Hence, it is realised that better understanding of weather and its variability and the overall effect on different processes contributing to agricultural production is necessary to adopt appropriate cropping systems, soil and water management practices. It is also necessary to know the weather conditions likely to prevail in order to bring in greater stability and the overall growth in the trend line of agricultural production. It is in this context that the importance of the science of agricultural meteorology can be realised for the benefit of the farming community. It has been said and rightly so that the Indian budget is a gamble in monsoon and 'Mr. Monsoon' is the Agriculture Minister of India, as stated by Mr. Chaturanan Mishra, one of the former Ministers for Agriculture in the Government of India, in a light vein.

1.5.3 Scope of Agricultural Meteorology

Agricultural meteorology is a multidisciplinary science. Though it is a branch of applied meteorology dealing with interactions between atmospheric variables and crop husbandry as well as animal husbandry, it interacts with the agricultural disciplines viz. Agronomy and Horticulture, Soil Science and Agricultural Chemistry, Agricultural Physics, Agricultural Engineering, Plant Physiology, Entomology, Plant Pathology and other allied agricultural sciences (see Fig.1.5). Hence, agricultural meteorology puts the science of meteorology to the service of agriculture to help farmers use their environment for producing more and more agricultural output in terms of quantity as well as quality. The scope of agricultural meteorology could also be explained through different ways and means as indicated below:

- To characterize climatic resources of a given area for effective crop planning.
- To evolve weather based effective farm operations.
- To study crop weather relationships in all important crops and forecast crop yields based on agroclimatic and spectral indices using remote sensing.
- To study the relationship between weather factors and incidence of pests and diseases of various crops and forewarn the incidence and spread.

- To delineate climatic/agroecological/agroclimatic zones for defining agroclimatic analogues so as to make effective and fast transfer of technology for improving crop yields.
- To prepare crop weather diagrams and crop weather calendars.
- To develop and revalidate crop growth simulation models for assessing and obtaining potential yields in different agroclimatic zones.
- To monitor agricultural droughts on crop-wise basis for effective drought management.
- To understand the impact of climatic variability or climate change on agricultural production including quality and project future crop trends.
- To develop pro-active measures against agrometeorological risks.
- To develop weather-based agroadvisories to sustain crop production utilizing various types of weather forecast including seasonal climate forecasts.

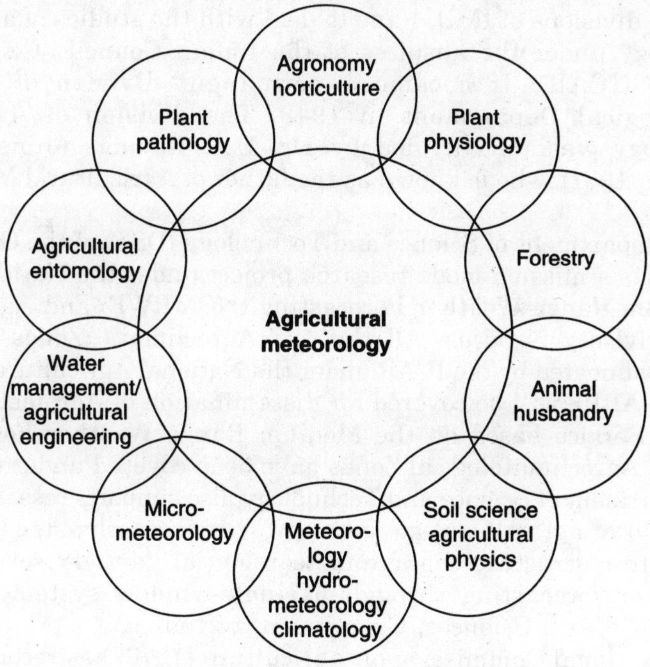

Fig. 1.5 Agricultural meteorology—a multidisciplinary science.

In the strict sense, the science of agricultural meteorology is different to that of agrometeorology, though both are the same from operational point of view. The science of agricultural meteorology deals with the application of meteorological skills on crops and animals while agrometeorology deals with crop husbandry only (see Fig.1.1).

1.6 STATUS OF METEOROLOGY AND AGRICULTURAL METEOROLOGY AT NATIONAL AND GLOBAL LEVEL

The India Meteorological Department (IMD), Government of India, is the unique institution in the country which caters to weather services. It was established in 1875. The headquarters of the IMD are located in New Delhi. The IMD headquarters are known as Mausam Bhavan. Its sole role in weather forecasting is dealt in detail in one of the chapters of this textbook entitled *Weather Forecasting*. The National Data Centre (NDC) was created at IMD, Pune in 1977, where large series of climatological data are maintained in computerised form from 1875 onwards. The Indian Institute of Tropical Meteorology (IITM), located in Pune, is another premier institution in the field of meteorological research in the tropics.

A division of Agricultural Meteorology was also created in 1932 as one of the divisions of IMD, Pune to deal with the studies of agricultural meteorology under the auspices of the Indian Council of Agricultural Research (ICAR). It became a permanent division of the India Meteorological Department in 1943. The division of Agricultural Meteorology was initially led by Dr. L.A. Ramdas (June 3, 1900–January 1, 1971), who is known as the father of Agricultural Meteorology in India.

The Department of Science and Technology (DST) of the Government of India has a mission mode research project under the National Centre for Medium Range Weather Forecasting (NCMRWF) and Agrometeorological Advisory Services. All the 127 Agroclimatic Zones across the country, delineated by the ICAR under the National Agricultural Research Project (NARP), will be covered for dissemination of Agrometeorological Advisory Service based on the Medium Range Weather Forecast. Till 2006, 107 Agroclimatological Zones have been covered under the project. The Department of Science and Technology also supports research projects in the field of agricultural meteorology. Very recently, steps have been initiated to restructure the agrometeorological advisory services based on weather forecasting through a single-window system under the Ministry of Earth Sciences, Government of India.

The National Commission on Agriculture (1976) has recommended a separate Department of Agrometeorology in all the State Agricultural Universities and the ICAR Research Institutes to strengthen teaching, research and extension in agricultural meteorology and to train the technical personnel. At present, a few State Agricultural Universities, namely Anand Agricultural University, Anand (Gujarat); Punjab Agricultural University, Ludhiana; CCS Haryana Agricultural University, Hissar; Kerala Agricultural University, Vellanikkara; Mahatma Phule Krishi Vidhyapeeth, Pune; Indira Gandhi Agricultural University, Raipur;

Assam Agricultural University, Jorhat; N.D. University of Agriculture and Technology, Kumarganj, Faizabad; and Tamil Nadu Agricultural University, Coimbatore have started P.G. programmes in Agricultural Meteorology.

In addition to the above, the Division of Agricultural Physics, IARI New Delhi, offers courses in Agricultural Meteorology at P.G. and Ph.D. levels, and it has a strong multidisciplinary team to carry out research in agricultural meteorology under the Division of Agricultural Physics. A few more State Agricultural Universities are likely to start P.G. programmes in the field of agricultural meteorology to meet the growing demand for technical personnel. Besides, almost all State Agricultural Universities have separate units in the field of agricultural meteorology.

The ICAR started a separate "All-India Co-ordinated Research Project on Agrometeorology" (AICRP on Agrometeorology) in 1985 under Dryland Agriculture and attached to the Central Research Institute for Dryland Agriculture (CRIDA), located in Hyderabad. Many State Agricultural Universities across the country participate in its co-ordinating research programmes on different crops and carry out need based research in agrometeorology. The ICAR also supports research projects in agricultural meteorology.

Research in agrometeorology was initiated on crop-wise basis for the first time in India in the 1970s at International Crops Research Institute for Semi-Arid Tropics (ICRISAT), located in Hyderabad. This is one of the premier international institutes where thrust is given on climate-based research as the name of the institute itself indicates.

The Andhra University, Waltair, and the Cochin University of Science and Technology, Cochin are the two premier universities where teaching programmes are undertaken at the Masters degree level in the field of Meteorology/Atmospheric Sciences. The Department of Meteorology and Oceanography, Andhra University also offers a two-year postgraduate course in the field of meteorology with agricultural meteorology as a special subject.

The World Meteorological Organisation (WMO), located in Geneva (Switzerland), is the nodal organization to develop the science of meteorology all over the world. The World Meteorological Day falls on 23rd March, which is celebrated by meteorologists all over the world every year, focussing annual themes related to the science and its impact on society. The WMO works together with the member countries in monitoring, understanding and predicting weather. It also looks at the development of science of agricultural meteorology as one of its major activities. The Commisssion for Agricultural Meteorology (CAgM) is one of the eight technical commissions of the WMO, which provides technical assistance to the member countries in improving agrometeorological services. The WMO also organizes frequent international conferences on latest issues related to meteorology/climatology/agricultural meteorology to create awareness among the member countries. India is one of the member countries. The latest focus all over the world is on climate change and its impact on various sectors by 2100.

REFERENCES/SUGGESTED READINGS

Ahrens, C.D., *Meteorology Today—An Introduction to Weather, Climate and the Environment*, 6th ed., Brooks/Cole, California, 2000, p. 528.

Ghadekar, A.E., *Meteorology*, Agromet Publishers, Nagpur, 2001, p. 251.

Gopalaswamy, N., *Agricultural Meteorology*, Rawat Publications, Jaipur, 1994, p. 153.

Kakde, J.R., *Agricultural Climatology*, Metropolitan Book Company, New Delhi, p. 387.

Lutgens, F.K. and Tarbuck, E.J., *The Atmosphere*, 7th ed., Prentice-Hall Inc., London, 1998, p. 434.

Menon, P.K., *Our Weather*, National Book Trust, New Delhi, 1989, p. 349.

National Commission on Agriculture, *Climate and Agriculture*, Abridged Report: 4, Ministry of Agriculture and Irrigation, Department of Agriculture, Government of India, New Delhi, 1976, pp.152–236.

Petterson, S., *Introduction to Meteorology*, McGraw-Hill, New York, 1958, p. 327.

Rao, G.S.L.H.V.P., *Agricultural Meteorology*, 2nd ed., Kerala Agricultural University, Thrissur, India, 2005, p. 326.

Riehl, H., *Introduction to the Atmosphere*, McGraw-Hill, New York, 1965, p. 365.

Trewartha, G.T., *An Introduction to Climate*, McGraw-Hill, New York, 1954, p. 395.

QUESTIONS

1. Define meteorology and agricultural meteorology.
2. Distinguish between
   (a) Weather and climate
   (b) Meteorology and agricultural meteorology
   (c) Agricultural meteorology and agrometeorology.
3. Write short notes on:
   (a) Greenhouse effect
   (b) Ozone
   (c) Branches of meteorology
   (d) Importance of agricultural meteorology
   (e) Scope of agricultural meteorology
   (f) Aerosol cooling
   (g) Scales of climate

(h) Status of meteorology in India
(i) Status of agricultural meteorology in India.

4. Expand the following:
(a) IMD
(b) NCMRWF
(c) IPCC
(d) WMO
(e) ICRISAT
(f) IITM

5. Fill in the blanks:
(a) ——— is the father of agricultural meteorology in India.
(b) The World Meteorological Day falls on ———.
(c) India Meteorological Department was established in ——— year.
(d) ——— is known as statistical meteorology.
(e) In ———, composition of gases in atmosphere is not uniform.
(f) In ———, composition of gases in atmosphere is uniform.
(g) ——— is the projection of CO_2 by 2050 A.D.
(h) ——— is responsible for regional cooling effect.
(i) ——— is responsible for ozone depletion.
(j) In ———, the division of agricultural meteorology was established by the IMD.
(k) In ———, AICRP on agrometeorology was established by the ICAR.

6. State true or false:
(a) Indian foodgrains production is related to all-India drought.
(b) Today's climate is pleasant.
(c) Atmospheric window is related to shortwave radiation.
(d) IITM is involved in research on tropical meteorology.
(e) Weather Forecasting Bulletins are issued by IITM.
(f) Numerical Weather Prediction is dealt in the field of synoptic meteorology.
(g) Agricultural Meteorology is not a multidisciplinary science.
(h) CO_2 increase is not responsible for global warming.
(i) Atmospheric water vapour is more above 10 km from the earth's surface.
(j) 3% of CO_2 by volume is present in atmosphere.

7. Elucidate the following:
(a) Composition of atmosphere.
(b) Importance and scope of agricultural meteorology.

2 RADIATION

Radiation received from the sun is the main source of the earth's atmospheric energy and ultimately it is the fuel of the atmosphere. The atmosphere acts mostly as a medium which is—transparent to the sun's radiation and opaque to the earth's radiation, and this is the driving force of atmosphere. Variations in radiation received over different regions of the globe due to its position with reference to the sun, together with the features of earth, produce weather and climate. Hence, the basic concepts of radiation are very important to understand the behaviour of the atmosphere. Just as water flows from a level of high to low and air moves from high atmospheric pressure to low atmospheric pressure, heat flow also occurs from a body of higher temperature to that of lower temperature. The radiation emitted by a body depends on its temperature and is known as *thermal radiation*. The thermal radiation from the sun is the ultimate thermal energy, creating contrast in surface temperature over different regions of the globe. It amounts to variations in atmospheric pressure, leading to movement of air from one region to another. In the process a huge amount of heat is transported from the equator to the poles, and thus weather and climate are the resultant atmospheric systems.

2.1 TRANSFER OF HEAT

Any body, which has a temperature, transfers heat in different forms. Three processes viz. conduction, convection and radiation are involved in heat flow or heat transfer. Every hot object radiates heat.

2.1.1 Conduction

The transfer of heat by matter through its adjacent layers from its hotter to colder parts, without any motion of the matter itself, is called conduction. For example, if one end of an iron bar is put in furnace, it is by conduction that the other end receives heat. Heat is received when hot objects are touched and is lost with cold ones. This is another example of conduction of heat.

2.1.2 Convection

The transfer of heat through the mass movement of hot fluid matter, either liquid or gas, is called convection. Steam flow from hot water bodies is an example of convection. A layer of steam is noticed over large water bodies during winter. This is another example of convection.

2.1.3 Radiation

The direct transfer of heat from a hot body to a cold body without the aid of any material medium is called radiation. The direct transfer of heat from furnace to our body is an example of radiation. Radiation is one of the heat processes which take place from the sun to the earth's surface directly without any material medium. It may be noted that radiation is not heat. Only when radiation is absorbed by the surface of a body, heat is produced.

All the above three processes are seen between the sun, earth and the atmosphere. For example, soil heat flow is through conduction, heat flow in the atmosphere is through convection, and heat flow from the sun to the earth is due to radiation (see Fig. 2.1).

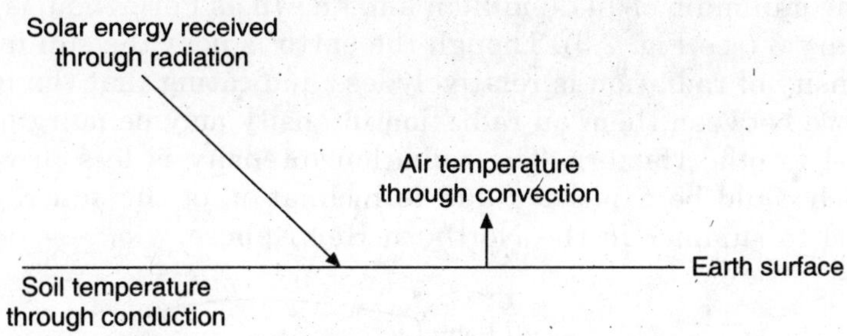

Fig. 2.1 Heat processes between the earth and the atmosphere.

2.2 SEASONS

The seasons form due to the revolution of the earth around the sun. A complete revolution around the sun takes one year. When the Northern

Hemisphere experiences six months summer, the Southern Hemisphere experiences winter; and the opposite happens for the remaining six months due to position of the earth with reference to the sun. The inclination of earth's axis makes the length of day and night is different in different seasons. The rotation of earth on its own axis makes the difference in time from one place to another. It takes 24 hours to complete one rotation of the earth on its own axis. Since the earth is spherical in shape, always half of the globe is only covered under sunlight and the other half is in dark. Half the sphere under sunlight experiences day while the dark portion experiences night (see Fig. 2.2).

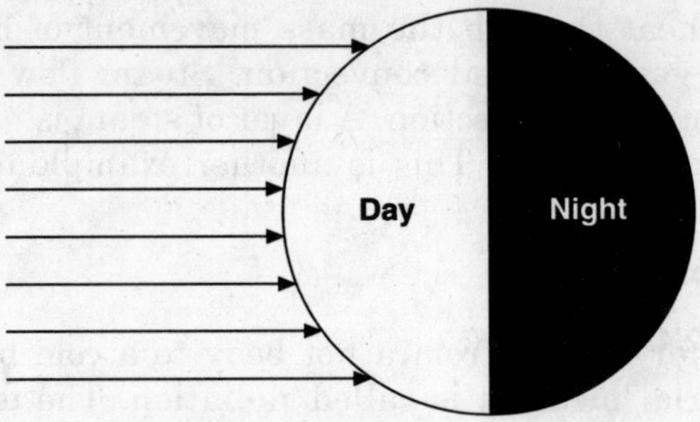

Fig. 2.2 Day and night.

2.2.1 Aphelion and Perihelion

The elliptical earth's orbit around the sun ensures that the distance between them is not uniform. The average distance between the earth and the sun is 149.5 million km. The distance between the earth and the sun reaches a maximum of 152 million km, known as Aphelion on July 4, and the minimum of 147.6 million km, known as Perihelion, is reached on January 3 (see Fig. 2.3). Though the earth is near the sun in winter, the intensity of radiation is relatively less, indicating that the influence of distance between them on radiation intensity may be marginal when compared to other factors. The radiation intensity is less in winter, a fact which could be explained due to inclination of the sun rays when compared to summer in the Northern Hemisphere.

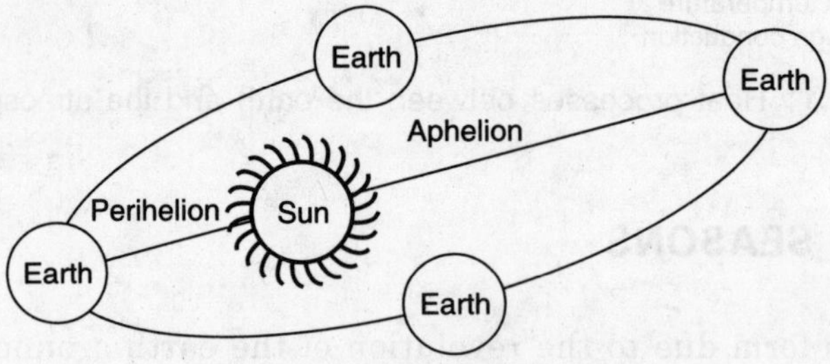

Fig. 2.3 Perihelion and Aphelion.

2.2.2 Equinox

Equinox means *equal night*. In other words, the duration of day and night is equal. It happens twice a year during March and September. If it is on 21st March, it is known as *spring equinox* or *vernal equinox*, while it is called *autumnal equinox* if it is on 23rd September. On the above dates the noon sun rays reach over the equator straight (see Fig. 2.4).

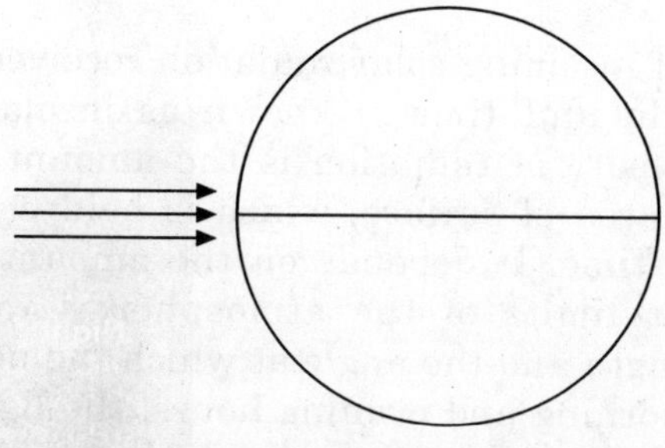

Fig. 2.4 Equinox (March 21st and September 23rd).

2.2.3 Solstice

Solstice means 'the sun stands' or that the position of the earth with reference to the sun is such that the sun rays reach straight over the Tropic of Cancer and the Tropic of Capricorn, and make the day the longest. It happens once in a year in the Northern Hemisphere as well as in the Southern Hemisphere. They are known as *summer and winter solstices.*

Summer solstice (21st June)—the sun is directly overhead at the Tropic of Cancer, making it the longest day in the Northern Hemisphere and the shortest day in the Southern Hemisphere (see Fig. 2.5(a)).

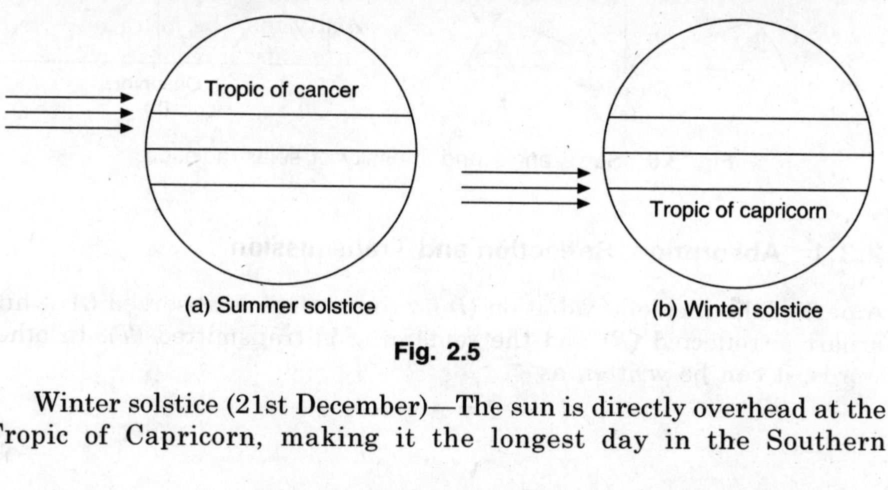

(a) Summer solstice (b) Winter solstice

Fig. 2.5

Winter solstice (21st December)—The sun is directly overhead at the Tropic of Capricorn, making it the longest day in the Southern

Hemisphere and the shortest day in the Northern Hemisphere (see Fig. 2.5(b)).

On 21st June, the Arctic has 24 hours of daylight whereas the Antarctic has 24 hours of darkness. On 21st December, the opposite happens.

2.3 INSOLATION

The total amount of incoming solar radiation received on a unit area of horizontal surface in unit time is known as insolation. Radiant flux density or the intensity of radiation is the amount of radiant energy received on a unit area of surface, which is held perpendicular to the solar beam in unit time. It depends on the amount of solar radiation reaching the outer limits of the atmosphere, transparency of the atmosphere, day length and the angle at which the noon sun rays strike the earth. In the morning and evening hours, the light intensity is less due to inclined angle as it travels a greater distance in the atmosphere when compared to the noon sun's rays. In the process, it losses more solar energy in the atmosphere before it reaches to the earth's surface (see Figs. 2.6(a) and (b)). In the case of inclined sun angle, it not only loses solar energy in travel through the atmosphere, but also covers more surface area on the earth surface. It leads to lower temperature. That is why higher temperature is noticed over the tropics and lower temperature over the higher latitudes. Again, the solar radiation at the outer atmosphere depends on the energy output of the sun, distance from the earth to the sun and interstellar dust in the solar system.

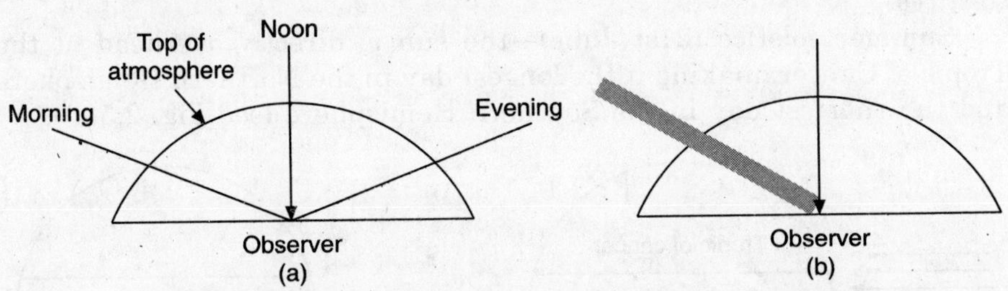

Fig. 2.6 Sun's angle and intensity of solar radiation.

2.3.1 Absorption, Reflection and Transmission

A part of the incident radiation (I) on the surface is absorbed (A) while a part is reflected (R) and the remaining is transmitted (T). In other words, it can be written as

$$I = A + R + T$$

If the simple equation given in the previous line is divided by incident radiation I, It can be written as follows:

$$\frac{I}{I} = \frac{A}{I} + \frac{R}{I} + \frac{T}{I}$$

The same can be put in the form of $1 = a + r + t$, where a is absorptivity, r is reflectivity and t is transmittivity. All the three become unity for any given amount of incident radiation on a given surface (see Fig. 2.7). From the above equation, all the three terms can be defined as follows.

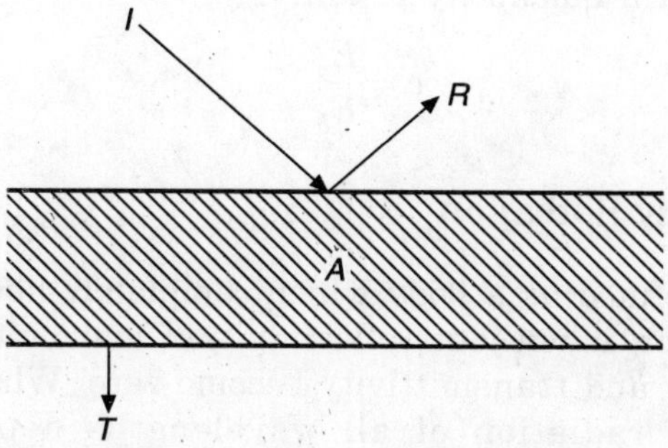

Fig. 2.7 Reflection, absorption and transmission.

2.3.2 Absorptivity (*a*)

Absorptivity of a substance is defined as the ratio of the amount of radiant energy absorbed to the total amount incident upon that substance. The absorptivity of a blackbody is unity. Natural bodies like the sun and the earth are nearly perfect blackbodies. Earth too behaves as a grey body; but in certain wavelength bands, it may be effectively a blackbody.

2.3.3 Reflectivity (*r*)

Reflectivity is defined as the ratio of the radiant energy reflected to the total incident radiation upon that surface. If it is expressed in percentage, it becomes *albedo*. A certain part of the energy received from the sun is reflected back to space from the earth. It is termed the *earth's albedo* if the ratio of the reflected radiation to the incident radiation expressed in percentage. Similarly, if it is from a cloud, it is known as the *cloud's albedo*. The cloud's albedo is more than the earth's albedo since the cloud looks more white. All the white surfaces reflect the maximum, while it is less in case of black surfaces and it is between black and white in case of grey objects. For example, the albedo in case of snow varies between 70% and 90% depending upon the freshness while it is only 10% in case of forests and 20% in case of bare soils.

2.3.4 Transmittivity (*t*)

Transmittivity is defined as the ratio of transmitted radiation to the total incident radiation upon that surface.

2.3.5 Emissivity

Emissivity (*e*) is defined as the ratio of the radiant energy emitted by a given surface (E_s) to the total heat energy emitted by a blackbody (E_b). The 'emissivity' of a blackbody is unity.

$$e = \frac{E_s}{E_b}$$

2.3.6 Blackbody Radiation

A blackbody is defined as a body which completely absorbs all the heat radiation falling on it without reflecting and transmitting any of it. It means reflectivity and transmittivity become zero. When such a body is heated, it emits radiation of all wavelengths depending upon its temperature. In nature, the sun and the earth are near perfect blackbodies.

2.4 RADIATION LAWS

The direct transfer of heat, from the sun to the earth, through the space and atmosphere, indicates that radiation of heat from one place to other occurs in the form of electromagnetic waves in the same manner and with the same speed as light. The wavelength of electromagnetic radiation is given by the equation (see Fig. 2.8)

$$\lambda = \frac{C}{\nu}$$

where λ is the wavelength, ν is the frequency which means number of vibrations or cycles per second, and C is the constant equal to the velocity of light, 3×10^{10} cm/sec.

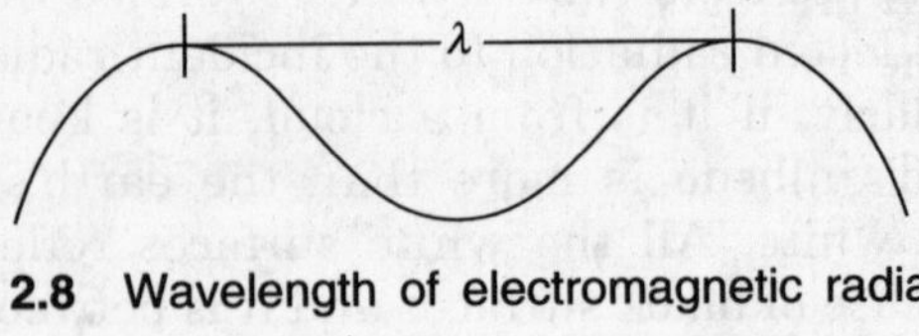

Fig. 2.8 Wavelength of electromagnetic radiation.

2.4.1 Kirchhoff's Law

The law states that the ratio of emitted energy to the absorbed energy is a function of temperature (*T*) and wavelength (λ).

$$\frac{E_{em}}{E_{ab}} = f(\lambda, T)$$

where E_{em} = radiant energy emitted and E_{ab} = radiant energy absorbed.

If $E_{ab} = 1$, the body is termed 'black', thus blackbodies possess high emissivity also.

If $0 < E_{ab} < 1$, the body is termed a *greybody*.

In other words, the law states that a perfect absorber of radiation is a perfect emitter and it is unity in the case of a blackbody.

2.4.2 Planck's Law

Planck introduced the particle concept. It is stated that electromagnetic radiation consists of a stream or flow of particles, or *quanta*, with each quantum having an energy content E determined by

$$E = h\,\nu$$

where h is Planck's constant = 6.62×10^{-34} Js (joule sec) and ν is the frequency.

It indicates that the electromagnetic radiation with greater frequency (shorter wavelength) contains more energy content of the quantum. Hence the ultraviolet radiation is more energetic and harmful than quanta of red light, as the wavelength of ultraviolet radiation is shorter.

2.4.3 Stefan-Boltzmann's Law

Stefan showed experimentally and Boltzmann demonstrated on theoretical grounds that the radiation emittance of a body (E) is a function of the fourth power of its absolute temperature (emissivity = 1 for blackbodies). Therefore, $E = \sigma T^4$ for blackbodies, where T is the absolute temperature and σ is the Stefan-Boltzmann's constant.

$$\sigma = 5.673 \times 10^{-8}\ \text{Wm}^{-2}\ \text{K}^{-4}$$

2.4.4 Wien's Displacement Law

The wavelength of maximum emission (λ_{max}) is related to the absolute temperature of the radiating body by

$$\lambda_{max} = \frac{\text{Constant}}{T\text{(in Kelvin)}} = \frac{2897}{T}\ \mu$$

The sun may be understood to behave as a blackbody whose surface temperature is 6000 K. By Wien's law,

$$\lambda_{max}(\text{sun}) = \frac{2897}{6000} = 0.483\ \mu$$

If the mean surface temperature of earth is assumed to be 300 K,

$$\lambda_{max}(\text{earth}) = \frac{2897}{300} = 9.66\ \mu$$

It shows that the sun emits maximum solar energy at 0.483 μ while the earth does so at 9.66 μ.

2.4.5 Solar Constant

The amount of solar radiation that is received in a minute, at the top of the atmosphere on a unit area of surface, which is held perpendicular to the beam when the earth is at its mean distance from the sun, is known as the solar constant. This occurs in April and October. The value of solar constant is estimated to be 1.94 cal/cm^2/min or (1.94 ly/min) = 1367 ± 7 Wm^{-2}.

The solar constant is not a true constant but fluctuates as much as ± 3.5 % about its mean value depending upon the distance of the earth from the sun. The maximum value of 2 langley per minute is recorded in January and the minimum (1.88 cal/cm^2/min or 1.88 langley/minute) by the end of June or the beginning of July.

2.5 SHORT AND LONGWAVE RADIATION

Solar energy covers a wide spectrum of electromagnetic radiation ranging from electric waves (wavelength of several hundred metres) to the ultraviolet and cosmic radiation of extremely short wavelength and high energy, capable of destroying all forms of life. The wavelength of solar radiation is so small and inconvenient to express in mm. Hence, smaller units such as micron (μ), millimicron (mμ) and angstrom (Å) are used. The conversion of the same in length units (m) is as follows:

$$1\ \mu = 10^{-6}\ \text{m}$$
$$1\ \text{m}\mu = 10^{-9}\ \text{m} = 1\ \eta\text{m}$$
$$1\ \text{Å} = 10^{-10}\ \text{m}$$

The energy for various purposes in the earth atmospheric system comes from the sun having surface temperature of 6000 K. The radiation emitted from the sun is known as *shortwave radiation*, generally called *solar radiation*. It ranges from 0.15 to 4.0 μ with a peak at 0.483 μ.

The solar energy received over the earth surface is maximum in visible radiation and near infrared radiation. About 80% is received in the above spectral region. About 10% each is covered towards ultraviolet radiation and far infrared radiation, respectively.

The *terrestrial radiation* is the radiation emitted by the earth. The range of terrestrial radiation (longwave radiation) spectrum ranges from 4 to 80 μ with a peak of 9.66 μ at an effective surface temperature of about 300 K. The longwave radiation is also called *nocturnal radiation* since it is the major radiate source of energy at night.

2.5.1 Electromagnetic Spectrum

It is an array of distribution of electromagnetic radiation in different wavelengths (see Fig. 2.9) which can be shown as follows:

Cosmic rays	Gamma rays	X-rays	Ultraviolet radiation	Visible radiation	Infrared radiation	Radio waves	Electric waves

$<10^{-10}\mu$ $10^{-8}\mu$ $10^{-4}\mu$ 0.4μ 0.76μ $3 \times 10^{3}\mu$ $>3 \times 10^{10}\mu$

Fig. 2.9 Electromagnetic spectrum.

The visible radiation between 0.4 and 0.76 μ is very important to plants. We see each other and all the objects through visible radiation. It is the light which is very important to plants in their photosynthetic process. Within the visible radiation, VIBGYOR is very important for plants, which is distributed in different wavelengths, namely, Red and Orange (0.6 to 0.7 μ), Yellow and Green (0.5 to 0.6 μ) and Blue and Violet (0.4 to 0.5 μ).

2.5.2 Depletion of Solar Radiation

The solar radiation reaching the ground is depleted due to the presence of air pollutants in addition to the angle at which the sun rays reach the ground. The reflection, absorption and transmission vary when the sun rays pass through the atmosphere based on its constituents. Water vapour is the one which strongly absorbs solar radiation when compared to other gases. Clouds reflect solar radiation to a considerable extent. That is why the solar radiation reaching the ground is relatively low when the sky is overcast (see Fig. 2.10). If the sky is covered with cumulus clouds, the depletion of solar radiation is more due to high reflection when compared to that of cirrus clouds covering the sky. Depending upon the state of atmosphere and the sun angle, the earth receives normally 50–70% of the solar radiation received on the top of the atmosphere. This could be understood well from the radiation balance discussed in the following section. Ozone also plays a major role in depletion of solar radiation depending upon its density and its variation with time.

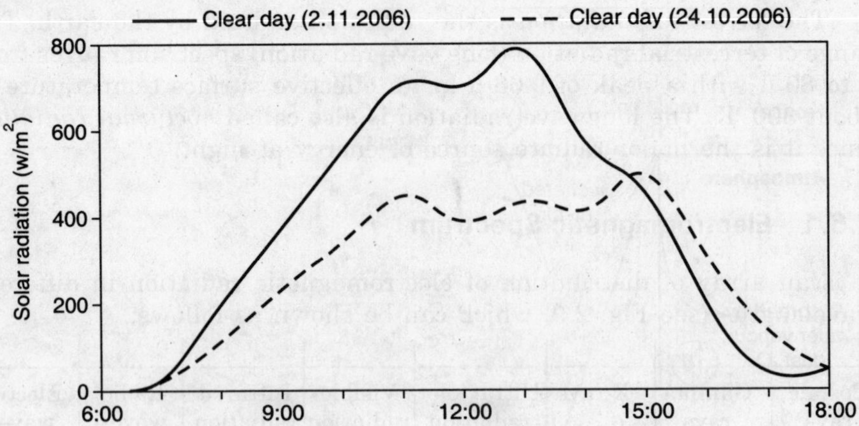

Fig. 2.10 Influence of clouds on solar radiation.

2.6 RADIATION BALANCE

The difference between all incoming and outgoing radiations at the earth's surface and in and above the atmosphere is known as radiation balance (R_B).

R_B is equal to incoming – outgoing radiation. It can be seen between the earth, atmosphere and space.

Based on certain measurements and observations carried out by worldwide weather bodies, it has been estimated that if the incoming solar radiation (insolation) is equal to 100 units at the outer atmosphere, 30 units are returned to space through reflection from the earth's surface (4 units), reflection from clouds (20 units) and back scattered by air (6 units) without heating the atmosphere. Out of 70 units remaining, 16 units are absorbed by water vapour, dust and ozone, and 3 units are absorbed by clouds. The remaining 51 units are transmitted through the atmosphere and absorbed at the ocean and land surface. Out of the 51 units absorbed at the earth's surface, 23 units are used for latent heat flux and 7 units are used for sensible heat flux and the earth emits 21 units in the form of longwave radiation. Out of 21 units emitted by the earth, 15 units are absorbed by water vapour and carbon dioxide and 6 units are lost to space. The net emission to space in the form of longwave radiation by water vapour and carbon dioxide is 38 units and 26 units by clouds, so as to balance the 100 units of incoming solar radiation received at the outer atmosphere as illustrated in Fig. 2.11. The positive sign is for incoming solar radiation and negative for outgoing radiation lost to space.

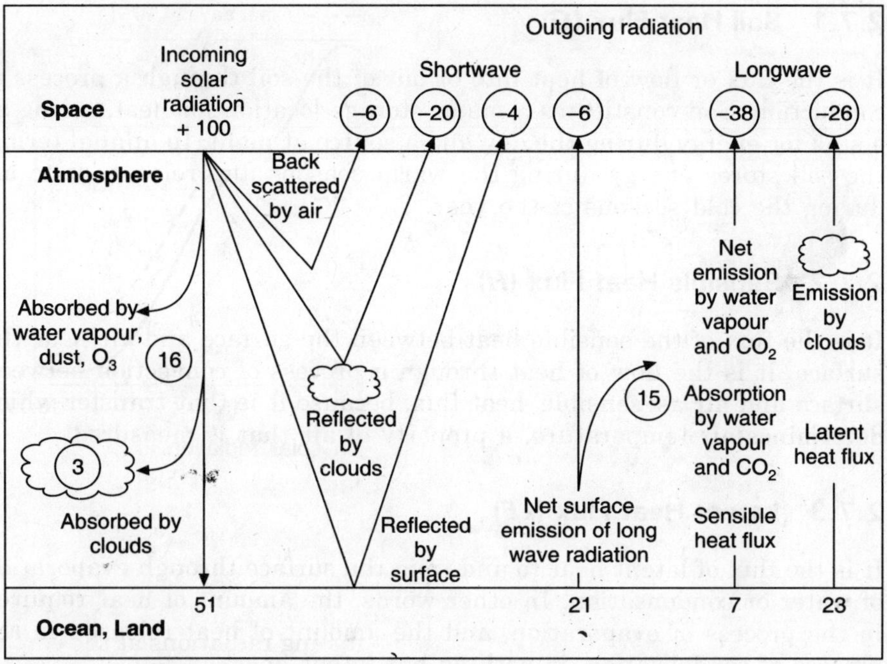

(Source: US, GARP, 1975)

Fig. 2.11 The mean annual radiation and heat balance of the atmosphere relative to 100 units of incoming solar radiation.

2.7 ENERGY BALANCE OR HEAT BALANCE

The net radiation is the difference between total incoming and outgoing radiations and is a measure of the energy available at the ground surface. It is the energy available at the earth's surface to drive the processes of evaporation, air and soil heat fluxes, as well as other smaller energy consuming processes such as photosynthesis and respiration. The net radiation over a crop surface is as follows:

$$R_n = G + H + LE + PS + M$$

where R_n is net radiation, G is surface soil heat flux, H is sensible heat flux, LE is latent heat flux, PS and M are energy fixed in plants by photosynthesis and energy involved in respiration, respectively. The PS and M are assumed negligible due to their minor contribution (about 1–2% of R_n). The net radiation is the basic source of energy for evapotranspiration (LE), heating the air (H) and soil (S) and other miscellaneous processes (M), including photosynthesis. If water is not limiting, most of the R_n is used for evapotranspiration. Thus air temperature increases when water is limited as the energy that is supposed to evapotranspiring water is used to heat air.

2.7.1 Soil Heat Flux (*G*)

It is the flux or flow of heat into or out of the soil through a process of conduction. Soil constitutes a major storage location for heat, acting as a sink for energy during the day and a source at night. In annual terms, the soil stores energy during the warm seasons and releases it to air during the cold seasons of the year.

2.7.2 Sensible Heat Flux (*H*)

It is the flux of the sensible heat between the surface and air near the surface. It is the flow of heat through a process of convection between surface and air as 'sensible' heat flux, because it is that transfer which determines air temperature, a property of air that is measured.

2.7.3 Latent Heat Flux (*LE*)

It is the flux of latent heat to and from the surface through evaporation of water or condensation. In other words, the amount of heat required in the process of evaporation, and the amount of heat released in the process of condensation, is nothing but latent heat.

2.8 RESPONSE OF PLANTS TO RADIATION

The visible radiation is very important in photosynthetic mechanism of plants, which is a very complex process. The visible radiation is known as photosynthetically active radiation (PAR), as it is essential for plants to produce carbohydrates, ultimately of biomass. The raw material like carbon dioxide and water in the presence of sunlight, plants produce carbohydrates in the form of biomass or dry matter.

$$CO_2 + 2H_2O \xrightarrow[\text{green plants}]{\text{sunlight}} CH_2O + O_2 + H_2O$$

In the visible radiation, a strong photosynthetic activity takes place in the blue-violet (0.4 to 0.5 μ) and in the orange-red (0.6 to 0.7 μ) region. The green-yellow (0.5 to 0.6 μ) is the region in which low photosynthetic activity takes place. The absorption by plants is nil to minimum depending upon the type of leaves between 0.7 and 0.8 μ (see Fig. 2.12).

Plants grow healthy under optimum sunlight if water and nutrients are unlimited. If the sun loving plants are grown in shade, they grow taller, their photosynthetic efficiency is less and thereby there is low crop biomass.

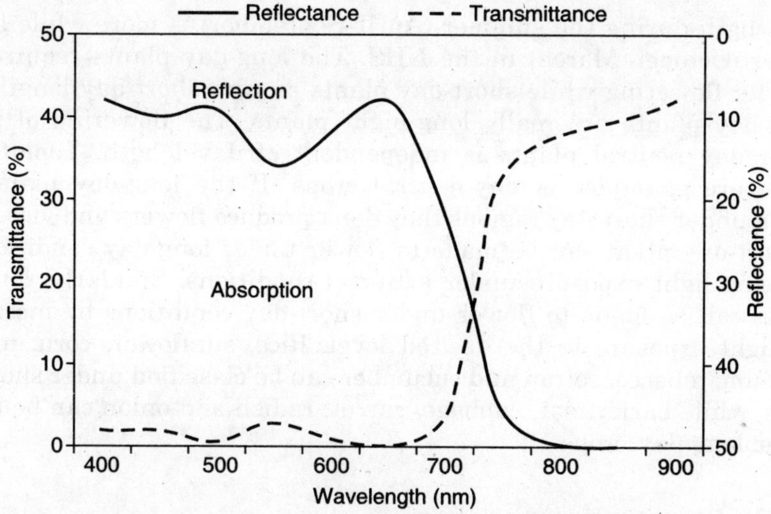

Fig. 2.12 Spectral response of a coconut leaf.

2.8.1 Photoperiodism

The response of plants to day length is known as photoperiodism. Plant species are grouped into long-day, short-day and neutral plants based on response of plants to day length. The day length varies from latitude to latitude depending upon the season (see Table 2.1).

Table 2.1 Length of day in northern latitudes (hours and minutes in the middle of month)

Month	*0*	*10**	*30*	*60*	*90*
January	12:07	11:45	10:24	06:38	00:00
February	12:07	11:50	11:10	09:11	00:00
March	12:07	12:10	11:57	11:41	00:00
April	12:07	12:20	12:53	14:31	24:00
May	12:07	12:30	13:38	17:04	24:00
June	12:07	12:30	14:04	18:49	24:00
July	12:07	12:30	13:56	17:31	24:00
August	12:07	12:15	13:16	15:46	24:00
September	12:07	12:15	12:23	13:00	24:00
October	12:07	12:10	11:28	10:11	00:00
November	12:07	11:45	10:40	07:37	00:00
December	12:07	11:30	10:14	05:54	00:00

*Observed values by the author at Thrissur (10°32′ N and 76°13′ E)

It is relatively uniform over the tropics while significant variation is noticed in day length when one moves north or south of the equator. The day length during the summer (April–September) is more while less in winter (October–March) in the NHS. The long-day plants require long days for flowering while short-day plants require short day length. The short-day plants are really long-night plants. The flowering of plants under day neutral plants is independent of day length. Tomato and maize are examples of day-neutral crops. If the long-day plants are grown under short-day regions they don't produce flowers and vice versa. A short-day plant can be made to flower under long-day conditions by reducing light exposure under artificial conditions. Similarly, long-day plants can be made to flower under short-day conditions by increasing the light exposure to the desired level. Rice, sunflower, corn, millets, sorghum, tobacco, cotton and cucumber can be classified under short-day crops, while barley, oat, cabbage, carrot, radish and onion can be placed under long-day crops.

2.8.2 Phototropism

Plant species also show phototropism, which is nothing but the growth of plants in the direction of sunlight. It is also known as *positive phototropism*. The classical example is sunflower. The flowers of sunflower move along with the direction of sunlight. In the case of coconut also, it takes several bends in search of sunlight if continuous shade falls on coconut trees. The above situation is noticed in homestead gardens. The northernmost crop rows in coconut plantations are seen bent in search of sunlight. The delay in biotic events is also noticed in fruit crops like cashew and mango towards the east and west quadrants of the north and produce less in the Northern Hemisphere, as sunlight is always in the south quadrant of a tree. In the case of banana, bunch emergence is always in the east and west of the south quadrant, showing positive phototropism. Sometimes, the high intensity of radiation may cause negative phototropism. If the plant growth is away from the direction of sunlight, it is known as *photonasty* or *negative phototropism*.

2.8.3 Photosensitivity

Local cultivars mostly are photosensitive, which means *season-bound*. Whenever crops are sown or transplanted they blossom only in a particular period of time or season. A local cultivar named Allikkannan in northern districts of Kerala is a classical example for photosensitivity. It cannot be grown round the year as it responds only to the first crop season (*kharif*). Mayflower is another classical example of photosensitive plants. Photosensitivity is an adverse character in an intensive cropping system, where crops are to be taken up round-the-year. Hence, photoinsensitive varieties play a major role in crop production as they can be grown throughout the year.

2.8.4 Photoinsensitivity

All the varieties normally are classified under the photoinsensitive category, unless otherwise mentioned. Whenever crops are sown or transplanted, they blossom according to the growth period. It means they can be grown round-the-year and achieve more crop production. All rice varieties are photoinsensitive.

2.8.5 Radiation Distribution in Plant Canopy

The radiation distribution in plant canopy depends upon the three aspects, viz. intensity of radiation—that is, the amount of radiant energy received on a unit surface area in a unit time, the spectral distribution of radiation, which governs the photochemical process of photosynthesis and the radiation distribution in time and is important for photoperiodic phenomenon. The above three aspects are again determined by several factors such as transmissibility of the leaf, leaf arrangement and inclination, plant density, plant height and angle of the sun.

Radiation distribution in coconut

The percentage decline in solar radiation varied between 29.9 and 31.9 at 10 feet level (inside coconut) in summer and southwest monsoon, respectively. The percentage decline of net radiation at coconut canopy to the solar radiation received just above the coconut varied between 26.5 and 33.1 during summer and southwest monsoon respectively. It also indicated that the percentage decline in net radiation was high during the southwest monsoon when compared to that of summer (see Table 2.2).

Table 2.2 Radiation distribution in coconut

Solar and net radiation	*Summer* (Wm^{-2})	*Southwest monsoon* (Wm^{-2})	*% decline*
Solar radiation above coconut (56′ height)	618.0	444.6	28.0
Solar radiation inside coconut (10′ height)	433.4 (29.9)	302.6 (31.9)	30.1
Net radiation at coconut canopy level (38′ height)	454.0 (26.5)	297.4 (33.1)	34.4

Figures in parenthesis indicate percentage decline to solar radiation above coconut (56′ height).

The high percentage decline in net radiation during southwest monsoon due to the fact that the emitted energy is relatively less in wet environment when compared to that of dry environment. The potential photosynthesis may be reduced during the rainy season as the decline in solar radiation as well as in net radiation was around 30%. In fact, the depletion in solar radiation and net radiation was around 51% inside the coconut garden during the southwest monsoon when considered the amount of solar radiation received just above the coconut garden during the summer. De Wit (1959) has reported that there could be 50% of reduction in potential photosynthesis when light intensity get reduced to only 20% of the intensities on clear days in the tropical areas at 10°N. It is clearly seen that the low light intensity during the monsoon is one of the crop production constraints.

Vertical profiles of radiation in coconut

The net radiation at the coconut canopy followed the same profile of the solar radiation received just above the coconut canopy in both summer and southwest monsoon. The different between solar and net radiation was higher in the forenoon when compared to that of afternoon and reach very close to each other before sunset (see Figs. 2.13(a) and (b)). This might be due to domination of emitted energy from the canopy as well as the ground surfaces while the depletion of solar radiation takes place.

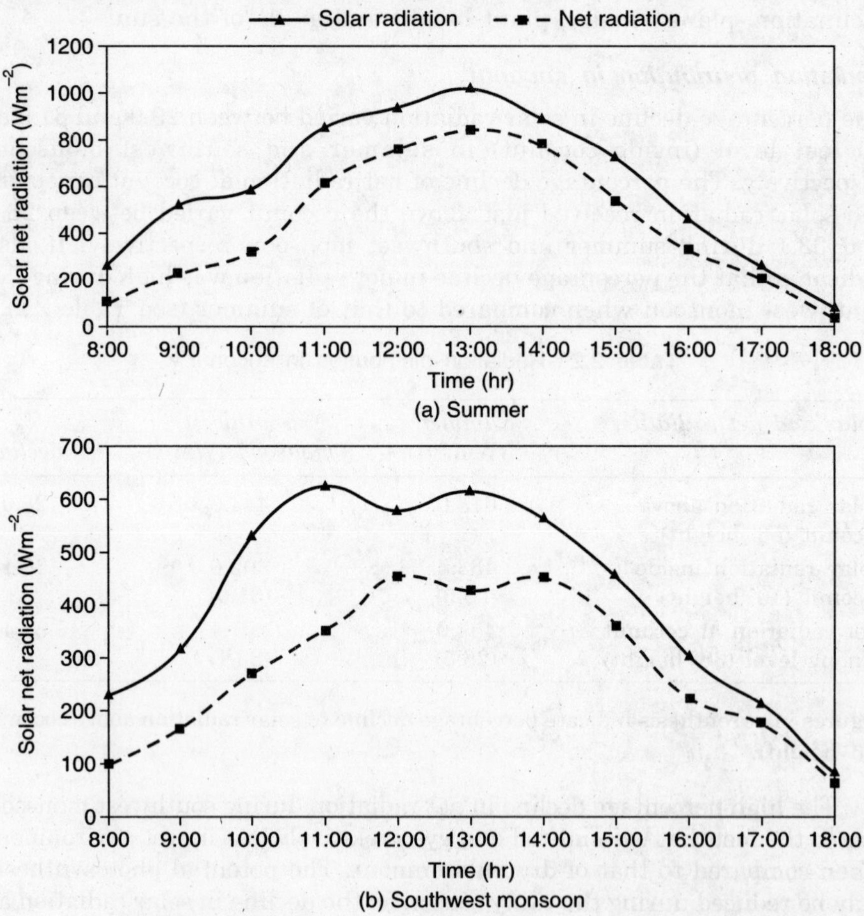

Fig. 2.13 Solar and the radiation (Wm^{-2}) profiles at RARS. Pilicode (12°12′ N and 75°10′ E).

Relationship between sunshine and solar radiation

The relationship between the bright sunshine hours and the solar and net radiation is given in Table 2.3. It helps in estimating solar radiation in and above coconut, as well as net radiation at the coconut canopy based on empirical relationship between the number of bright sunshine hours and incoming solar radiation. Such equations can be developed in different agroclimatic zones for each crop as they are location- and crop-specific.

Table 2.3 Relationship between bright sunshine hours and solar and net radiation in coconut garden

Location	*Coconut*	'r'
RARS, Pilicode,	$R_{nc} = 0.786\ R_{st} - 39.17$	0.96
Kasaragod district, Kerala	$R_{nc} = 26.39\ n + 168.52$	0.93
(12°12′N and 75°10′E)	$R_{si} = 0.774\ R_{st} - 39.17$	0.99
	$R_{si} = 29.68n + 161.41$	0.96
	$R_{st} = 33.78n + 261.58$	0.98

R_{st} : Direct solar radiation above coconut canopy (56′ height)
R_{si} : Direct solar radiation inside coconut (10′ height)
R_{nc} : Net radiation at the coconut canopy (38′ height)
n : Number of bright sunshine hours per day

2.9 INSTRUMENTS

The following instruments can be used to measure radiation:

2.9.1 Sunshine Recorder

The sunshine recorder (shown in Fig. 2.14) is a convex glass sphere mounted on a brass bowl with three sets of grooves for holding the sunshine recorder cards. The convex glass sphere burns a trace on the card when exposed to the sun. The length of burning trace is the measure of the duration of bright sunshine. It is measured using a plastic scale provided along with the equipment. The duration of sunshine is relatively less in cloudy weather while more in clear skies. There are three types of sunshine cards viz. long curved for summer, short curved for winter and straight card at the equinoxes.

2.9.2 Lux Meter

The lux meter, shown in Fig. 2.15, measures the light illumination. It is used to measure the shade falling on crops in different homestead gardens. The shade depends upon the solar angle and the crop stand under which the test crop is grown. The shade is more in forenoon and afternoon

(40–80 %) and less(20–40%) during the noon hours. It is also dependent on cloudy weather. It is used to define shade loving crops. It is expressed in percentage for comparison.

Fig. 2.14 Sunshine recorder.

Fig. 2.15 Lux meter.

2.9.3 Quantum Sensor

The quantum sensor, shown in Fig. 2.16, is the instrument, which measures the intensity of photosynthetically active radiation (PAR) in the 400–700 ηm wavelength range. It gives an output of quantum flux rather than energy flux as quantum response is more appropriate in photosynthesis studies. It measures in moles.

2.9.4 Tube Solarimeter

It is the instrument which measures the intensity of incoming solar radiation (see Fig. 2.17). This instrument is mostly used to measure solar radiation where the distribution of radiation is not uniform. In crop micrometeorology under different cropping systems, where normally the distribution of solar radiation is not uniform, tube solarimeters are used. It measures in Wm^{-2}.

Fig. 2.16 Quantum sensor.

Fig. 2.17 Tube solarimeter.

2.9.5 Angstorm Phyrheliometer

It measures the intensity of direct component of solar irradiation in $cal/cm^2/min$ or Wm^{-2}.

2.9.6 Pyranometer

It measures the global solar radiation both direct and diffuse radiations. It can be used as an Albedometer if it is exposed to the earth's surface. It measures in Wm^{-2}.

2.9.7 Net Radiometer

It measures the difference between the upward and downward total radiant energy directly. It measures in Wm^{-2}.

2.9.8 UV Biometer

The UV biometer (shown in Fig. 2.18) is the instrument which measures the filtered UV-B radiation of 280–320 ηm in terms of Minimum Erythema Dose, reaching the ground. The biological effectiveness of the UV irradiation is measured in MED/HR (Minimum Erythema Dose per Hour). One MED/HR would cause minimal redness of the average skin after one hour irradiation. In India, very few locations have the facility to measure UV-B radiation.

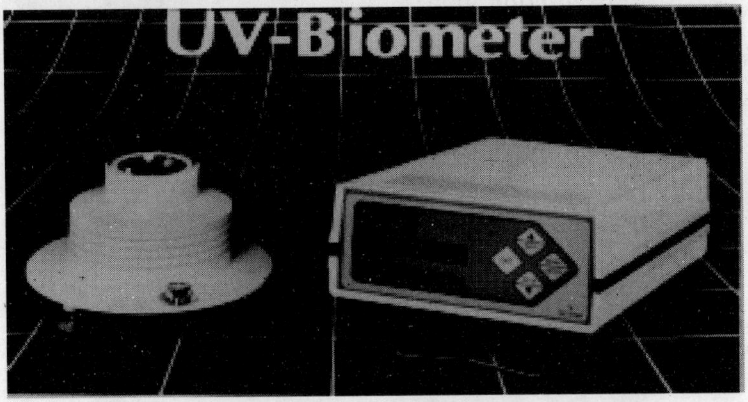

Fig. 2.18 UV biometer.

2.9.9 Total Ozone Mapping Spectrometer (TOMS)

It is the instrument used in satellite to measure the ozone levels in the atmosphere over the earth's surface.

2.9.10 Solar Backscatter Ultra Violet (SBUV) Instrument

It is another instrument used to measure the total ozone levels.

REFERENCES/SUGGESTED READINGS

Ahrens, C.D., *Meteorology Today—An Introduction to Weather, Climate and the Environment*, 6th ed., Brooks/Cole, California, 2000, p. 528.

Cosgrove, B., *Weather*, Dorling Kindersley, London, 1991, p. 64.

De Wit, C.T., 'Potential photosynthesis of crop surfaces', *J. Agric. Sci.*, 7: 141–149, 1959.

Ghadekar, A.E., *Meteorology*, Agromet Publishers, Nagpur, 2001, p. 251.

Gopalaswamy, N., *Agricultural Meteorology*, Rawat Publications, Jaipur, 1994, p. 153.

Lal, M. and Singh, J.K., 'Global warming and monsoon climate', *Mausam,* 52: 245–262, 2001.

Lutgens, F.K. and Tarbuck, E.J., *The Atmosphere*, Prentice-Hall, London, 1998, p. 434.

Mavi, H.S., *Introduction to Agricultural Meteorology*, Oxford & IBH, New Delhi, 1994, p. 281.

Pant, G.B. and Kumar, K.R., *Climates of South Asia*, John Wiley & Sons, Chichester, 1997, p. 320.

Petterson, S., *Introduction to Meteorology*, McGraw-Hill, NewYork, 1958, p. 327.

Riehl, H., *Introduction to the Atmosphere*, McGraw-Hill, New York, 1965, p. 365.

Rosenburg, N.J., *Microclimate: The Biological Environment*, John Wiley and Sons, New York, 1974, p. 300.

Trewartha, G.T., *An Introduction to Climate*, McGraw-Hill, New York, 1954, p. 395.

QUESTIONS

1. Define the following:
- (a) Solar constant
- (b) Blackbody radiation
- (c) Albedo
- (d) Emissivity
- (e) Reflectivity
- (f) Transmissivity
- (g) Absorptivity

2. Distinguish between
- (a) Aphelion and Perihelion
- (b) Shortwave and longwave radiation
- (c) Reflectivity and Albedo
- (d) Solstice and Equinox
- (e) Photoperiodism and Phototropism
- (f) Photosensitivity and Photo-insensitivity
- (g) Long-day and short-day plants

3. Write short notes:
- (a) Transfer of heat
- (b) Radiation laws
- (c) Radiation balance
- (d) Energy or heat balance

(e) Response of plants to radiation
(f) Depletion of solar radiation
(g) Radiation instruments

4. Fill in the blanks:
(a) ———is known as the fuel of atmosphere.
(b) At 300 K, λ_{max} is equal to ——— μ.
(c) λ_{max} = ———.
(d) ——— is the instrument used to measure UV radiation.
(e) ——— is the instrument used to measure illumination.
(f) A perfect absorber is the perfect emitter in case of ———.
(g) At ——— and ——— ranges absorption of solar radiation is maximum in leaf.
(h) $R_n = G + H +$ ———.
(i) ——— % is of incoming solar radiation is reflected by the earth surface.
(j) ——— is known as photosynthetic active radiation.

5. Match the following:

(a) Visible radiation	>3 × 10[10] μ
(b) Ultraviolet radiation	0.4–0.76 μ
(c) Radiowaves	0.4–10[–4] μ
(d) Electric waves	3 × 10[3]–3 × 10[10] μ
(e) Infrared radiation	10[–4]–10[–8] μ
(f) X-rays	0.76–3 × 10[3] μ

6. Elucidate the following:
(a) Radiations and heat balance between the earth, atmosphere and space
(b) Radiation distribution in crops
(c) Response of plants to solar radiation

3 TEMPERATURE

Temperature is a measure of the hotness of a body. If a water vessel is kept on flame and we monitor the state of water by dipping our fingers, we first feel mild warmth, then warmth and then heat. A stage comes further when we cannot measure hotness through our fingers if the water vessel continued to be on flame. It means something flows to water from flame, which is nothing but heat flow. In other words, heat flow results change in temperature, which can be measured by a device known as *thermometer*. The heat flow in unit time is known as *flux*. If heat flow is within the ground, it is known as ground heat flux, and it is sensible heat flux if it is from ground to air, just above the ground. The sensible heat is a part of the available heat energy at the earth surface and acts as a heat source to the atmosphere to raise its temperature. Heat does not rise. It makes the air layer just above the ground warmer and lighter when comes to contact with warm ground. The lighter air having low density rises if surrounded by air of higher density. Hence, the earth is the heat source to the atmosphere and not the sun, as the atmosphere is a transparent medium to incoming solar radiation.

3.1 STRATIFICATION OF ATMOSPHERE

The atmosphere has several properties with reference to incoming solar radiation and outgoing earth radiation, as seen in radiation balance. The properties vary depending upon atmospheric constituents with height and time. Though the properties of atmosphere are studied at the high school level, it is necessary to recollect them here since these are the causal elements for stratification of atmosphere. The properties of atmosphere are as follows:

(a) It is not visible, but it is felt if in motion.
(b) It has no smell or taste or colour.
(c) It occupies space and has mass.
(d) It conducts sound and supports combustion.
(e) It exerts pressure.
(f) It is transparent or translucent, and acts as a filter to almost all incoming solar radiation.
(g) It acts opaque to longwave radiation emitted from the earth.
(h) It diffuses heat and gases.
(i) It is denser over the earth surface and becomes thinner at high altitudes.

Due to various properties of the atmosphere that are mentioned above, there is a definite pattern in change of atmospheric temperature and pressure with height. Depending upon the variation in thermal property of atmosphere with height, the entire atmosphere is divided into different layers, which is nothing but stratification of atmosphere. It is also known as the thermal structure of atmosphere (see Fig. 3.1).

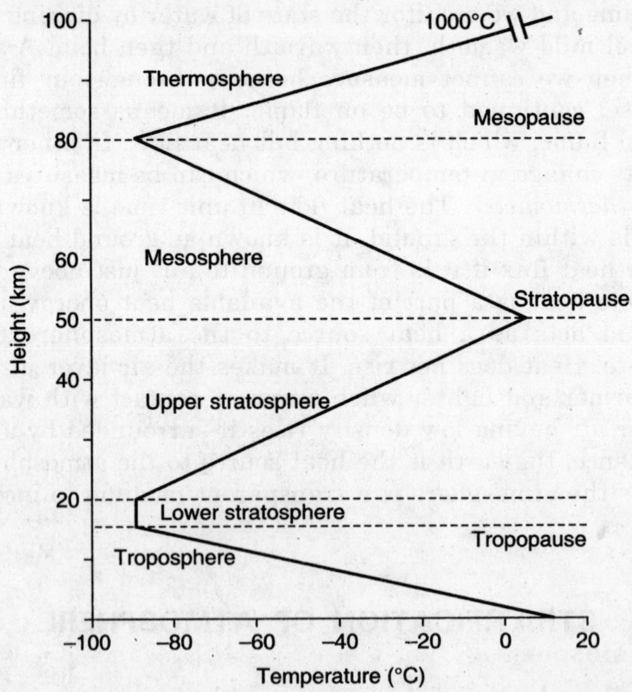

Fig. 3.1 Vertical distribution of temperature in the atmosphere.

3.1.1 Troposphere

'Tropo' is a Greek word meaning *vertical mixing*. A region where vertical mixing of air is predominant is known as troposphere, which is nearest to the earth's surface and has a more or less uniform decrease of temperature with increase in height. This is the region where weather phenomena is very well-marked and has aptly been described as the *weather-making layer of the atmosphere.* The top of the troposphere, where the decrease of temperature with height ceases, is the 'tropopause'. The height of the tropopause varies with latitude, having 16 km over the equator and 8 km over the poles (see Fig. 3.2). The height of tropopause also varies with seasons. The average temperature at the tropopause over the equator is the order of –80°C while it is –40°C over poles. The decrease in temperature with height is at the rate of 6.5°C/km in troposphere, known as the *normal lapse rate* or *environmental lapse rate.* There is an abrupt change in air temperature at the level of tropopause. The principal characteristics of troposphere are as follows:

(a) More or less uniform decrease in temperature with height
(b) Increasing wind speed with height
(c) Considerable moisture in the lower levels

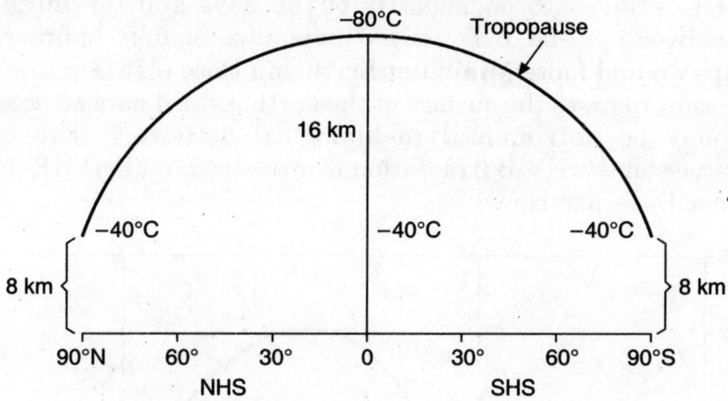

Fig. 3.2 Height of tropopause with latitude.

Appreciable vertical air movement, formation of clouds, and in general, the atmospheric phenomena are collectively called the *weather*. Within the troposphere, there are three layers viz. laminar or surface layer, frictional layer and free atmosphere. The above layers exist up to a height of 1000 metres. These are together also known as *Atmospheric Boundary Layer*, which is very important for micro- and phyto-meteorological studies.

3.1.2 Stratosphere

The stratosphere could be divided into two layers,viz. lower and upper stratospheres. The isothermal character (i.e., temperature is constant with height) of the lower stratosphere terminates at a height of approximately 20 km. It is replaced by a very gradual temperature increase with height. This region is called *upper stratosphere*. The stratosphere (both lower and upper) is just above the tropopause, extending to a height of 50 km where the stratopause occurs. The temperature in this region is more or less constant to start with, but increases with height in the upper stratosphere, reaching about earth's temperature at the stratopause. The increase in temperature is attributed to the absorption of UV radiation by the ozone layer, which has the maximum concentration at about 25 km. The air density is so much less that even limited absorption of solar radiation by the atmospheric constituents, notably ozone, produces a temperature increase. Unlike the troposphere, the stratosphere contains little or no water vapour. The human-made interventions in the industrial development leads to ozone depletion; thereby filtered UV radiation reaches the ground, resulting in various human, animal and crop diseases. The ozone loss has the potential to increase the incidence of skin cancer and cataracts in humans, harm some crops and interfere with marine life. The diurnal variation of UV-B radiation recorded at Vellanikkara (10°31′N) showed that the UV filtered radiation (>1MED) reaches the ground surface between 1000 hr and 1530 hr (see Fig. 3.3) on majortiy of the days and the duration of it varies between 3 and 6 hours. A maximum of four hours radiation, reaching two and more Minimum Erythema Dose (MED–a unit of UV-B filtered radiation), to the surface of the earth, is also noticed occasionally, which may be detrimental to biological activities. The biological effectiveness of the UV-B irradiation is measured in MED/HR (Minimum Erythema Dose per Hour).

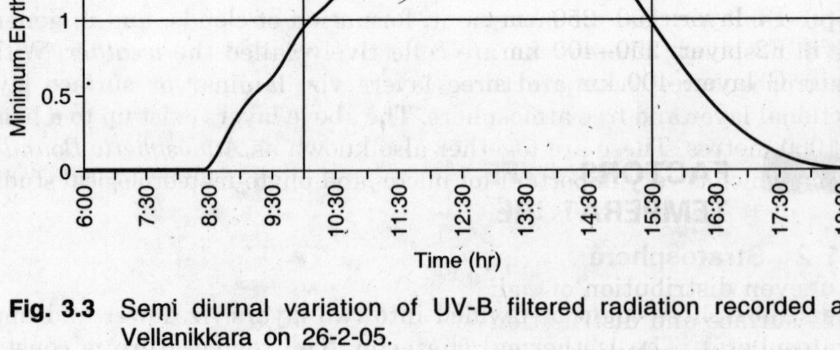

Fig. 3.3 Semi diurnal variation of UV-B filtered radiation recorded at Vellanikkara on 26-2-05.

3.1.3 Mesosphere

The mesosphere starts at the stratopause, roughly from 50 km to 80 km where the temperature again decreases with height.The coldest region in the atmosphere with temperature reaching the lowest value of nearly –95°C is at the mesopause at a height of about 80 km. The temperature decreases with height in this sphere, as the ozone in the stratosphere acts as a heat source; below and above this, the temperature decreases.

3.1.4 Thermosphere

The thermosphere is just above the mesopause (80 km) where the temperature increases again with height. The temperature in the thermosphere is of the order of 1000°C. By about 400 km, the density is so low that some of the gas molecules escape the earth's gravitational pull and fly into space. Within the thermosphere, three more layers viz. ionosphere, exosphere and heliosphere are noticed. In these regions the gas molecules (H and He) are in a highly ionised state.

3.1.5 Ionosphere

The ionosphere in atmosphere is seen at about 80 km onwards from the earth's surface. In this layer, the ionisation of the atmosphere begins to occur and extends up to 400 km, above which it is known as the *exosphere.* The first knowledge about the ionisation of atmosphere at such heights was known by means of radiowaves. Radiowaves are reflected in the ionosphere where ultraviolet radiations split up N_2 and O_2 into free positively charged ions of O^+ and N_2O^+. The characteristics of these layers are very important in radio communication. The ionosphere consists of the following ionised layers:

(i) D layer: 60–90 km
(ii) E layer: 90–130 km
(iii) Sporadic E layer: 130 km
(iv) E2 layer: 150 km
(v) F1 layer: 150–250 km
(vi) F2 layer: 250–400 km
(vii) G layer: 400 km and above.

3.2 FACTORS AFFECTING SURFACE AIR TEMPERATURE

The uneven distribution of radiation, the physiographic features of the earth's surface and distribution of water and land on the earth surface influence the distribution of global surface temperature. Of the factors effecting surface air temperature, latitude is the most important factor, followed by altitude and distribution of land and ocean surfaces. Mountains also play a role in deciding surface air temperature and water vapour between windward and leeward sides. The average global temperature is said to be constant at 15°C, provided the release of greenhouse gases into the atmosphere due to human-made interventions is controlled. The major factors are, namely, latitude, altitude, distribution of land and water, nearness to sea, topography (mountains), ocean currents, prevailing winds and clouds. All these influence the earth's temperature.

3.2.1 Latitude

The rotation and the revolution of the earth as well as the inclination of its axis control the duration and intensity of radiation. The surface air temperature at a particular location depends upon the total amount of radiation received by it. The amount of total radiation received at the equator and at the lower latitudes is more intense when compared to that of higher latitudes due to earth's position with reference to sun's rays. Due to the above reason, the surface air temperature decreases with latitude from the equator. For example, the mean monthly temperature at New York (42°N) is low round the year when compared to Singapore (1°N). The mean annual temperature range is high at higher latitudes when compared to equator, where the temperature is uniform throughout the year (Figs. 3.4(a) and (b)). The mean annual temperature at New York is around 12°C while it is 27°C at Singapore. Similar is the case within the country when compared between Thiruvananthapuram (8°29′N) and New Delhi (28°36′N). Thiruvananthapuram records a mean annual temperature of around 27°C while it is around 24 °C at New Delhi. The difference in temperature between latitudes is explained due to weak

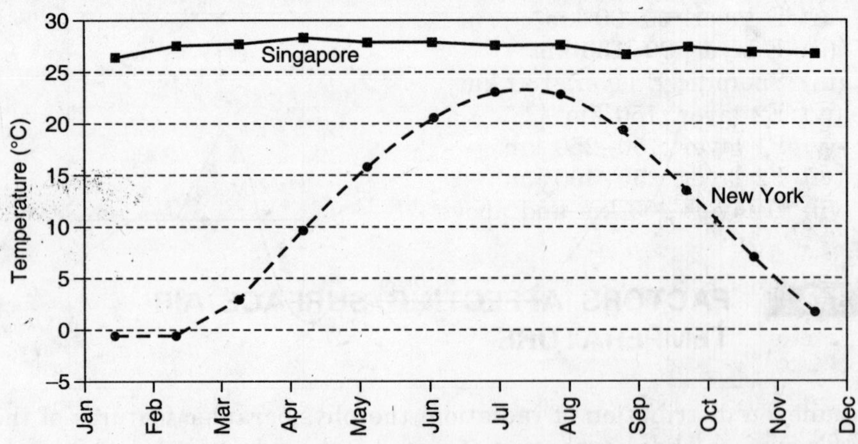

Fig. 3.4(a) Effect of latitude on mean monthly distribution of surface air temperature.

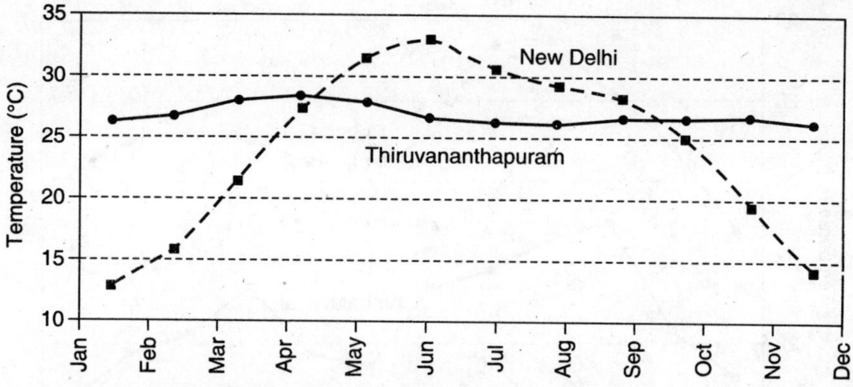

Fig. 3.4(b) Effect of latitude on mean monthly distribution of surface air temperature.

solar radiation as inclined sun's rays reach the earth surface at higher latitudes. The time of occurrence of maximum monthly mean temperature and minimum monthly mean temperature also depends on latitude of a place. For example, the coldest month is January in the northern regions of India while it is December in the south. Similarly, the warmest month is May in the south and June in the north across the country.

3.2.2 Altitude

The surface air temperature decreases with increasing altitude from the mean sea level as the density of air decreases. Since the density of air is less at higher altitudes, the absorbing capacity of air is relatively less with reference to earth's longwave radiation. This explains why the surface air temperature at higher altitudes is relatively low, though it is relatively nearer to the sun's rays. The following are the further reasons for decreasing temperature at high altitudes:

(i) The source of heat lies at the surface of the earth. High densities, high water vapour content and more dust particles near the surface layers absorb more earth's emitted radiation.

(ii) In the troposphere, the temperature decreases with height at the rate of 6.5°C/km.

The mean monthly surface air temperature over Ambalavayal at 1000 m above mean sea level (AMSL) is low round the year when compared to Pilicode (13 m AMSL). Similar is the case between Thrissur (25 m AMSL) and Pampadumpara (1000 m AMSL). The altitudinal difference in surface air temperature at these locations is due to the reasons explained above (Figs. 3.5(a) and (b)). A sudden drop in surface air temperature is noticed at all the four locations during the monsoon season (June–September). It is one of the characteristic features in the humid tropics like Kerala, where rainfall is high and relatively continuous.

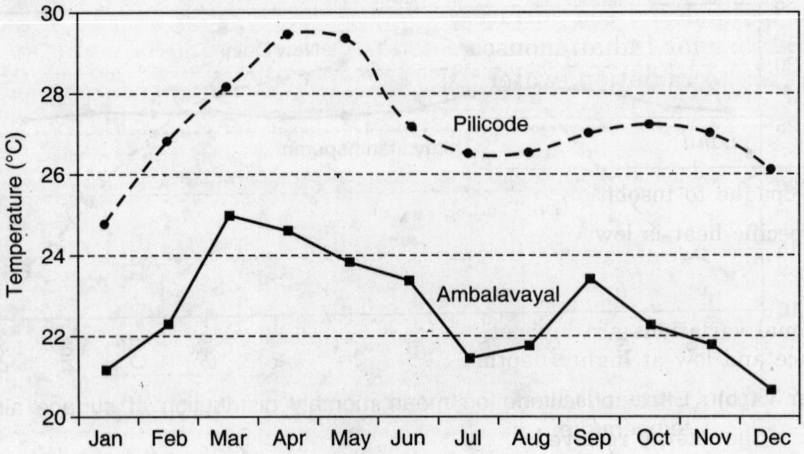

Fig. 3.5(a) Effect of altitude on monthly distribution of surface air temperature.

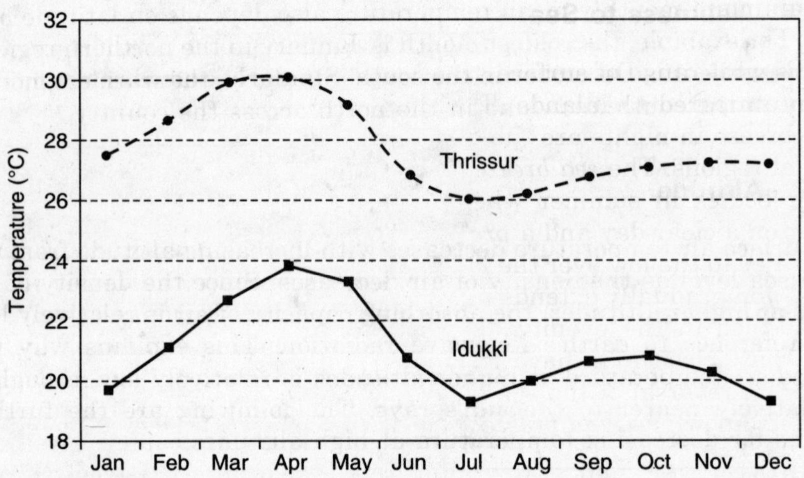

Fig. 3.5(b) Effect of altitude on monthly distribution of surface air temperature.

The maximum temperature suddenly falls due to heavy clouds and minimum temperature rises due to low re-radiation cooling from the wet surface of the earth. It reveals that the altitudinal effect on surface air temperature is significant across Kerala due to its unique topographical features. It is true in the north-eastern states also, when topography plays a predominant role in crop distribution.

3.2.3 Distribution of Land and Water

Land and water surfaces react differently to the insolation. Because of the great contrasts between land and water surfaces, their capacity for heating the atmosphere varies. Variations in air temperature are much greater over the land than over the water. The differential heating process between land and sea surfaces are due to their properties. It is one of the reasons for Indian monsoon. The properties of land and water with reference to radiation, water vapour and temperature are as follows:

Land	*Water*
It is opaque to insolation.	It is transparent to insolation.
Its specific heat is low.	Its specific heat is high (three times greater when compared to land).
Seasonal variations are high at surface and low at higher depths.	Seasonal variations are low.
Water vapour content is less.	Water vapour content is high.
Most of the energy received is lost through evaporation.	Energy loss is less through evaporation.

3.2.4 Nearness to Sea

The diurnal range of surface air temperature is less in coastal locations when compared to inlands. The sea breeze during daytime as well as land breeze in nights are the common phenomena in summer along the coastal regions. The *sea breeze* develops along sea coasts or large inland water bodies in summer when the land heats much faster than the water on a clear day and a pressure gradient is directed from high over the water to the low over the land. It begins offshore in the late morning hours and gradually extends inland to decrease afternoon temperatures. It subsides in the evening. Because of the sea breeze, areas on the immediate coast may have lower temperatures than a few kilometres inland. A reversal of the sea breeze may occur at night, a phenomenon which is known as *land breeze*. For example, the surface air temperature

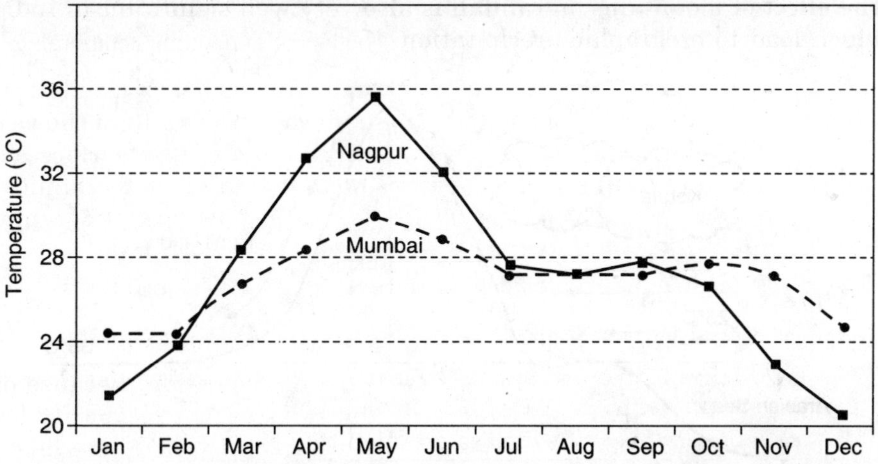

Fig. 3.6 Temperature difference between coastal and inland climates.

at Mumbai (coastal) varies between 24°C and 29°C while it is between 22°C and 35°C at Nagpur (inland). It indicated that the annual mean temperature range is high over the inland stations when compared to coastal stations. It is due to the presence of water vapour content in the atmosphere, which is relatively high along the coast and regulates the surface air temperature, thereby the annual mean temperature range (the difference between maximum monthly mean and minimum monthly mean temperatures) is less at Mumbai (5°C) while it is higher at 13°C over Nagpur (see Fig. 3.6). The maximum monthly mean temperature is noticed in May at both the locations while it is minimum in December. The difference in temperatures between the two locations is attributes to continentality factor.

3.2.5 Mountains

Air at the top of mountain makes little contact with the ground and is therefore cold, while in the valley at the foothills it makes a great deal of contact and is therefore warm. That is, the lower region of the earth's atmosphere is relatively warmer when compared to hillocks. Vertical displacements of air currents can be seen if air from the sea meets an obstacle along the way in the form of a mountain chain (see Fig. 3.7). As the air mass ascends, its temperature falls with height due to adiabatic cooling and thus condensation of water vapour, which leads to formation of clouds and precipitation. At the same time, it creates high temperature and air becomes dry while descending along the mountain slopes in the leeward side. Hence, mountains play significant role in temperature variation. It is one of the reasons why Kerala's low surface air temperature and humid air lead to high surface air temperature and dry weather over Tamil Nadu during monsoon, which falls in the leeward side and gets very low amount of rainfall. The reason for temperature variation due to mountains is also explained in the Chapter 7 on precipitation. The effect of mountains on rainfall is also very well significant in India, which lead to orographic precipitation.

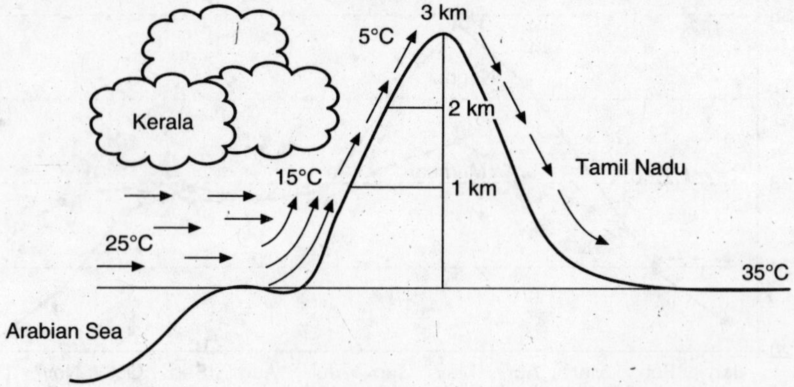

Fig. 3.7 Influence of mountains on surface air temperature.

3.2.6 Ocean Currents

The energy received over the ocean surface is carried away by the ocean currents from the warm areas to cool areas. This results in temperature contrast between the equator and polar regions. The warm currents always move from the equator to poles. The occurrence of El Nino is due to change in sea surface temperature between two oceanic regions over the globe.

3.2.7 Prevailing Winds

Winds can moderate the surface temperature of the continents and oceans. In the absence of winds, we feel warm in hot climates. At the same time, the weather is pleasant if wind blows.

3.2.8 Cloudiness

The amount of cloudiness affects the temperature of the earth's surface and the atmosphere. A thick cloud reduces the amount of insolation received at a particular place and thus the daytime temperature is low. At the same time, the lower layers in the atmosphere absorb earth's radiation. This results in increasing atmospheric temperature during night. That is why cloudy nights are warmer. This is common in the humid tropical climates.

3.3 TEMPERATURE INVERSION

If air temperature increases with height, it is known as temperature inversion. Sometimes, temperature inversion happens due to the following reasons:

(a) Earth's surface losses heat more than it gains.
(b) Cold air from hill tops and slopes tend to flow downward and replaced by warm air.
(c) Cold air masses replace warm air masses when air masses with different temperatures come together.

In crops also, temperature inversion is noticed due to active surface of crop during its maximum growth. The cropped surface becomes secondary heat source during daytime and this results in high temperature at the crop canopy. This phenomenon mostly depends on the nature of crops.

3.4 DIURNAL AND MONTHLY VARIATIONS OF SURFACE AIR TEMPERATURE

The minimum air temperature occurs at about sunrise, after which there is a constant rise till it reaches the maximum level. The maximum air temperature is recorded between 1300 hr and 1400 hr although the maximum solar radiation level is attained at noon. A steady fall in temperature till sunset is noticed after it attains the maximum. Thus the daily march displays one maximum and one minimum (see Fig.3.8). The difference between the two is called the *diurnal range of air temperature.* The diurnal range of air temperature is more on clear days, while cloudy weather sharply reduces daily amplitudes. The diurnal range of temperature is also influenced by soils and their coverage in addition to seasons. Addition of daily maximum and minimum temperature divided by two is nothing but daily mean or average temperature. That is,

Daily mean temperature

$$= \frac{\text{Daily maximum temperature + Daily minimum temperature}}{2}$$

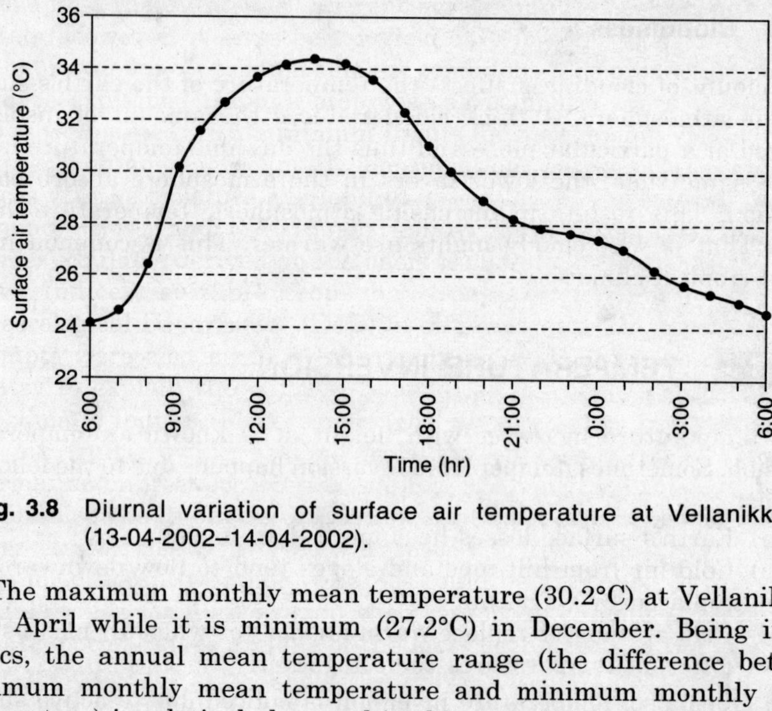

Fig. 3.8 Diurnal variation of surface air temperature at Vellanikkara (13-04-2002–14-04-2002).

The maximum monthly mean temperature (30.2°C) at Vellanikkara is in April while it is minimum (27.2°C) in December. Being in the tropics, the annual mean temperature range (the difference between maximum monthly mean temperature and minimum monthly mean temperature) is relatively less and in this case, it is only 3°C. However, it is relatively more when compared to that of the equatorial belt, as seen in the case of Singapore and New York. It reveals that the annual mean temperature is low with increase in latitude.

3.5 ISOTHERMAL CHARTS

The spatial or horizontal distribution of surface air temperature on a regional map can be brought out by drawing lines connecting places with equal values of air temperature. Such lines are called *isotherms.* The charts showing isotherms of different locations over a region or state or country or world are known as isothermal charts. From the isothermal charts, one can know the surface air temperature of a given region. In one such country, the data compiled by the weather bodies (see Fig. 3.9) shows the variation in mean annual surface air temperature from 20°C in the extreme north to 27°C in the south, excluding the hilly areas. The figure also clearly indicates that the temperature distribution is influenced by latitude. The mean monthly temperature over India in January varies between 5°C in extreme north and 25°C in the extreme south. In contrast, the mean monthly temperature during July is high in north.It varies between 25 and 30°C except in hill areas across the country.

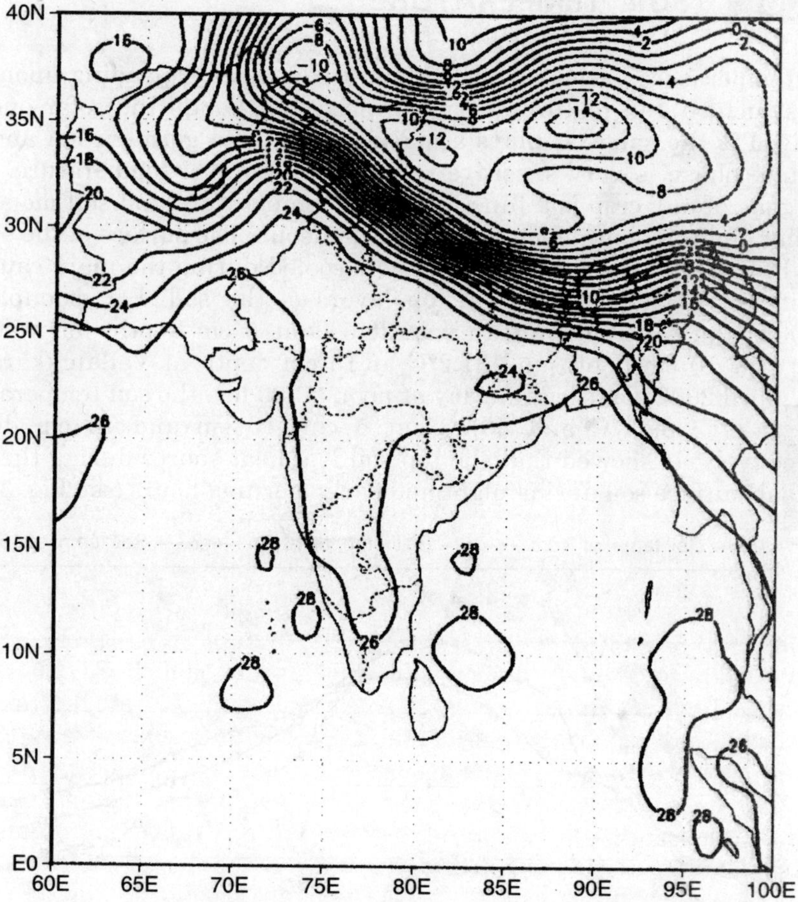

Fig. 3.9 Mean annual surface air temperature.

The global isotherms generally run east-west along latitude parallels following the amount of solar radiation received. In general, the highest temperatures are in the tropical interiors and the lowest temperatures in the polar interiors. The annual temperature range is higher (10 to –40°C) at higher latitudes and significantly low at equator (25–27°C). In January, the temperature range is between 10 and –41°C and in July, it is between 36 and –1°C.

From the isothermal charts, it is revealed that the mean annual temperature over coastal belt of Kerala varies from 25.5 to 27.5°C while the central part and the Palakkad gap experiences between 27.5 and 29.5°C. In extreme east and hill areas, the mean annual temperature is between 17.5 and 21.5°C. In remaining parts of Kerala, it varies between 21.5 and 25.5°C.

3.6 SOIL TEMPERATURE

Soil temperature varies among different soil types depending upon the soil structure and texture at a given location though the solar energy received is the same. It plays a major role in seed germination and its juvenile phase is very sensitive to variations in soil temperature. The root zone of any crop is a function of soil temperature and soil moisture if other environmental factors are homogeneous. The surface of the earth gets heated up during the day and gets cooled during the night causing significant diurnal changes in top layers of the soil. For example at 0800 hr, the soil temperature was 28.5°C at 5 cm depth, followed by 29.9°C at 10 cm depth and 31.2°C at 15 cm depth at Vellanikkara on 13th April 2002. On the same day at noon (1300 hr), the soil temperature was 44.4°C, 39.7°C and 36.7°C at 5 cm, 10 cm and 15 cm depth respectively. It showed that the top soil is a heat source during the day and sub-surface soil during night and early morning hours (see Fig. 3.10).

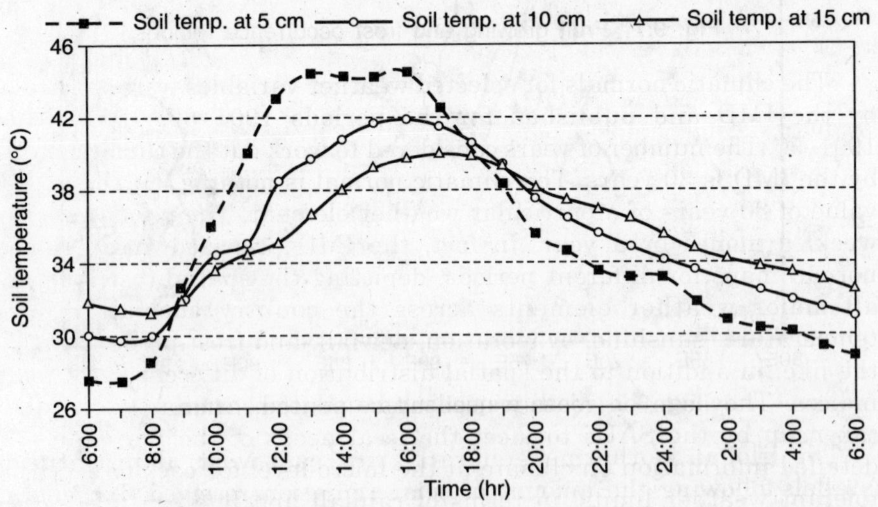

Fig. 3.10 Hourly soil temperature at different depths (13-04-2002–14-04-2002) at Vellanikkara.

There is also a lag time between the maximum and minimum soil temperature with depth. The maximum soil temperature was 44.4°C (5 cm), 42°C (10 cm) and 40°C (15 cm) at 1300 hr, 1600 hr and 1700 hr, respectively. The minimum soil temperature was 27.5°C (5 cm), 29.8°C (10 cm) and 31.2°C (15 cm) at 0600 hr, 0700 hr and 0800 hr respectively. The amplitude in diurnal soil temperature was high (27.5 to 44.4°C) at the surface and low (31.2 to 40.0°C) at relatively higher depths (15 cm). At higher depths, the diurnal amplitude is relatively marginal while significant difference is noticed seasonally (see Fig. 3.11). A gradual increase in soil temperature was noticed at all the depths (2.5 to 90 cm) from the surface during the four seasons, viz., monsoon, post monsoon, winter and summer.

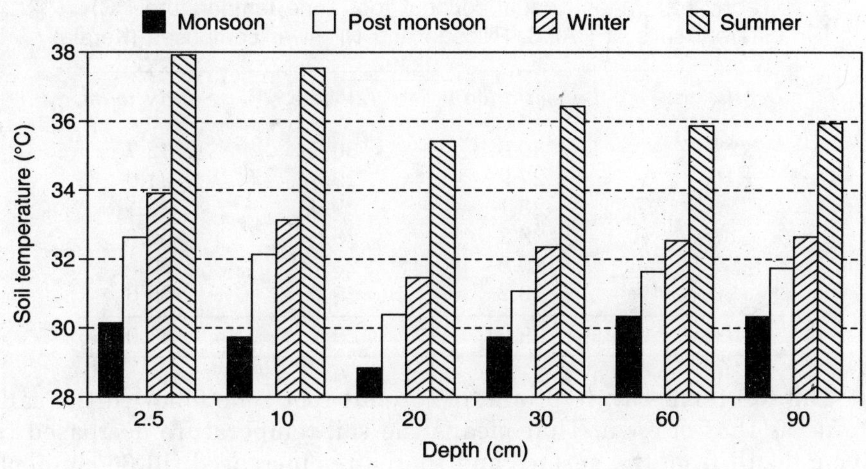

Fig. 3.11 Seasonal variation in soil temperature at Pilicode in different detphs.

3.6.1 Thermal Structure of Soils

The range in soil temperature was high (31.5–33.6°C) in open when compared to that of coconut root zone (29.0–30.7°C). The difference in soil temperature between open and coconut root zone varied from 2.5°C to 3.8°C. It is quite interesting to note that the difference in soil temperature between coconut and open was higher at 10 cm depth (see Table 3.1).

The soil temperature profile as a whole showed an increase of 2.9°C in open when compared to that of coconut root zone. It also revealed that the coconut root zone temperature in laterite is intermediary when compared to that of the coconut root zone in lateritic loam and sandy loam (see Table 3.2).

Table 3.1 Mean annual (1990–93) soil temperature (°C) in coconut root zone and open at different depths under laterite soils at RARS, Pilicode (Kerala)

Depth (cm)	*Open*	*Coconut*	*Difference*
2.5	33.6	30.7	2.9
10	33.1	29.3	3.8
20	31.5	29.0	2.5
30	32.4	29.6	2.8
60	32.6	29.7	2.9
90	32.6	29.7	2.9
Mean	32.6	29.7	2.9

Table 3.2 Mean annual coconut root zone temperature (°C) in different soils at RARS, Pilicode and Nileswar campuses (Kerala)

Depth (cm)	*Lateritic loam*	*Laterite*	*Sandy loam*
2.5	30.6	30.7	32.7
10	27.9	29.3	31.0
20	28.1	29.0	30.1
30	29.2	29.6	30.6
60	29.2	29.7	30.9
90	29.2	29.7	30.6
Mean	29.0	29.7	31.0

The soil temperature profile in coconut root zone followed the same profile as that of open. That means the soil temperature decreased till 20 cm depth from the surface and thereafter increased till 60 cm depth and followed constantly further. It appears that a sort of troposphere and stratospheric thermal structure is followed in soils also like in atmosphere (see Fig. 3.12(a) and (b)).

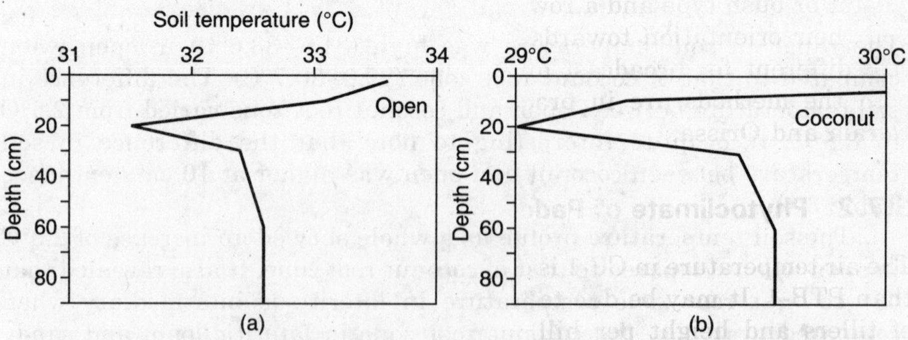

Fig. 3.12 Mean annual soil temperature (°C) in coconut rot zone and in open at different depths under laterite soil at RARS, Pilicode.

Though the difference in soil temperature is significant and varies widely among different soils, its direct effects on root zone spread and its thickness are yet to be understood, since it can be regulated or manipulated to a considerable extent through irrigation scheduling. Nevertheless, the root zone temperature may have direct adverse effect on plant growth, development and yield under rainfed conditions. Studies in this direction are scant in India.

3.7 VEGETATION AND AIR TEMPERATURE

Natural vegetation or crop cover has considerable influence on the distribution of temperature in the layer of air near the soil. If the ground surface is totally or partially covered by crops during its maximum growth stage, the incoming solar radiation is intercepted due to their canopy and the amount received at the ground surface is relatively less. The crop surface itself acts as an active surface just like the ground surface, the temperature profile is changed within the crop surface, and it is different from the thermal structure in the open, very near the ground. A temperature inversion is noticed in crops, as the canopy acts as a secondary heat source. In daytime the temperature of the plant cover is higher, and lower at night when compared to that of the surrounding air. The temperature of plant cover depends on the height of the plant, the stand density and the solar elevation. If tall trees they cover the ground, the lowest summer temperature occurs at the surface of the soil during the day.

3.7.1 Phytoclimate

It is a modified form of microclimate produced due to influence of crops. It is controlled by the structure of plant cover viz. plant height, number of leaves or tillers, leaf area, the distribution of plants (pointed crop, cluster or bush type and a row stand) and spacing given between plants and their orientation towards sunlight. Of course, the phytoclimate is still different for broadcast and transplanted crops. In case of paddy, both the methods are in practice in some paddy growing states like Kerala and Orissa.

3.7.2 Phytoclimate of Paddy

The air temperature in CC-1 is comparatively more throughout the canopy than PTB-4. It may be due to the fact that the growth of CC-I (number of tillers and height per hill) may be less, which allows more direct radiation to the surface when compared to that of PTB-4. The decrease in vapour pressure with height in the crop indicates that water vapour flux is consistently upward within the canopy and more stable than in open. The above phenomenon is specific in case of paddy, as it is grown under puddled conditions.

3.7.3 Phytoclimate of Coconut

The vapour pressure is always high inside coconut when compared to that outside. Higher air temperatures may be noticed in coconut in morning hours while less in the afternoon when compared to that outside. A second maximum can be seen in temperature and vapour pressure at canopy level in coconut when compared to that in the open. It is true for almost all the crops.

3.7.4 Phytoclimate of Sugarcane

Temperature inversion (increases with height) has been noticed inside sugarcane crop till 12' height during its maximum growth, as the canopy acts as the active surface. The concentration of moisture in the sugarcane field also showed a secondary maximum at about 12' height from the surface, the height at which maximum foliage is seen and therefore maximum transpiration occurs (IMD, 1957).

3.8 AIR TEMPERATURE AND CROPS

The surface air temperature is one of the important variables, which influences all stages of crop during its growth, development and reproductive phase. A short-duration crop becomes medium or long duration crop depending upon its environmental temperature under which it is grown. Most of the crops have upper and lower limits of temperature, below or above which they may not come up. Also, the crop productivity is related to temperature if other crop-related environmental factors are not limiting. The influence of temperature on various phases of crops could be explained better through concepts like cardinal temperature points, Hopkins bioclimatic law, thermoperiodism and growing degree days, which are explained in brief in preceding sections.

3.8.1 Cardinal Temperature Points

Every physiological plant process undergoes a well-defined range of limits of temperatures. There must be at least a minimum essential temperature to the initiation of activity; the activity will proceed at the highest rate when the optimum temperature is reached; eventually activity will come to a close at the maximum temperature point. These three points are known as the *cardinal temperature points*. They may vary widely with the age or stage of development of the plant, and they vary considerably for different species. Wilsie (1962) gave a table of cardinal temperature points for germination of crop seeds (based on data collected by Haberlandt and reported by Graphe) which provides some idea of the differences among crops as to their temperature requirements (see Table 3.3).

Table 3.3 Cardinal temperature points for germination of certain crop seeds

Crop	Cardinal points in degree Fahrenheit at indicated temperatures			Days required for germination			
	Minimum	*Optimum*	*Maximum*	40°F	54°F	60°F	66°F
Red clover	34	86	99	7.5	3.0	1.75	1.0
Alfalfa	34	86	99	6.0	3.75	2.75	2.0
Hemp	34–36	95	113	3.0	2.0	1.0	1.0
Peas	34–36	86	95	5.0	3.0	1.75	1.75
Rye	34–36	77	86	4.0	2.5	1.0	1.0
Vetch	34–36	86	95	6.0	5.0	2.0	2.0
Flax	35–37	77	86	8.0	4.5	2.0	2.0
Timothy	37–39	79	86	–	6.5	3.25	3.0
Wheat	39–40	77	86–90	6.0	3.0	2.0	1.75
Barley	39–40	68	82–86	6.0	3.0	2.0	1.75
Oats	39–41	77	86	7.0	3.75	2.75	2.0
Carrot	39–41	77	86	–	6.75	4.25	3.25
Sugar beet	39–41	77	82–86	22.0	9.0	3.75	3.75
Lentils	39–41	86	97	6.0	4.0	2.0	1.75
Maize	46–50	89–95	104–111	–	11.25	3.25	3.0
Sorghum	46–50	89–95	104	–	11.25	4.75	4.0
Rice	50–53	86–90	97–101	–	–	–	–
Tobacco	55–57	82	95	–	–	.0	6.25
Pumpkin	54	89–93	104	–	–	10.75	4.0
Melon	54–59	95	104	–	–	15.0	17.0

3.8.2 Hopkins' Bioclimatic Law

Hopkins (1938) attempted to express the importance of latitude, longitude and altitude in the distribution and rate of development of plants by means of a bioclimatic law. It may be stated as follows: "A biotic event in North America will, in general, show a lag of four days for each degree of latitude, five degrees of longitude and 400 feet of altitude, northward, eastward and upward in spring and early summer". In case of mango, the India Meteorological Department also established that there is a delay in flowering from south to north of India, which generally followed Hopkins' bioclimatic law (IMD, 1957). In the case of cashew also, all the biotic events of cashew in the three test varieties delayed with increase in latitude and altitude from south to north and northeast of Kerala. It is evident that the delay in biotic events such as bud break, flowering, harvest and the total crop duration of cashew depend on latitude and altitude if the distribution of rainfall is uniform and genotype is alike under better crop management situation in rainfed conditions (Rao, 2002). However, there is a difference in number of days delayed in cashew flowering at each degree of north latitude while the influence of altitude on time of cashew flowering is similar as stated by Hopkins (see Fig. 3.13). It also revealed that the influence of latitude and altitude reflected on the surface air temperature, which led to delay in biotic events of cashew. Such studies, if undertaken in fruit crops, will be of immense use in selecting suitable variety, manipulation of flowering and prophylactic crop protection measures.

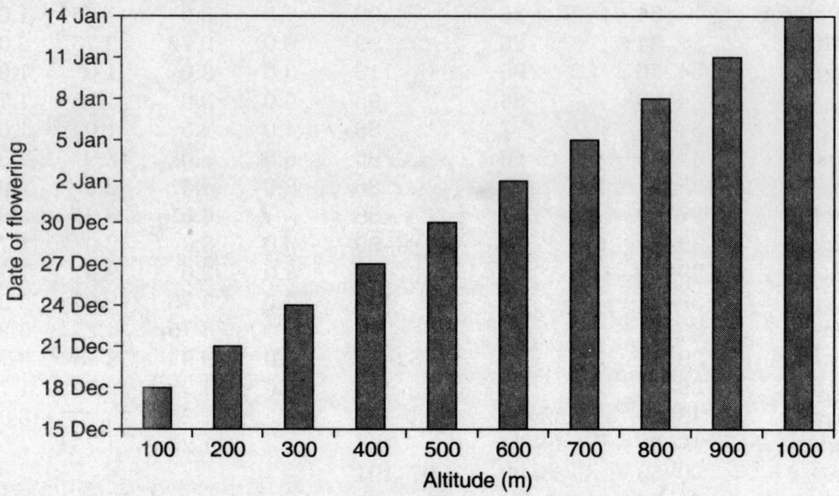

Fig. 3.13 Influence of altitude on cashew flowering.

3.8.3 Thermoperiodism

The response of plants to diurnal or seasonal fluctuation in temperature is known as thermoperiodism. A number of physiological processes, including germination, stem elongation, fruiting, floral development and increase in frost-hardiness may proceed most satisfactorily under a rhythm of alternating temperatures. The tomato is found to be sensitive to night temperatures. The optimum temperature may be 26–30°C in case of young seedlings. As the plant grows older, it favours lower temperature and the optimum night temperature may be a range of 13–20°C, depending on the variety. The fruit set is abundant with day light temperature of 26.5°C and night temperature of 15–20°C. A high optimum daytime temperature is effective because it increases the photosynthetic rate. Tomatoes grow mainly at night and with temperatures above 18°C. Translocation of sugars becomes a limiting factor. Actually, little fruit sets in the field with night temperatures above 22°C or below 10°C. In southern California, in winter and early spring, night temperatures are often below 10°C, but late afternoon temperature may be between 15 and 20°C. But covering plants at 3 pm, an optimum dark (night) temperature exists for a few hours and fruit set is improved considerably. In India too, more tomatoes could be produced per plant and fruit size is high if harvested during winter, when compared to that of other warm seasons during which temperature may be a limiting factor.

3.8.4 Growing Degree Days (GDD)

The concepts of growing degree days, heat unit and thermal unit was developed assuming that there is a direct and linear relationship between crop growth and temperature. The assumption was that the growth of plants is dependent on the total amount of heat to which it is subjected during its lifetime. The heat unit or growing degree days are defined as follows:

$$GDD = [(T_{max} + T_{min})/2] - T_t$$

where, T_{max} is the maximum temperature in °C, T_{min} is the minimum temperature in °C and T_t is minimum threshold or base temperature.

The minimum threshold temperature is the temperature below which no growth takes place. It varies from 4.5 to 12.5°C for different crops. It is 5°C for rice, 8°C for millets (jowar, ragi and bajra), 10°C for maize, 14–15°C for cotton and 13°C for coconut and cashew. The degree days concept can be used for forecasting the duration of crop and its harvest dates. The crop yields also could be predicted, based on crop-specific thermal units.

In the case of okra, the fruit yield is directly related to the number of heat units. The fruit yield is high if the number of heat units is high irrespective of time of sowing and vice versa (see Fig. 3.14). It clearly indicates that okra may produce better fruit yield if sowing is taken up from February to April and from September to October. However, more studies are needed in this direction at different locations before to draw such conclusions. The cashew yield and kernel weight are influenced by

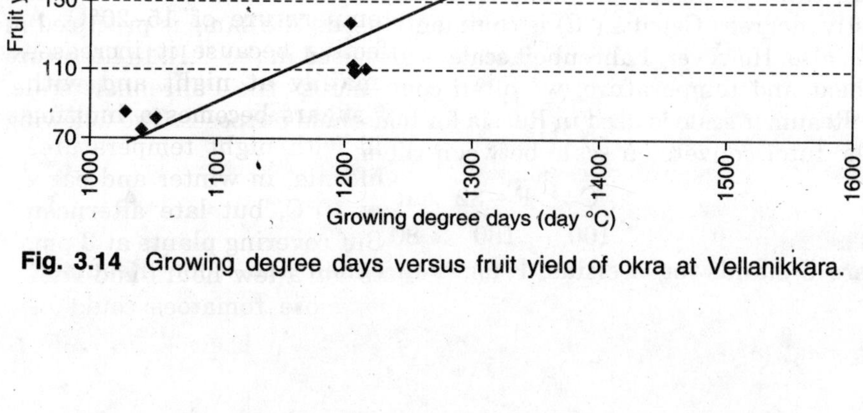

Fig. 3.14 Growing degree days versus fruit yield of okra at Vellanikkara.

the number of heat units if other meteorological factors are not in hindrance with the nut development and yield (see Fig. 3.15). The early varieties need less number of heat units while more in the case of late varieties in cashew. The second phase of nut development in coconut is sensitive to heat units, since it is the critical phase which finally decides the nut size in coconut. If the number of heat units during the second phase of nut development (4–7 months old nut) is more than 2100 day °C, the final nut size may be diminished. It is one of the seasons for low coconut size in non-traditional areas like in Tamil Nadu, Karnataka and Andhra Pradesh. It clearly indicates that the number of heat units influence crop duration and yield to a large extent. The genetic coefficients are derived for each crop on the basis of thermal units, and used in crop growth simulation models which are widely operational for decision making.

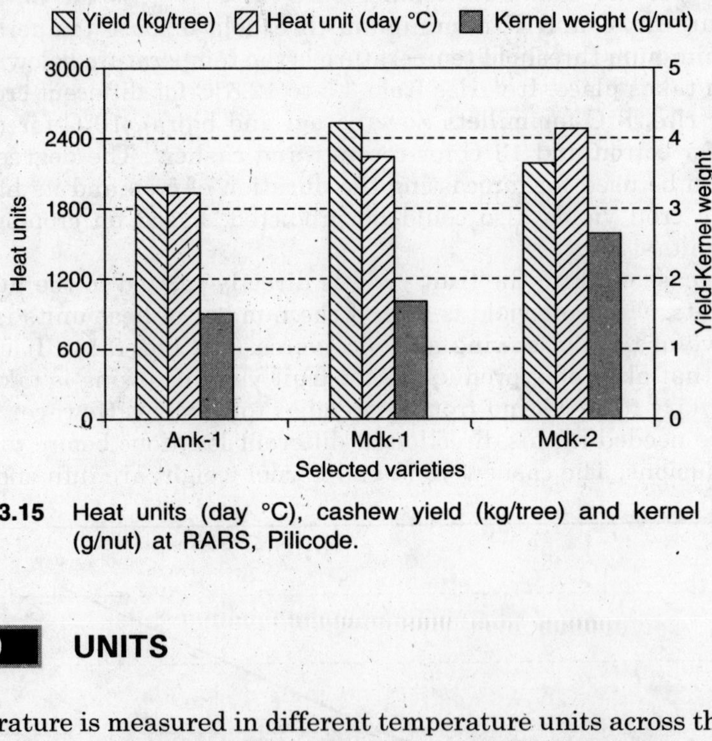

Fig. 3.15 Heat units (day °C), cashew yield (kg/tree) and kernel weight (g/nut) at RARS, Pilicode.

3.9 UNITS

Temperature is measured in different temperature units across the world. Mostly, degrees Celsius (°C) is commonly used. The same is practiced in India also. However, Fahrenheit scale is also used in the U.K. In literature on heat and temperature, we often come across the Fahrenheit scale. The Reaumer scale is used in Russia for household purposes. The following is the inter-conversion scale between them:

$$\frac{C}{100} = \frac{F - 32}{180} = \frac{R}{80}$$

where C stands for °Celsius, F for °Fahrenheit and R for °Reaumer.

Degrees Celsius is same as degrees Centigrade. The melting point of ice is taken as 0 in Celsius scale and Reaumer scale while 32 in Fahrenheit scale. The boiling point of water is taken as 100, 212 and 80 in Celsius, Fahrenheit and Reaumer scales, respectively. The scales are equally graduated between melting and boiling points. In case of Celsius, it is between 0 and 100 while it is between 32 and 212 in case of Fahrenheit. It is between 0 and 80 in case of Reaumer. That is how the inter-conversion scale is obtained. The inter-conversion scale between Celsius and Fahrenheit could be written as follows:

$$C = (F - 32) \times \frac{5}{9}$$

Both scales give equal value at –40°C; –40°C = –40°F. In absolute scale, 0°C becomes 273 A; it means 20°C is equal to 20 + 273 = 293 A. In thermodynamic scale, the absolute scale is known as the Kelvin scale. Hence, degrees Absolute (A) is equal to degrees Kelvin (K).

3.10 INSTRUMENTS

The instruments used for measuring temperature are as follows:

3.10.1 Single Stevenson Screen

It is a wooden box specifically designed for housing 4 thermometers which record the surface air temperature in °C (see Fig. 3.16). They are maximum, minimum, dry and wet bulb thermometers. The box has a provision inside to place the thermometers separately such that each thermometer has its own position in the screen. The wooden box is placed at 4 feet height from the surface. It has uniform design of double

Fig. 3.16 Single Stevenson screen.

louvre from all the 4 sides. This provision admits uniform flow of air to the thermometer bulbs. The box is white painted and also has double roof. The air space between the double roof and white painting prevent direct heating from the intensity of sunlight. It has 4 wooden legs to support the box. The legs are fixed in the soil such that the concrete material used for fixing legs should not be exposed and seen from outside.

3.10.2 Maximum Thermometer

The maximum thermometer is a mercury-in-glass thermometer which has a constriction (small bend) just below the lower graduation (see Fig. 3.17). It is placed on the upper platform of the screen horizontally which is earmarked for the maximum thermometer. It should be rested at an angle of 2° with the bulb to the lower side. Once the mercury in thermometer reaches its maximum level with increase in air temperature, the mercury does not come back due to the constriction when surface air temperature starts falling. The purpose of the constriction is to allow mercury from the bulb when temperature rises and not to allow mercury down to bulb when temperature decreases. After taking the reading, the thermometer should be shaken gently till it reaches the reading level of dry bulb thermometer by positioning it vertically in the right hand. That is why the maximum thermometer gives the maximum temperature recorded during the last 24 hours. The maximum temperature is generally obtained between 1300 and 1500 hours depending upon season. The principle of the maximum thermometer is similar to that of the clinical thermometer.

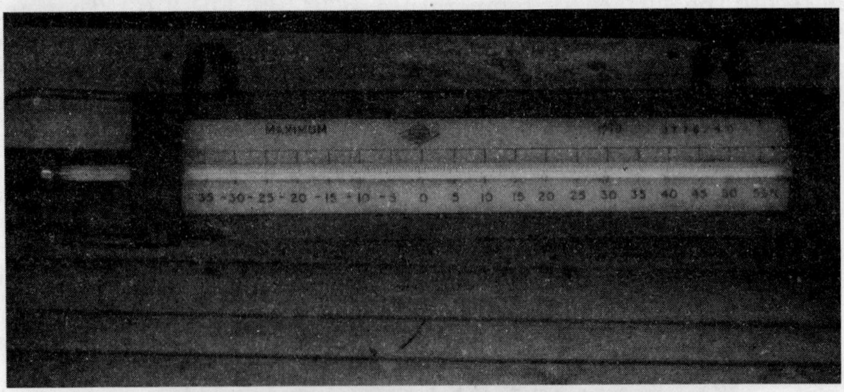

Fig. 3.17 Maximum thermometer.

3.10.3 Minimum Thermometer

The minimum thermometer is placed on the lower platform of the screen just below the maximum thermometer (see Fig. 3.18). It should be rested at an angle of 2°, having the bulb on the lower side as in the case of maximum thermometer. The main difference between the maximum and minimum thermometers is that in the former, mercury is used and in the latter, alcohol. It is a thermometer with lightweight movable index in its capillary tube. When temperature rises the alcohol in capillary tube alone expands, leaving the movable index (blue glass pin). The alcohol pushes the index back towards the bulb due to its surface tension when temperature decreases. The right side of the position of the glass index is the reading of the minimum thermometer. After taking the reading of the minimum thermometer, the minimum thermometer should be tilted towards right side such that the glass pin touches the surface of the alcohol column. After adjustments, the minimum thermometer reading and the dry bulb thermometer reading should be the same within the accuracy of ± 0.5°C. The minimum thermometer gives the minimum temperature recorded during the last 24 hours in °C. It generally occurs just before sunrise.

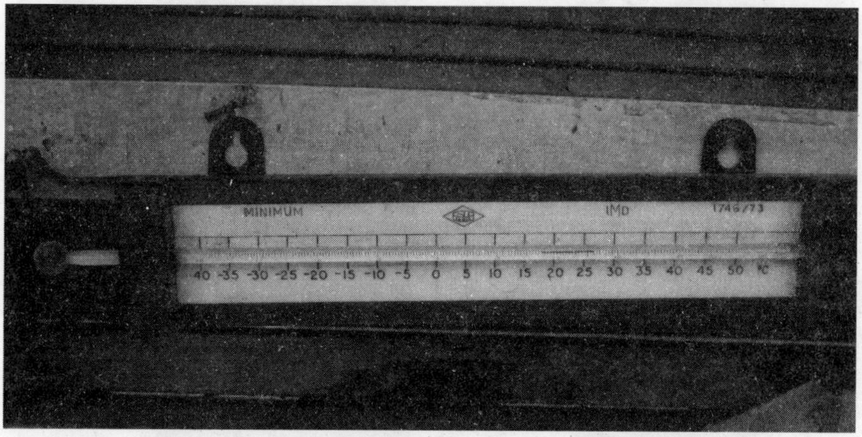

Fig. 3.18 Minimum thermometer.

3.10.4 Dry Bulb Thermometer

The dry bulb thermometer consists of a bulb at one end and is connected with a tube in which the mercury is filled (see Fig. 3.19). It is placed in the Stevenson screen vertically towards left side of the inside box.It gives instantaneous air temperature in °C. Air temperature at a given point of time can be obtained by reading this thermometer.

3.10.5 Wet Bulb Thermometer

The dry bulb thermometer together with wet bulb thermometer is known as the *station hygrometer*. This is the best instrument to measure relative humidity. The construction of wet bulb thermometer is same as that of dry bulb thermometer, except that the bulb of the thermometer is wrapped with a muslin cloth using cotton thread (see Fig. 3.20). The other end of the strings of the cotton thread is dipped in a small and narrow mouth glass bottle having distilled water. This facility is provided to see that the bulb of the wet bulb thermometer is always wet due to the capillary action of the cotton thread dipped in water. It is also placed on a vertical platform seen on the right side in the box.

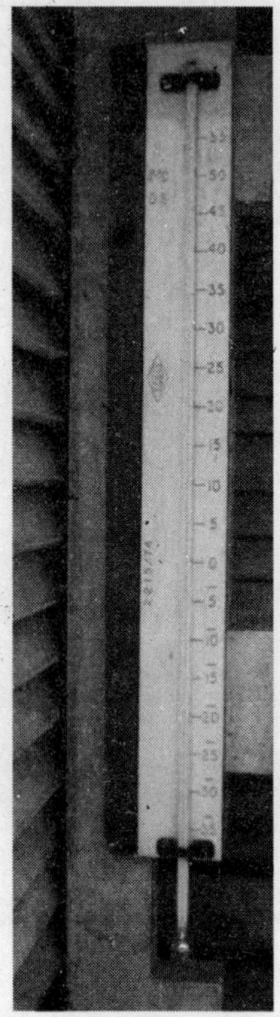

Fig. 3.19 Dry bulb thermometer.

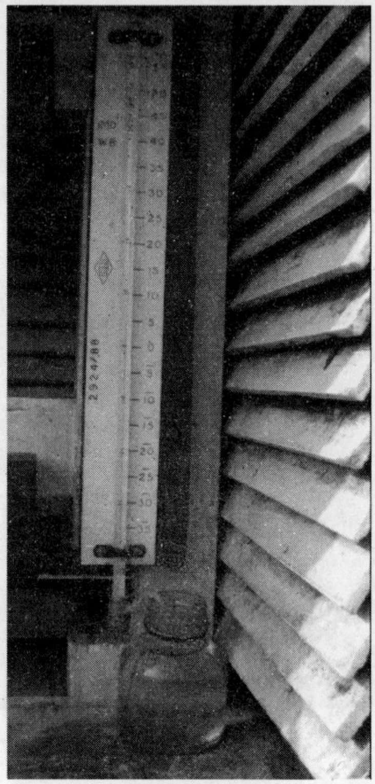

Fig. 3.20 Wet bulb thermometer.

The dry bulb and the wet bulb readings are noted down to obtain relative humidity directly using hygrometric table. The relative humidity is recorded in percentage.

3.10.6 Infrared Thermometer

The state-of-the-art in infrared technology is applied to non-contact surface thermometry. It is a non-contact type of thermometer which is different from conventional types of thermometer. It is a useful tool in agricultural sciences for irrigation scheduling, crop yield predictions and insect pest and disease forecasting. The infrared thermometer measures the temperature of the surface which is exposed to it (see Fig. 3.21). For example, if the thermometer is positioned towards ground it gives surface temperature, and crop temperature if it is exposed to crops. The instrument gives differential temperature (DT), which is nothing but the difference between crop temperature and its ambient temperature if the thermometer is exposed to crop surface. From the above, the crop water stress index (CWSI) could be worked out. This can be used as a tool for irrigation scheduling. AG multimeter is another type of infrared thermometer, which is also used to measure differential temperature in addition to direct solar radiation and vapour pressure deficit. The instruments look alike (of gun type).

Fig. 3.21 Infrared thermometer.

3.10.7 Assmann Psychrometer

It is also known as *aspiration psycrometer*. It is a portable instrument housing dry and wet bulb thermometers and a small clockwork ventilating fan (see Fig. 3.22). Thermometers are protected from direct radiation by suitable covering with highly polished material. It is used to obtain vertical temperature and humidity profiles (micrometeorological observations). The levels of measurement are at the surface 1, 2, 4, 8 and 12 feet from the surface. A micrometeorological tower is provided for this purpose.

3.10.8 Psychron

It is another device to measure temperature and relative humidity based on the concept of dry bulb and wet bulb thermometers (Psychrometry). The ventilation to the thermometers is provided through a battery system, which provides power supply for ventilation, unlike in case of the Assmann psycrometer where winding mechanism is provided for ventilation (see Fig. 3.23). Psychron is a better device to monitor air temperature and humidity very near to the ground at different heights as well as in crops.

Fig. 3.22 Assmann psychrometer.

Fig. 3.23 Psychron.

3.10.9 Six's Maximum and Minimum Thermometer

This instrument was introduced in India by the British. Even now the U-shaped instrument is seen in many agricultural farms and research stations (see Fig. 3.24). Using this instrument, maximum and minimum temperatures could be measured. The instrument uses alcohol as the expanding substance. After attaining maximum and minimum temperatures, the readings are noted down and the instrument is reset by pulling down the indices to the mercury surfaces by means of a magnet. The only disadvantage with this instrument is that alcohol wets the glass tube and is not a good conductor. It cannot be easily seen and has to be coloured.

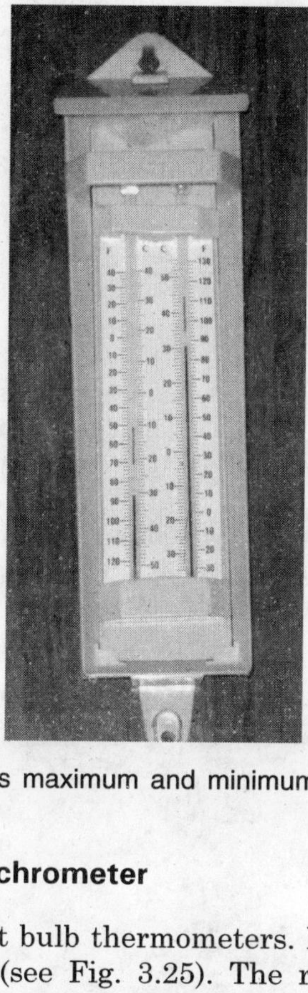

Fig. 3.24 Six's maximum and minimum thermometer.

3.10.10 Whirling Psychrometer

It consists of dry and wet bulb thermometers. It is handled by rotating firmly with a handgrip (see Fig. 3.25). The rotation is continued for three minutes and then the readings are taken. It will be repeated for another three minutes, so that both the readings should be the same. From the wet bulb and dry bulb readings, the relative humidity can be obtained directly using the hygrometric tables. This instrument is used for micrometeorological studies. However, Assmann psychrometer and Psychron are the better devices for micro meteorological studies to measure vertical temperature and relative humidity when compared to whirling psychrometer, thermograph, hygrograph, thermohygrograph and barograph.

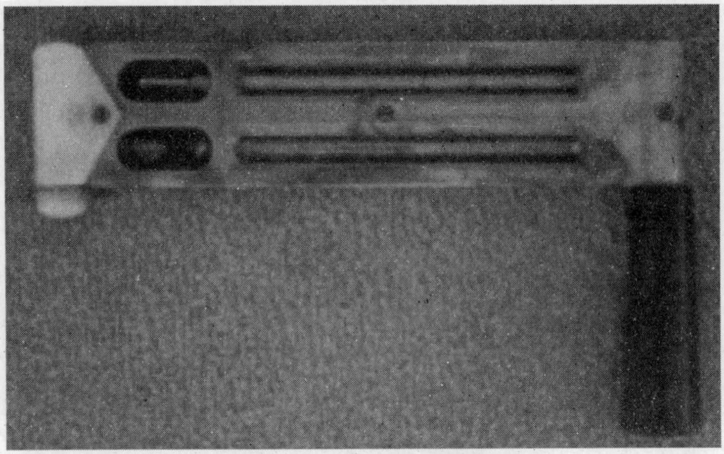

Fig. 3.25 Whirling psychrometer.

3.10.11 Double Stevenson Screen

The design of the Double Stevenson screen is exactly the same as the Single Stevenson screen except that it is double the size (see Fig. 3.26). Self-recording instruments like thermograph, hygrograph, thermo-hygrograph and barograph are placed inside the box to record temperature, relative humidity, temperature and relative humidity and atmospheric pressure, respectively. The equipment gives continuous record for 24 hours if it is daily, for a week if it is weekly, and for a month if it is monthly.

Fig. 3.26 Double Stevenson screen.

3.10.12 Thermograph

This instrument is used for recording continuous air temperature of 24 hours. It consists of a metallic sensitive element called Helix which is a bimetallic spring which deforms proportional to temperature changes and amplified by mechanical levers (see Fig. 3.27). A pen attached to them record temperature variations on a mechanical clock drum which makes a complete circle within 24 hours. The chart has reading of hours on x-axis and of temperature in °C vertically (y-axis).

Fig. 3.27 Thermograph.

3.10.13 Grass Minimum Thermometer

This is also known as *terrestrial radiation thermometer*. This is kept on a grass surface, which is very close to the ground (2.5 cm height of the grass tips from the surface is regularly maintained). Two such grass plots are maintained. In the centre of the grass plot, V-shaped wooden platforms having height of 2.5 cm from the surface are erected to place the grass minimum thermometer (see Fig. 3.28). The principle of grass minimum thermometer is similar to that of the sunset minimum thermometer placed inside the Stevenson screen. Readings are taken just before sunrise. It provides the lowest temperature attained at the ground surface by the exposed vegetation surface. It is used from October to March period only, when frost occurrence liability is expected.

3.10.14 Soil Thermometer

A soil thermometer contains mercury in a glass thermometer (see Fig. 3.29). There is a bend of 120° just above the bulb, with the rest of the stem being straight. Thus, when the bulb rests horizontally on the soil at the correct depth, the stem is inclined at 120° from the ground to facilitate the recording on the scale. The graduation begins at a distance of 6.5, 17.5 and 35 cm from the bulb in the 5, 15 and 30 cm depth of the soil thermometers respectively. These soil depths are known as *standard soil depths*, which are within the plant root zone with considerable temperature variations. Iron stands are used as the platform for soil thermometers. The thermometers used beyond 30 cm depth are known as *earth thermometers*. They are used for specific purposes. The temperature can be read with an accuracy of 0.1°C with eye estimation.

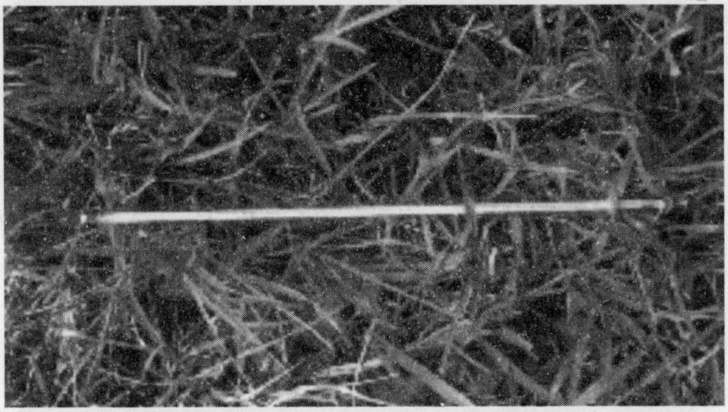

Fig. 3.28 Grass minimum thermometer.

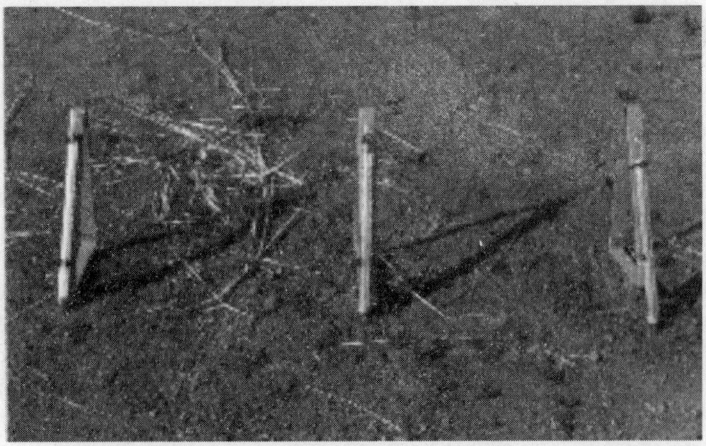

Fig. 3.29 Soil thermometer.

REFERENCES/SUGGESTED READINGS

Cosgrove, B., *Weather*, Dorling Kindersley, London, 1991, p. 64.

Hopkins, A.D., *Bioclimates, a Science of Life and Climate Relations*, United States Department of Agriculture Misc. Publications, 1938, p. 280.

IMD, *Agricultural Meteorology in India Meteorological Department (1932–57)*, Silver Jubilee of the Agricultural Meteorology Division, India Meteorological Department, Yeravada Prison Press, Poona, India, 1957, p. 42.

Middleton, W.E.K., *Meteorological Instruments*, Meteorological Service of Canada, Toronto, 1941, p. 276.

Pant, G.B., Singh, N. and Sonatakke, N.A., *Agrometeorological Data, Monitoring and Management*, Indian Institute of Tropical Meteorology, Pune, India, 1998, p. 463.

Petterson, S., *Introduction to Meteorology*, McGraw-Hill, New York, 1958, p. 327.

Rao, G.S.L.H.V.P., *Climate and Cashew*, Kerala Agricultural University, Thrissur, India, 2002, p. 100.

Riehl, H., *Introduction to the Atmosphere*, McGraw-Hill, New York, 1965, p. 365.

Rosenburg, N.J., *Microclimate: The Biological Environment*, John Wiley and Sons, New York, 1974, p. 300.

Sacheti, A.K., *Agricultural Meteorology: Instructional-cum-Practical Manual*, National Council of Educational Research and Training, New Delhi, 1985, p. 53.

Trewartha, G.T., *An Introduction to Climate*, McGraw-Hill, New York, 1954, p. 395.

Vitkevich, V.I., *Agricultural Meteorology*, Israel programme for scientific translators, Jerusalem, 1963, p. 320.

Wilsie, C.P., *Crop Adaptation and Distribution*, Eurasia Publishing, New Delhi, 1962, p. 448.

QUESTIONS

1. Write short notes on the following:

(a) Troposphere	(b) Stratosphere
(c) Isothermal charts	(d) Phytoclimate
(e) Soil temperature	(f) Sea and land breeze
(g) Cardinal temperatures	(h) Thermoperiodism
(i) Hopkins bioclimatic law	(j) Temperature units
(k) Growing degree days	(l) Temperature instruments

2. Fill in the blanks:
   (a) ———— is known as the weather making layer.
   (b) The height of the tropopause is —— at the equator and—— at the poles.
   (c) ———————— is the temperature at top of the troposphere.
   (d) Ozone is maximum at ———— height.
   (e) The number of days delayed in flowering by ———— days for every 1° N latitude.
   (f) ———— postulated bioclimatic law in ———— year.
   (g) ———— modified bioclimatic law in cashew.
   (h) ———— is known as the lapse rate.
   (i) ———— is the coldest layer in lowest atmosphere.
3. State true or false:
   (a) Temperature increases with height in troposphere.
   (b) Isotherm is noticed in lower stratosphere.
   (c) Tropopause is the height at which coldest temperature is noticed.
   (d) Temperature inversion is noticed in crops during daytime.
   (e) The principle in clinical thermometer and maximum thermometer is the same.
   (f) Minimum and grass minimum thermometers are one and the same in principle.
   (g) Thermograph is kept in single Stevenson screen.
   (h) Stevenson screen is erected at 3 feet height from the ground.
   (i) Assmann psychrometer is used for micrometeorological studies.
   (j) The annual mean temperature range is high in inland stations.
   (k) Temperate climate is not seen in high ranges near the equator.
   (l) Sea breeze is noticed in the morning hours along the coast.
   (m) Degrees Celsius and degrees Fahrenheit are not the same at any temperature.
   (n) Low temperature over the windward side becomes high temperature over the leeward side.
4. Elucidate the following:
   (a) Factors affecting surface air temperature.
   (b) Stratification of atmosphere.
   (c) Temporal and spatial distribution of surface air temperature.
   (d) Thermal distribution of soils.

4 GENERAL CIRCULATION

Atmospheric pressure is an important weather element in weather forecasting. It may not directly influence crop production. Nevertheless, the partial atmospheric pressure exerted due to water vapour influences crop evapotranspiration to a great extent. This is dependent on vapour pressure deficit just above the crop. The difference in atmospheric pressure results in horizontal motion of air, which is nothing but wind. Wind also plays a significant role in crop evapotranspiration. More than that, the strong winds depending on pressure gradient lead to mechanical damage in crops. Seasonal and biennial crops like tapioca, banana and drumstick are sensitive even to moderate winds. Therefore, it is very important to learn about the general circulation of atmosphere. It is influenced by the standard atmospheric pressure and wind systems across the globe.

The atmosphere surrounding the earth is thin as compared to the size of the earth and is often compared to the skin of a tomato or apple. However, its mass is considerably big and is capable of supporting all the life we know on our planet. Earth's atmosphere has mass and gets compressed under its own weight, making it dense at lower levels and progressively thin at higher levels. The earth's gaseous atmosphere has weight and thereby exerts pressure at the earth's surface, which was first demonstrated by Torricelli (an Italian scientist) in 1643. He showed that the atmospheric pressure could support a column of mercury, 76 cm high, in a mercurial barometer (see Fig 4.1). As the total column of atmosphere at the mean sea level is high, the atmospheric pressure is also high. If that is the case, the atmospheric pressure is still higher in mines as the height of atmosphere is furthermore when compared to the mean sea level. The total weight of a vertical column of air extending up to the outer limit of the atmosphere, supported by a unit area of the earth's surface is called the atmospheric pressure. In simple terms, it is defined as the force per unit area. Buys Ballot and William Ferrel established in 1857 that surface wind flows from 'high' to 'low' pressure and is deflected by a force called the *Coriolis force*, caused by the earth's rotation around its own axis. In the Northern Hemisphere, air pressure is lower on the left than on the right if we stand with our back to the wind. In contrast, lower atmospheric pressure will be on our right and higher pressure on our left in the Southern Hemisphere. This is known as the *Buys Ballot law*. Buys Ballot was a Dutch meteorologist who postulated the above theory in 1857.

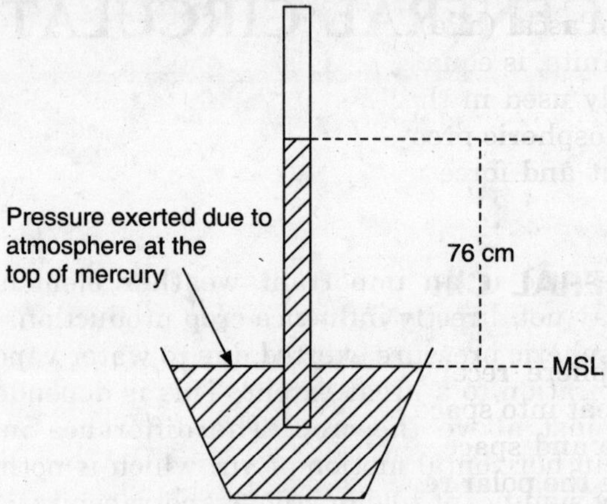

Fig. 4.1 Mercury barometer.

4.1 UNITS OF ATMOSPHERIC PRESSURE

The atmospheric pressure can be found out on the basis of the height of mercury as shown by Toricelli by setting up a simple mercury barometer. At the mean sea level, the vertical height of mercury measures 76 cm or 760 mm as against the total atmosphere supported in one square centimetre, extending from the earth's surface to the top of the atmosphere as shown in Fig. 4.1. It is known as the *standard atmospheric pressure*.

The atmospheric pressure can be measured by weight also. It could be obtained by simply multiplying the height of mercury (76 cm) with its density (13.6 g/cm^3), which is

$$76 \text{ cm} \times 13.6 \text{ g/cm}^3 = 1033.3 \text{ g/cm}^2$$

It means that the weight of a unit column of air over one square centimetre surface area extending up to the top of the atmosphere weighs just more than 1 kilogram. The pressure exerted due to the above weight of air over 1 square centimetre is known as 1 *standard atmosphere* in terms of weight units.

The atmospheric measure can be measured in absolute units of force, that is, in dynes/cm^2, as follows:

$$P = h\rho g = 76 \times 13.6 \times 980.6 = 10{,}13{,}250 \text{ dynes/cm}^2$$

where P is atmospheric pressure, h is the height of mercury in cm, ρ is the density of mercury in grams per cm^3 and g is the gravitational unit of force in cm/sec^2.

1 million dynes/cm^2 is equal to 1 bar, and 1 bar is equal to 1000 millibar. This can be taken as equivalent to 1013.25 mb. 1 millibar is equal to 1 hectoPascal (hPa). Hence 1 standard atmosphere, in terms of absolute force units, is equal to 1013.25 hPa. The unit of atmospheric pressure commonly used in the literature now is hectoPascals (hPa).

Thus, the atmospheric pressure can be measured in terms of height of mercury, weight and force units.

4.2 GENERAL CIRCULATION

Earth and atmosphere receive heat from the sun and re-radiate the same amount of heat into space as seen in the radiation balance between earth, atmosphere and space. The equatorial region absorbs more heat than it loses, while the polar regions give off more heat than they receive. This heat process creates contrast in surface air temperature between the above regions. The contrast in temperature also leads to difference in atmospheric pressure and thereby atmosphere is in horizontal motion, known as wind. The exchange of heat from warm to the cold regions takes place when atmosphere is in motion. The atmospheric motion or wind when summarised over the earth surface is known as the 'general circulation' of atmosphere. The general circulation of atmosphere is closely related to the standard atmospheric pressure and wind systems prevailing over the globe. Hence, the atmospheric pressure and wind systems are discussed under the general circulation of atmosphere.

4.3 VERTICAL PROFILE OF ATMOSPHERIC PRESSURE

The amount of pressure exerted by air at a particular point is determined by the height of air column, density and temperature. The atmosphere near the ground is naturally denser when compared at higher elevation, since the atmosphere is held to the earth by the force of gravity. This means that atmospheric pressure decreases with increasing elevation.

The atmospheric pressure decreases at a greater rate and is uniform near the earth's surface where pressure is relatively higher (see Fig. 4.2).

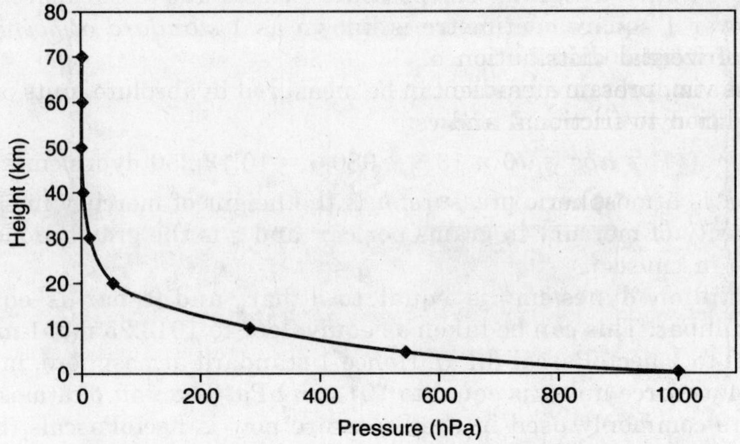

Fig. 4.2 Vertical distribution of atmospheric pressure.

In terms of international standard atmosphere, the rate of decrease of atmospheric pressure with height is as shown in Table 4.1.

Table 4.1 Rate of decrease of atmospheric pressure with height

Height (km)	*Atmospheric pressure* (hPa)	*Atmospheric pressure when compared to MSL* (%)
MSL	1013	–
5	540	53.3
10	265	26.1
15	121	11.9
20	55	5.3
25	25	2.3
30	12	1.1

It cleary indicates that about 50% of atmospheric pressure from the mean sea level decreases within 5 kilometres of height from the earth's surface. A schematic arrangement of atmospheric pressure with height is also depicted in Fig. 4.3. As the height of air column with elevation decreases, the atmospheric pressure diminishes. The atmospheric pressure in the lower atmosphere decreases at the rate of 2.54 cm in the first 270 m height of atmosphere from the ground; the same decrease is noticed in the next atmospheric layer of 300 m, while it is 330 m in the next atmospheric layer, which also indicates that the decrease in atmospheric pressure is very sharp very near to the ground where the density of atmosphere and the temperature are maximum (see Fig. 4.4).

4.4 HORIZONTAL DISTRIBUTION OF ATMOSPHERIC PRESSURE

The horizontal distribution of air pressure mainly depends upon two factors viz., pressure gradient and Coriolis forces due to earth's rotation in addition to frictional and centrifugal forces.

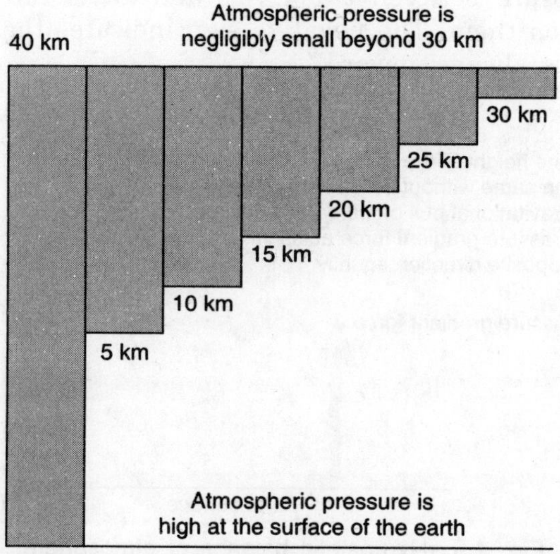

Fig. 4.3 Schematic arrangement of atmospheric pressure with height.

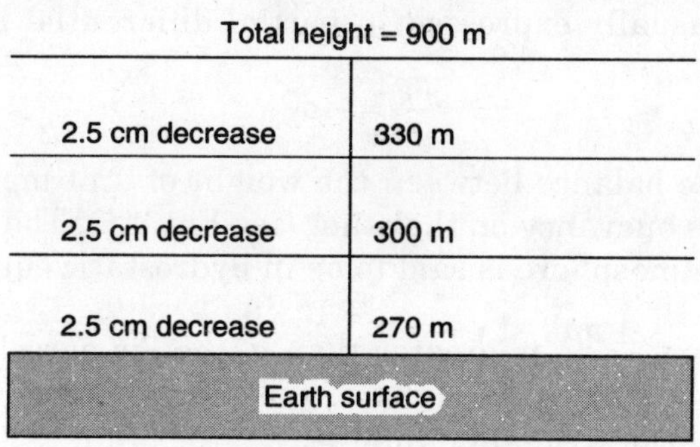

Fig. 4.4 Sharp decrease in atmospheric pressure with height very near the ground.

4.4.1 Hydrostatic Balance of Atmosphere

The hydrostatic balance of atmosphere keeps the atmosphere without collapse due to the gravitational pull and the pressure gradient force. They act equally in opposite directions (see Fig. 4.5). The pressure gradient force is the change in atmospheric pressure between two layers with difference in height. It acts as a driving force for air motion, which moves from high to low pressure. It can be expressed as

$$PG = -\frac{1}{\rho}\frac{dp}{dn}$$

where PG is the pressure gradient force, ρ is the density of air, dp is the change in pressure between two levels and dn is the perpendicular distance between them. The negative sign indicates that the pressure gradient force is always upward.

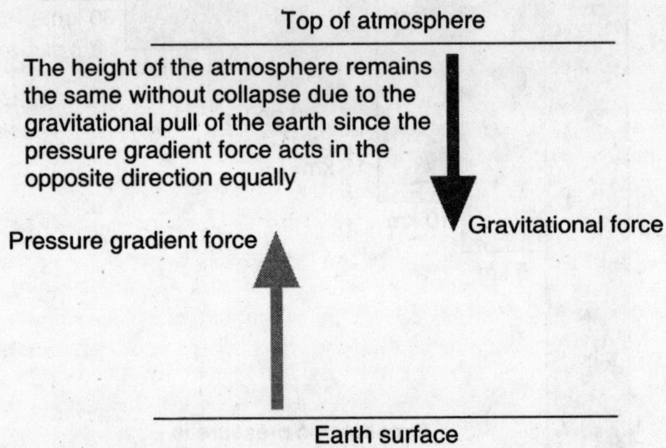

Fig. 4.5 Hydrostatic balance of atmosphere.

It can be written as $dp = -\rho g dn$, which is known as the *hydrostatic equation*. It is usually expressed in partial differential form as

$$\frac{dp}{dn} = -\rho g$$

It represents the balance between the weight of unit mass of air on the one hand and its buoyancy on the other (see Fig. 4.6). The above equation is valid if the astmosphere is said to be in hydrostatic equilibrium. If the term denoted by $-\frac{1}{\rho}\frac{dp}{dn}$ is greater than g, the air parcel will rise while it sinks if the term is smaller than g.

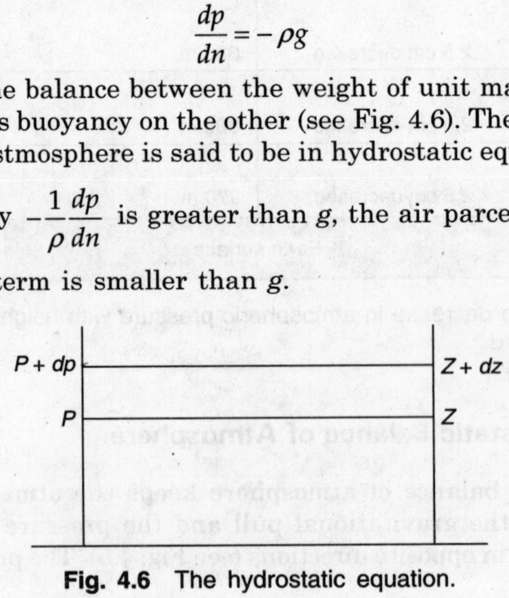

Fig. 4.6 The hydrostatic equation.

4.4.2 Coriolis Force

Coriolis force is not a force, but an effect due to the rotation of the earth. Air motion is deflected to the right of the direction of motion in the Northern Hemisphere and to the left in the Southern Hemisphere. This effect is named after G.D. Coriolis, who was a French mathematical physicist. He postulated the above effect in 1844. This can be expressed as

$$f = 2\omega V \sin \varphi$$

where

f = Coriolis force
ω = angular velocity of the earth
V = wind velocity
φ = latitude

This effect is zero at the equator and increases polewards as 50% at 30° latitude and 100 % near poles (see Fig. 4.7). The above figures of the effect of latitude could be obtained if we substitute φ with 0, 30 and 90 as sine 0 is zero, sine 30 is 0.5 and sine 90 is 1.

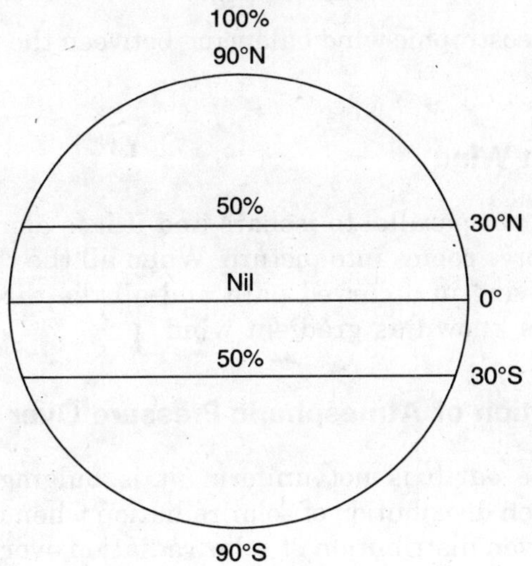

Fig. 4.7 Effect of Coriolis force with latitude.

4.4.3 Geostrophic Wind

If a parcel of air at rest is subjected to a pressure gradient force, it will initially commence to move directly from high to low pressure. As soon as it starts to move, it will be deflected to the right in the Northern Hemisphere. The faster it moves the greater will be the deflection force and the wind will flow parallel to the *isobars*, the lines on a map connecting various places with equal atmospheric pressure. When this occurs, the deflective and pressure gradient forces will be acting in opposite directions. When these two forces are equal in magnitude they balance each other. The resulting wind is called the *geostrophic wind*, which moves parallel to isobars (see Fig. 4.8).

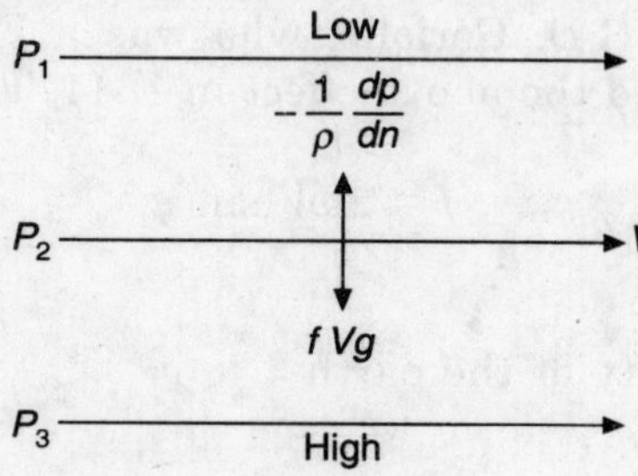

Fig. 4.8 Geostrophic wind-balance between two forces.

It can be expressed as

$$Vg = \frac{1}{2\rho\omega \sin \varphi} \frac{dp}{dn}$$

where *Vg* is the geostrophic wind balancing between the pressure gradient and Coriolis forces.

4.4.4 Gradient Wind

If the air flow is not parallel to isobars and if it is along a curved path, the centrifugal force comes into picture. When all the three forces act on an air parcel moving in a curved path and all the three forces balance one another, it is known as gradient wind.

4.4.5 Distribution of Atmospheric Pressure Over the Globe

The shape of the earth is not uniform; it is bulging at the equator, subjected to uneven distribution of solar radiation when it revolves around the sun. The uneven distribution of solar radiation over different regions of the globe leads to contrast in surface air temperature. This results in variations of surface atmospheric pressure systems, which are known as *standard atmospheric pressure systems* or *belts*. There are altogether seven alternating low and high pressure belts on the earth's surface. They are as follows:

1. Equatorial trough of low pressure (between 5°N and 5°S)
2. Subtropical high pressure belt (Northern Hemisphere)–(25°N and 35°N)
3. Subtropical high pressure belt (Southern Hemisphere)–(25°N and 35°N)

4. Subpolar low pressure belt (Northern Hemisphere)–(60°N and 70°N)
5. Subpolar low pressure belt (Southern Hemisphere)–(60°N and 70°N)
6. Polar high (Northern Hemisphere)
7. Polar high (Southern Hemisphere).

The equatorial region receives more solar radiation and thus the surface air temperature is high, which creates air near the ground is lighter when compared to higher latitudes. The above condition leads to low atmospheric pressure over the equatorial region while subtropical high pressure belts develop in both the hemispheres between 25 and 35° latitudes due to relatively low surface air temperature. It is due to low solar radiation received due to inclined sun's rays over the subtropical region when compared to the equatorial belt (see Fig. 4.9). Likewise, alternate low and high atmospheric pressure belt systems are developed across the globe from the equator to the poles.

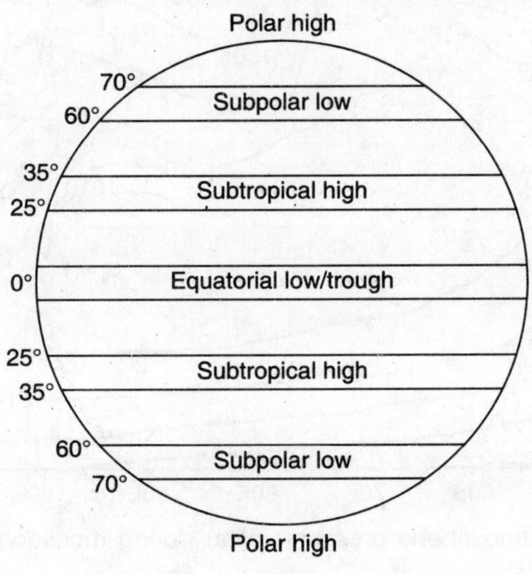

Fig. 4.9 Schematic presentation of standard atmospheric pressure system.

4.4.6 Distribution of Atmospheric Pressure Over India

The horizontal distribution of atmospheric pressure over India indicates that it varies from 1013 hPa in the south to 1019 hPa in the north in January, while it is from 1001 hPa in the south to 997 hPa in the north during July. These surface atmospheric pressure changes cause for both the monsoons viz. southwest monsoon and northeast monsoon during summer and winter, respectively. As per the data collected by the weather bodies, the atmospheric pressure variation during the southwest monsoon indicates that there is a pressure gradient of 8 hPa from south (1008 hPa) to north (1000 hPa) of India, which makes monsoon active (see Fig. 4.10). As stated before, *Isobar* is a line on a map connecting various places having equal atmospheric pressure. *Isallobar* is a map line connecting places of equal atmospheric pressure gradient. The difference in atmospheric pressure with horizontal distance is known as the *pressure gradient*. The pressure gradient is high if the difference in atmospheric pressure is more with less distance between the two locations.

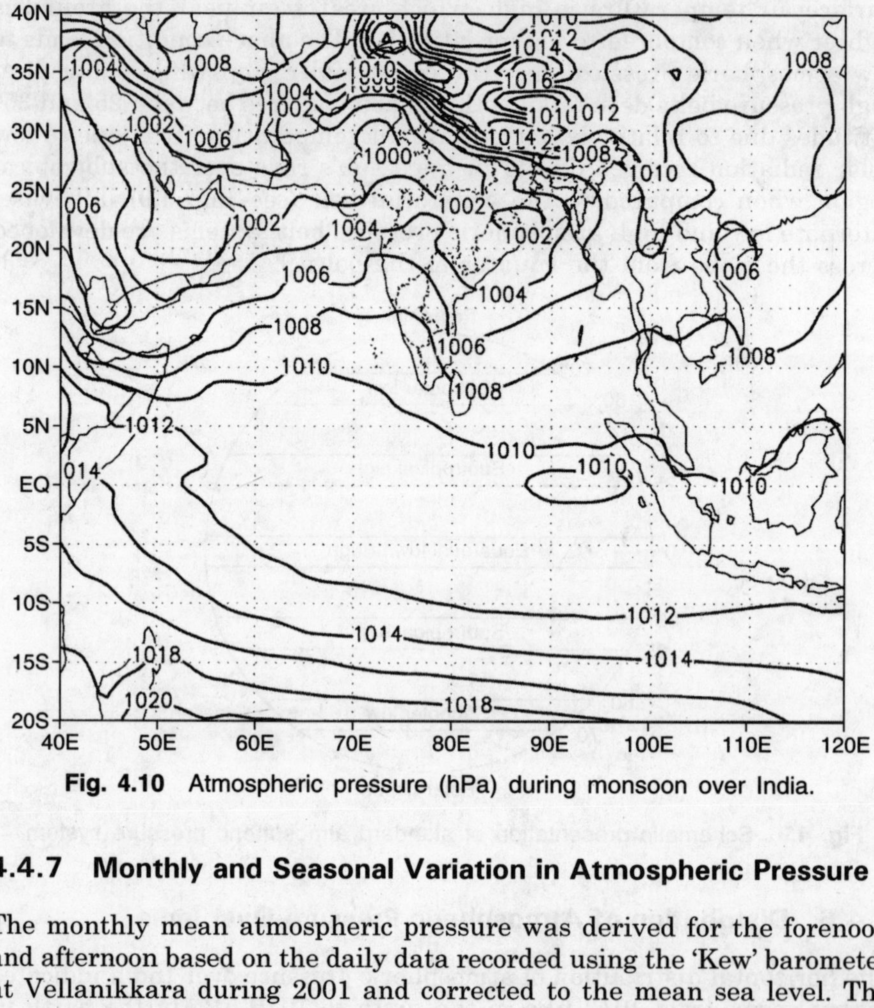

Fig. 4.10 Atmospheric pressure (hPa) during monsoon over India.

4.4.7 Monthly and Seasonal Variation in Atmospheric Pressure

The monthly mean atmospheric pressure was derived for the forenoon and afternoon based on the daily data recorded using the 'Kew' barometer at Vellanikkara during 2001 and corrected to the mean sea level. The annual range of atmospheric pressure in the forenoon varied between 1014.4 hPa (December) and 1009.5 hPa (June) and between 1010.5 hPa (December) and 1006.9 hPa (June) in the afternoon. It revealed that the maximum was recorded during December while the minimum was recorded in April, May and June. The low atmospheric pressure during summer leads to monsoonal wind over India. The atmospheric pressure range between morning and evening was high (4.5 hPa) during February while it was less (2.3 hPa) during July and August (see Table 4.2). The monthly mean atmospheric pressure varied between 1012.4 (December) and 1008.2 (June) over Vellanikkara at the mean sea level.

Table 4.2 Monthly mean sea level atmospheric pressure at Vellanikkara during 2001

Month	*Atmospheric pressure* (hPa) *at 0830* hr	*Atmospheric pressure* (hPa) *at 1730* hr	*Mean atmospheric pressure* (hPa)
January	1012.2	1008.0	1010.1
February	1012.4	1007.9	1010.1
March	1011.8	1007.4	1009.6
April	1010.3	1006.7	1008.5
May	1009.6	1006.9	1008.3
June	1009.5	1006.9	1008.2
July	1010.2	1007.9	1009.0
August	1010.7	1008.4	1009.5
September	1010.9	1007.6	1009.2
October	1010.8	1007.8	1009.3
November	1012.1	1008.2	1010.2
December	1014.4	1010.5	1012.4
Annual	1011.2	1007.9	1009.5

The atmospheric pressure is high (1010.8 hPa) during winter and low (1008.9 hPa) during summer, followed by post-monsoon season (1009.8 hPa), indicating that high surface air temperature during summer leads to low atmospheric pressure and vice versa during winter (see Fig. 4.11). The atmospheric pressure is low during summer and gradually increases during monsoon and post monsoon, reaching a peak in winter. The hourly distribution of atmospheric pressure indicates that during the morning hours, it shows high levels and later on decreases with temperature increase.

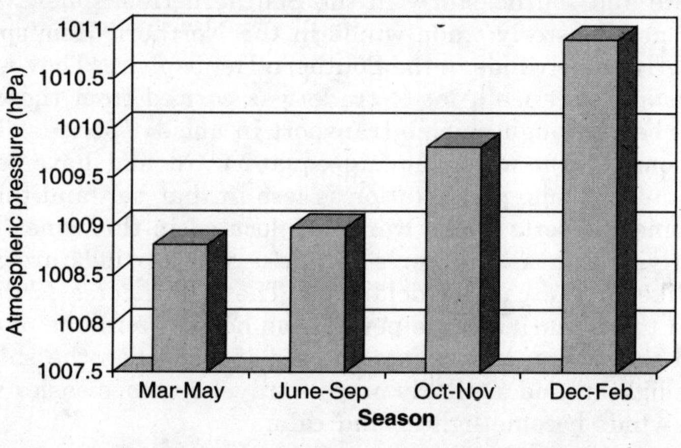

Fig. 4.11 Seasonal atmosphere pressure (hPa) at Vellanikkara during 2001.

4.5 WIND DISTRIBUTION

The horizontal movement of air is known as *wind*. Though the vertical movement of air is noticed, it is negligibly small when compared to horizontal movement as the height of atmosphere is only few kilometres. However, the vertical movement or uplift of air only causes for significant weather changes in terms of cloud formation and rain. The standard pressure patterns are accompanied by corresponding wind patterns, which can be divided into the following belts or zones:

1. Doldrums/Intertropical convergence zone (ITCZ)
2. Trade wind belt
3. Antitrade wind belt
4. Westerlies
5. Polar easterlies.

4.5.1 Doldrums

The average location of the doldrums belt is roughly between 5°N and 5°S where the winds are calm due to convergence of trade winds from both the hemispheres. This is also known as the *intertropical convergence zone.*

4.5.2 Trade Wind Belt

This belt is seen roughly from 5° to 30° latitude. The surface flow is equatorward, and the flow in upper troposphere is poleward. The trade winds originate due to pressure gradient from the subtropical belts of high pressure to the equatorial belt of low pressure (see Fig. 4.12). Instead of blowing direct from north towards equator, the wind is deflected due to Coriolis force and blows from northeasterly direction in the Northern Hemisphere and southeasterly in the Southern Hemisphere. They are known as northeasterly trade winds in the Northern Hemisphere and southeasterly trade winds in the Southern Hemisphere. They are known as *trade winds* because a lot of trade was carried from those belts to equatorial belt through marine transport in ancient periods. The trade winds become warm when moving equatorward and have less water vapour content. Thus, precipitation is less in that belt and that is why all the tropical deserts of the world are located in this zone. This zone is popularly known as the *horse latitudes* where winds are light and variable like in the case of doldrums. The name *horse latitudes* was popular in those marine trade periods, as horses were one of the trade items and they were thrown into the sea if and when the ship needed to become lighter, and similarly marine travel becomes easier whenever the trade winds become lighter and calm.

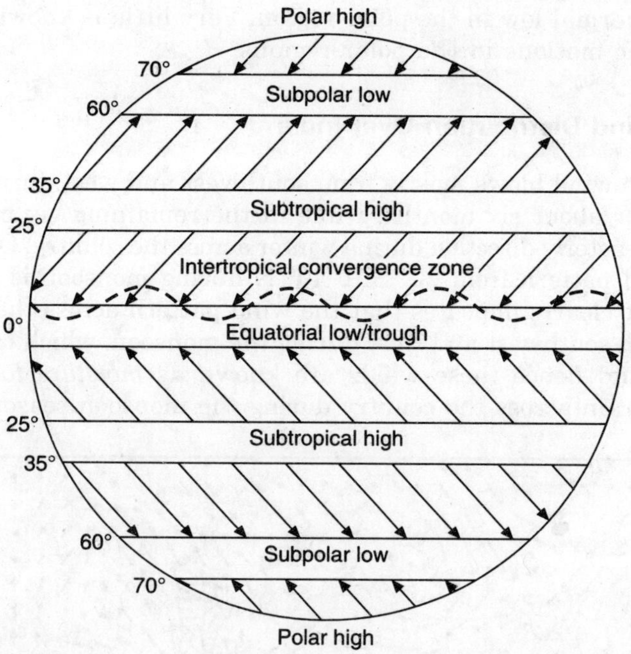

Fig. 4.12 Pressure and wind distribution over the globe.

4.5.3 Antitrade Wind Belt

In this wind pattern belt, the wind flow in upper troposphere is towards poles. The winds here are called *antitrade winds.*

4.5.4 Westerlies

This belt lies between 30° and 60° latitude. The westerlies are winds that move out from the subtropical high pressure belts. These winds, while moving into higher latitudes, are deflected and become the southwesterly and the northwesterly winds in the northern and southern hemispheres, respectively. They are more variable and intense compared to the trade winds. While moving towards higher latitudes, the westerlies become cool due to lower temperature and carry high water vapour content. The highest precipitation zones in the world are seen in this belt. This is the belt of cyclones due to westerly disturbances. It is one of the reasons for winter rains in northwest India. The western disturbances are popularly known as *Nor'westers* in India.

4.5.5 Polar Easterlies

The polar easterlies are those winds which move out of the Polar Highs towards the sub-polar low pressure belts. The Polar High pressure belt is due to thermal low in the polar region. Very little is known about the atmospheric motions in the polar regions.

4.5.6 Wind Distribution Over India

The surface wind blows mostly from southwest and west during summer monsoon for about six months, while in the remaining six months it is from northeasterly direction during winter across the country. One example of the wind pattern from 20° S to 40° N during monsoon is depicted in Fig. 4.13. It clearly indicates that the wind pattern across the country is mostly from southwest and west during the monsoon, which carries lot of moisture and hence these winds are known as *moisture laden winds.* They give rain across the country during the monsoon season.

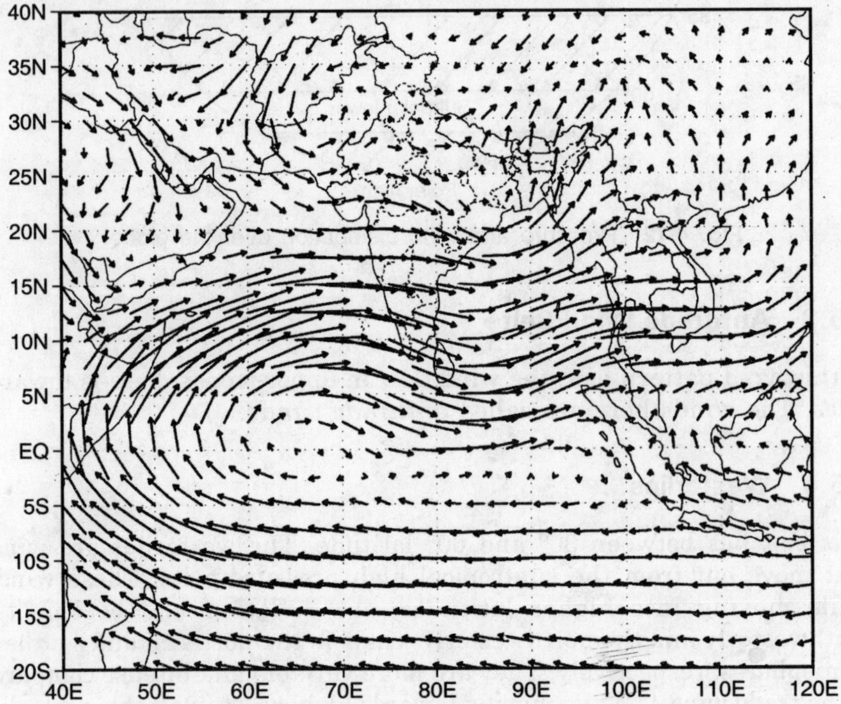

Fig. 4.13 Wind flow during monsoon period over India.

4.6 AIR MASSES AND FRONTS

An air mass is a thick and extensive portion of the atmosphere whose temperature and humidity characteristics are approximately homogeneous in a horizontal direction at different levels. If the air mass is said to be relatively uniform, there may not be much change in air temperature and humidity near the surface and they change very slowly over large areas in terms of 1000 km or more. Air masses can be classified on the basis of air temperature and humidity near the surface of the earth. If the surface air temperature is less, it is known as the *cold air mass,* and is called the *warm air mass* if its temperature is high. The source of cold air mass may be at higher latitudes or the polar region while the tropics for warm air mass. The cold air mass is known as the polar air mass (P) while the warm air mass is called the tropical air mass(T), based on the geographical origin. Similarly, the maritime and continental air masses are classified on the basis of humidity. The humidity is less in case of continental air mass while it is more in case of maritime air mass. Combining temperature and humidity based on geographical origin, the principal air masses are classified into 4 types, viz. Continental polar air mass (cP), Maritime polar air mass (mP), Continental tropical air mass (cT) and Maritime tropical air mass (mT).

The modification of air mass may be either thermodynamic or mechanical in origin. The thermodynamic modification results from the transfer of heat between the two air masses, which is common in the tropics. In valleys during winter, a temperature inversion is expected as the warm air mass at the ground level is replaced by cold air mass. It results in chilling of the surface air and increase in atmospheric stability. The mechanical modification of air mass is mostly seen in the temperate and high latitudes when there is a contrast in temperature between two air masses and cyclones are formed. These are different from the cyclonic processes and development in the tropics.

4.6.1 Fronts

A front is a zone of sharp changes in temperature between warm and cold air masses. Cyclones in high latitudes are born along a front between warm and cold air masses, which may be blowing in opposite directions. There are different types of front viz. cold front, warm front, occluded front and stationary front. They are plotted in different symbols on weather maps for analysis. A cold front is a front along which colder air replaces warmer air. A warm front is a front along which warmer air replaces colder air. If any air mass does not replace, it is known as stationary front. If a cold front overtakes warm front, it is known as occluded front. The occluded front is the last stage during the development of a cyclone.

4.6.2 Hadley Cell

The vertical circulation of the atmosphere between the tropics and subtropics is called the Hadley cell. Another vertical circulation in the tropical atmosphere is known as the Walker circulation. It develops due to difference in sea surface temperature between the east and west of the equatorial Pacific. There are two further cells of circulation at higher latitudes—the Ferrel cell and another one ending up at the poles. The details of these atmospheric circulations are not discussed here.

4.7 WIND ROSE

A wind rose is a pictorial representation of wind direction and wind speed. It gives an overall idea of the distribution of wind direction and wind at a given location for a given period of time. It is a very useful tool in windbreak and shelterbelt studies so as to reduce evapotranspiration losses, and thereby to increase crop yields. The wind damage due to high winds can be minimised and crops saved from mechanical damage in case of wind-sensitive crops like banana, tapioca, drumstick and like crops in wind-prone areas, provided planting is taken up on the basis of wind roses.

4.7.1 Construction of a Wind Rose

The monthly wind rose diagrams could be constructed based on daily wind speed and direction data collected using Anemometer and Wind vane, respectively. The wind direction can be recorded in 16 directions based on compass using Wind vane (see Wind vane under instruments). The wind direction and corresponding degree on compass are as follows:

N(360°), NNE(22.5°), NE(45°), ENE(67.5°)
E(90°), ESE(112.5°), SE(135°), SSE(157.5°)
S(180°), SSW(202.5°), SW(225°), WSW(247.5°)
W(270°), WNW(292.5°), NW(315°), NNW(337.5°)

In addition to the above 16 wind directions, 'C' and 'V' are also reported if the wind at the time of observation is 'Calm' and 'Variable' respectively.

To find out the prominence wind direction in a particular month, take daily wind speed and wind direction data of the selected month. The percentage occurrence of each direction can be obtained, based on the actual data of wind direction. Assign a suitable scale in centimetre, covering the whole range of wind directions prevailed during the month. For example, if the range of wind direction varies between 2% and 10%, and 2% is taken to be equal to one centimetre, 10% becomes 5 centimetres. The remaining wind directions vary between 1 and 5 centimetres. Then draw a small circle around a point, which is nothing but the location where the data are recorded. Using the protractor, earmark the angles for each direction and draw the scaled arms for all the directions. If there is 'C' and 'V', it is represented by simply writing the percentage of occurrence value inside the circle.

Wind speed can be shown inside the scaled arms of each wind direction drawn over the circle, with legend given for each class of wind speed.

For this, divide the wind speed ranges in a given direction to convenient class intervals and find out the frequency distribution in each interval of wind speed for each direction in terms of length units already obtained. Accordingly, the length of each class interval of wind speed in centimetres for a given direction can be demarcated by assigning legends. Such diagrams are known as 'wind roses'.

The wind roses were constructed based on the data collected at Vellanikkara for each month, representing different seasons, and depicted in Fig. 4.14.

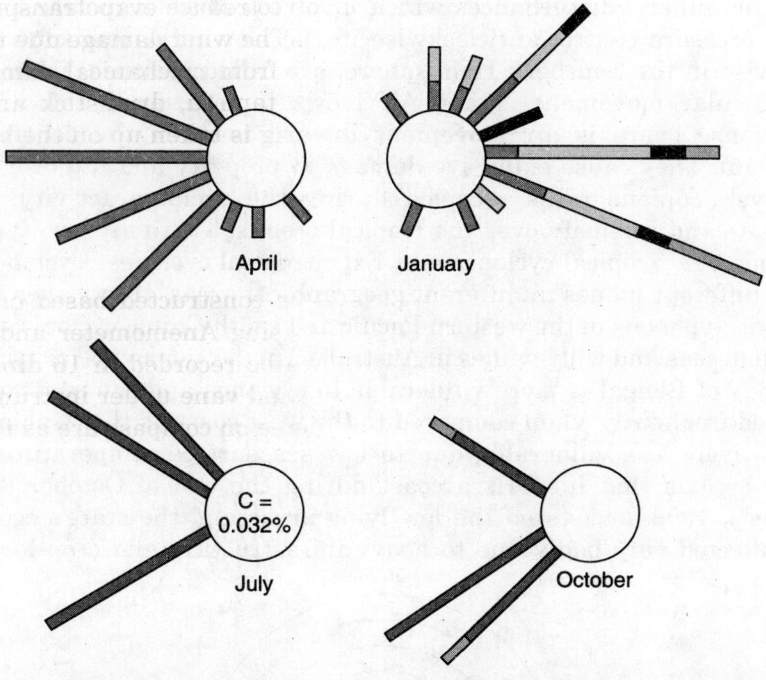

Fig. 4.14 Wind rose diagrams during selected months at Vellanikkara during 2002.

4.7.2 Windward and Leeward

Any region or location, if naturally positioned along the direction from which wind prevails, is called *windward*. The western part of Western Ghats (Kerala) is located windward during the southwest monsoon. Tamil Nadu is on the *leeward* side as monsoon wind is obstructed by the Western Ghats, which leads to low rain over Tamil Nadu. The reverse happens during northeast monsoon. Tamil Nadu is then on windward while Kerala is on leeward. Of course not all the regions of Kerala do not fall under leeward. If any place or region is away from wind direction due to its geographical position, it is known as leeward. Winds shift in counter clockwise direction are called 'backing' winds and winds shift in clockwise direction are called 'veering' winds in the Northern Hemisphere. Veering winds indicate fine weather and backing winds predict continued bad weather. The reverse is the wind direction in the Southern Hemisphere.

4.8 CYCLONES AND ANTICYCLONES

The atmospheric disturbances which involve a closed circulation about a low pressure centre, anticlockwise in the Northern Hemisphere and clockwise in the Southern Hemisphere, are called cyclones. It refers to a particular movement of air. Air blows in horizontal and vertical directions. There is air movement inward towards the centre (see Fig. 4.15). They cause extensive damage to property and life over land. However, copious rains received during the cyclonic activity. They originate and intensify over the tropical oceans. There are two types of cyclones, viz. Tropical cyclones and Extratropical cyclones. Cyclones are given different names in different geographical areas. Hurricanes in the Atlantic, typhoons in the western Pacific and south China, cyclonic storms in Indian seas and willy-willies in Australia are the synonyms for cyclones. The Bay of Bengal is more vulnerable to cyclones and the east coast is affected frequently when compared to the west coast as the Arabian sea is relatively less vulnerable due to low sea surface temperature. The super cyclone that hit Orissa coast during the end of October (27–30 October), 1999 devastated the low lying areas and the state's economy was affected very badly due to heavy infrastructure and crop losses.

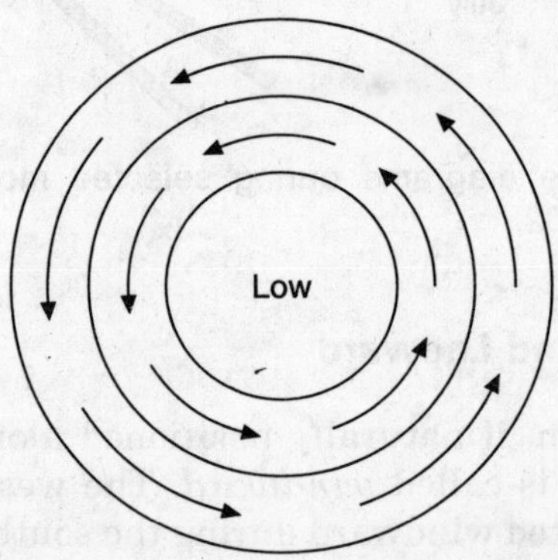

Fig. 4.15 Schematic presentation of winds in a cyclone system.

4.8.1 Classification of a Cyclonic System

A cyclonic system can be classified based on T-Number, where T stands for tropical storm. The T-Number is assigned to different classes of wind speed and the same is described as a cyclonic system, which begins with low pressure and ends up with super cyclone. The T-Numbers for storms are given in Table 4.3.

Table 4.3 T-Numbers for storms

T-Number	*Maximum wind speed in knots*	*Pressure drop in* hPa *in Indian seas*	*System Classification*
1.0	25	–	Depression
1.5	25	–	Depression
2.0	30	4.5	Deep Depression
2.5	35	6.1	Cyclone
3.0	40	10.0	Cyclone
3.5	55	15.0	Severe Cyclonic Storm
4.0	65	20.9	Very Severe Cyclonic Storm
4.5	77	29.4	Very Severe Cyclonic Storm
5.0	90	40.2	Very Severe Cyclonic Storm
5.5	102	51.6	Very Severe Cyclonic Storm
6.0	115	65.6	Very Severe Cyclonic Storm
6.5	127	80.0	Super Cyclonic Storm
7.0	140	97.2	Super Cyclonic Storm
7.5	155	119.1	Super Cyclonic Storm
8.0	170	143.3	Super Cyclonic Storm

An area where the atmospheric pressure is lower than the surroundings and enclosed by one or more isobars is designated as *low*, *depression* or *cyclone*, depending on the pressure deficit at the centre of the system. A system where the pressure is higher at the centre is termed a *high* or *anticyclone*. An outward extension from the low is a trough and that from a high is a ridge. An area between two highs and lows is a 'col'.

4.8.2 Anticyclones

Anticyclones play a more important role in weather phenomena at the surface of the earth, in comparison to the cyclones. The anticyclones are associated with fine weather. They are secondary atmospheric whirls, wherein a region of higher atmospheric pressure is located at the centre, and spiral air movement is outward from the central area in clockwise motion in the Northern Hemisphere and counterclockwise in the Southern Hemisphere (see Fig. 4.16). Heating up of air thrown off by the centrifugal force from the cyclonic whirls and the local cooling of air are responsible for this. The anticyclones are associated with scanty weather. Subsidence and divergent wind system within an anticyclone do not favour condensation and cloud formation. In winter, the cold anticyclones originating in the snow covered sub-polar regions always bring with them very low temperature. The middle latitude anticyclones always produce the lowest temperatures of the season. In summer, the stagnant type of warm anticyclones associated with the air of tropical or subtropical origin produce extremely high temperatures called heat waves. The presence of anticyclone at about 3 km in atmosphere over Rajasthan is one of the reasons for low rains during the monsoon.

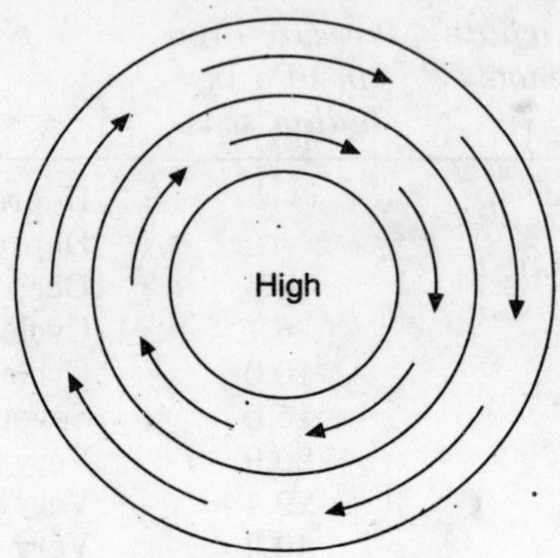

Fig. 4.16 Schematic presentation of winds in an anticyclone system.

4.9 BEAUFORT SCALE

In the absence of an anemometer, wind can be estimated reliably by its effect on objects on land and the state of the sea. The system was originally devised by Admiral Beaufort of the British Navy in 1805, for estimating wind at a height of 20 ft. above the ground without instruments. The Beaufort scale is given in Table 4.4. The Beaufort number is 2 and the mean wind speed is 5 nautical miles per hour if wind is felt on our face, while the wind speed is 60 knots if there is a widespread damage found during the storm.

Table 4.4 Beaufort scale of wind speed

Beaufort no.	*General description*	*Observations for estimate (i) over land / (ii) over sea*	*Wind (in Knots)*	*Speed (in kmph)*
			(Average in brackets)	
0	Calm	Smoke rises vertically	Less than 1	Less than 1
		Calm. Sea like mirror	(0)	(0)
1	Light air	Smoke bends from the vertical and drifts slowly with the winds, but wind vane not affected. Small ripples seen.	1–3 (2)	1–5 (3)

(Contd.)...

Table 4.4 Beaufort scale of wind speed *(Contd.)...*

Beaufort no.	*General description*	*Observations for estimate (i) over land (ii) over sea*	*Wind (in Knots)*	*Speed (in kmph)*
			(Average in brackets)	
2	Light breeze	Wind felt on face, leaves rustle, windvanes move. Small, short wavelets. Crests do not break.	4–6 (5)	6–11 (9)
3	Gentle breeze	Leaves and small twigs in constant motion; wind extends light flag. Large wavelets(crests begin to break), scattered white horses.	7–10 (9)	12–19 (16)
4	Moderate breeze	Raises dust and loose paper, small branches move. Small waves becoming longer; fairly frequent white horses.	11–16 (13)	20–28 (24)
5	Fresh breeze	Small trees begin to sway. Moderate waves taking a more pronounced long form, white horses with some spray seen.	17–21 (18)	29–31 (34)
6	Strong breeze	Large branches in motion, whistling heard in telegraphic wires. Large waves begin to form, white horses with some spray, extensively seen.	22–27 (24)	39–49 (44)
7	Near gale	Whole trees in motion. Rise in sea level seen, white foam from breaking waves carried along with the wind.	28–33 (30)	50–61 (55)
8	Gale	Twigs break off trees, difficulty experienced in walking against the wind. Moderately high waves of great length, edges of waves break, with spray seen along sea surface everywhere.	34–40 (37)	62–74 (68)

(Contd.)...

Table 4.4 Beaufort scale of wind speed *(Contd.)...*

Beaufort no.	*General description*	*Observations for estimate (i) over land (ii) over sea*	*Wind (in Knots) (Average in brackets)*	*Speed (in kmph)*
9	Strong gale	Slight structural damage occurs. High waves. Crests of waves with streaks of foam roll over. Spray reduces visibility.	41–47 (44)	75–88 (82)
10	Storm	Trees uprooted, considerable structural damage. Sea white with foam, high waves, heavy tumbling of sea, smoke like spray reducing visibility.	48–55 (53)	89–102 (96)
11	Violent storm	Widespread damage, very rarely experienced. Unusually high waves, small and medium size ships lost behind waves. Sea completely covered with white foam and froth. Poor visibility.	56–63 (60)	103–117 (110)
12	Hurricane	Widespread damage, shed roofs may be blown off, Electric and telegraph poles may fall. Sea-level visibility rises, sea completely white with driving spray. Very poor visibility.	64 and over	118 and over

4.10 INSTRUMENTS

The following instruments are used to obtain the atmospheric pressure, wind direction and wind speed at a particular place.

4.10.1 Aneroid Barometer

It consists of a thin shallow circular metal box from which all air has been removed (see Fig. 4.17). Both the top and bottom of the box are corrugated to make them sensitive to changes in atmospheric pressure. When the pressure of the air increases, the lid of the box is depressed and vice versa. This causes a pointer to move round a scale fixed in front of the instrument. The reading of this scale denotes the atmospheric pressure of that place. As there is no liquid in the instrument, it is easy to be carried about from place to place. No liquid is used in this type of barometer. Aneroid barometer was invented by Vidin in 1843.

Fig. 4.17 Aneroid barometer.

4.10.2 Kew Barometer

Its essential parts are (i) a glass tube (about 90 cm long) closed at the top and open below; (ii) a cup or cistern and (iii) a brass scale (see Fig. 4.18). The glass tube is filled with mercury and its open end is dipped in the mercury in the cistern, which prevents air from entering the tube. Above the column in the tube is an empty space and great care is to be taken to remove all air from this space, as its presence even in very minute quantities will vitiate the readings of the instrument. The mercury column in the tube is supported by the weight of the air on the surface of the mercury in the cistern. This forms the basic principle on which a barometer is constructed. The mercury level in the barometer tube rises or falls due to change in the atmospheric pressure. Another type of barometer is known as the Fortin barometer, which works on the same principle.

4.10.3 Barograph

It is an automatic recorder which measures atmospheric pressure continuously (see Fig. 4.19). It is kept in double Stevenson screen for recording surface atmospheric pressure. The principle is based on aneroid barometer.

Fig. 4.18 Kew barometer.

Fig. 4.19 Barograph.

4.10.4 Wind Vane

It is a horizontal metallic vane suspended on a frictionless vertical axis to show the direction from which the wind is coming (see Fig. 4.20). It is installed 10′ above the ground. The position of the arrowhead should be recorded at the time of observation. It gives the wind direction from which it blows towards the station.

Fig. 4.20 Wind vane.

The vane consists of two flat plates joined together at an angle. The rotating arm is mounted by means of ball bearing and brass sleeve on wind vane stand. The brass sleeve is fixed by means of two screws on iron rod of the stand.

Below the rotating arm, eight fixed direction arms are fixed on a brass bar for indicating main and intermediate directions.

Wind direction is expressed in degrees measured clockwise from the North. For reporting purposes, however, it is expressed on the compass scale from 0 to 360 degrees and the total direction as well as code numbers in different directions are indicated in Section 4.7.1.

4.10.5 Anemometer

The Robinson cup anemometer consists of a wheel with three semi-conical cups (see Fig. 4.21). The cups are attached to a centre spider by means of three arms made of brass. The spider is attached to a vertical steel spindle and the top of the spindle moves on small ball bearings while the bottom rests on a thrust bearing. The lower tip of spindle is enclosed in a bushing, which allows free rotation of the spindle. The rotation of the upright spindle is transferred, by means of a wheel and gear, to a counter having bold figures with a range from 00000 to 99999. The four white figures to the left give whole number in kilometres and the last figure in red gives values in tenths of a kilometre.

The reading of cup anemometer is taken at 0725 hours and 1425 hours daily. For finding the mean wind speed, the difference of two readings is taken for three minutes and worked out per hour. The wind speed is calculated in km/hr generally. To work out wind speed, an example is given below:

Initial reading of anemometer at the time of observation = 7966.4 km
Final reading of anemometer after three minutes = 7966.6 km
Difference in three minutes = 0.2 km
Wind speed in km/hr = 0.2 × 60/3 = 4 km
Wind speed at the time of observation is 4 km/hr.

Fig. 4.21 Robinson cup anemometer.

REFERENCES/SUGGESTED READINGS

Chritchfield, H.J., *General Climatology*, 4 ed., Prentice-Hall of India, New Delhi, 1994, p. 453.

Ian, J.P. and Mansur, M., *Will it Rain? The effect of the Southern Oscillation and EI Nino in Indonesia*, The State of Queensland, Department of Primary Industries, Brisbane, 2002, p. 50.

Koteswaram, P., 'Weather and crops', *Handbook of Agriculture*, 5th ed., Indian Council of Agricultural Research, New Delhi, 1980, pp. 1–20.

Kumar, K.R., *Climate Maps: NCEP/NCAR Reanalysis*, Personal Communication, Indian Institute of Tropical Meteorology, Pune, 2003, p. 12.

Lutgens, F.K. and Tarbuck, E.J., *The Atmosphere*, Prentice-Hall, London, 1998, p. 434.

Mallik, A.K., 'Weather and crops', *Handbook of Agriculture*, Indian Council of Agricultural Research, New Delhi, 1969, pp. 1–31.

Menon, P.K., *Our Weather*, National Book Trust, New Delhi, 1989, p. 349.

Murthy, R.K.V., *Terminology in Agricultural Meteorology*, Sri Venkateswara Publishers, Hyderabad, 1996, p. 230.

Murthy, R.K.V., *Basic Principles of Agricultural Meteorology*, BSPBS Publications, Hyderabad, 2002, p. 261.

NCMRWF, *Development on Agrometeorology in India*, Department of Science and Technology, New Delhi, 1990, p. 96.

Ramdas, L.A., 'Weather and crops', *Handbook of Agriculture*, Indian Council of Agricultural Research, New Delhi, 1961, pp. 1–31.

Ranganathan, C., *Cyclone Havoc and Other Weather Phenomena*, Tamil Nadu State for Science and Technology, Chennai, 2000, p. 115.

TNAU, *National Interactive Workshop on Cyclone Disaster Management*, Tamil Nadu Agricultural University, Coimbatore, 2002, p. 110.

QUESTIONS

1. Define the following:
   (a) General circulation
   (b) Atmospheric Pressure
   (c) Front
   (d) Wind rose
   (e) Buys Ballot law
   (f) Standard atmosphere.
2. Distinguish between:
   (a) Pressure gradient force and Coriolis force
   (b) Geostrophic wind and Gradient wind
   (c) Equatorial low and subtropical high
   (d) Wind vane and anemometer
   (e) Cyclones and anticyclones
   (f) Barometer and Barograph.
3. State true or false:
   (a) Coriolis effect is more at equator.
   (b) Geostrophic wind moves parallel to isobars.
   (c) Beaufort's scale is used to estimate wind direction.
   (d) T-number stands for anticyclone system.
   (e) Low pressure is seen in cyclonic system.
   (f) The direction of wind is deflected towards its right in Northern Hemisphere due to pressure gradient force.
   (g) The balance between pressure gradient force and Coriolis force is known as hydrostatic balance.
   (h) Wind rose is constructed using the difference in atmospheric pressure.
   (i) Anemometer is used for measuring wind direction.
   (j) Liquid is used in aneroid barometer.
   (k) Mercury level in the barometric tube rises or falls due to change in atmospheric temperature.
   (l) Subtropical high pressure system are noticed between 60 and 70° N & S latitudes.

4. Write short notes on:
   (a) Units of atmospheric pressure
   (b) Vertical structure of atmosphere
   (c) Air masses and fronts
   (d) Cyclones and anticyclones
   (e) Wind roses
   (f) Trade and anti trade winds
   (g) Beaufort scale
   (h) Wind instruments
   (i) Distribution of atmospheric pressure over India.
5. Elucidate pressure and wind systems.

5 ATMOSPHERIC MOISTURE

The amount of water vapour that is present in atmosphere is known as atmospheric moisture or humidity. It is mostly dependent on the source of water at the earth's surface and its microclimate very near the evaporating surface. It is said to be *moist air* if the water vapour content is relatively high, and *dry air* if the water vapour content in the atmosphere is low. The water vapour in atmosphere is less in arid climates like in Rajasthan and high in humid climates like in Kerala. It has greater significance in day-to-day weather changes since it is highly variable. Since water is seen in three forms viz. liquid, vapour and solid between the earth and atmosphere, it becomes complex to understand the dynamics of atmosphere and it is a difficult task for weathermen to predict weather. It is more so in the humid tropics. Humidity plays a major role in crop evapotranspiration and keeping the quality of crop produce. If the humidity is more in the atmosphere, crop evapotranspiration may be affected. The quality of crop produced in the greenhouse is better due to regulation of temperature and humidity. It is more so in case of vegetable cultivation inside the greenhouses. At the same time, high humidity is favourable for incidence of various crop pests and diseases. The incidence of pest and diseases is also comparatively less in the greenhouse cultivation, due to regulation of humid conditions. Humidity and temperature also play a key role in soil moisture and nutritional balance of crops and thereby in crop production. Like atmospheric temperature and pressure, the water vapour content also decreases with height, since the source of water is at the earth's surface. It is one of the reasons why all the clouds are seen only up to a maximum height of 8 to 10 kilometres. It is more than 1% near the earth's surface and reaches near 0 beyond 8 kilometres height in atmosphere. A sharp dicrease in water vapour is noticed like in case of atmospheric pressure up to the height of 6 kilometres from the earth's surface (see Fig. 5.1).

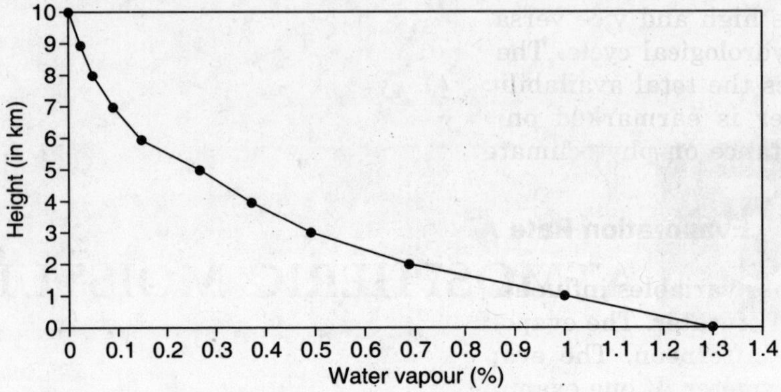

Fig. 5.1 Variation of water vapour (%) with height in atmosphere.

5.1 EVAPORATION

Evaporation is dealt with first in this chapter, as it is the process through which water vapour is present in atmosphere. Water bodies like oceans, rivers, canals, reservoirs, wells and ponds are the major sources of water for evaporation. It is a physical process which leads to movement of water from water and land surfaces to the atmosphere. The rate of evaporation is defined as the amount of water lost by evaporation from a surface unit area in unit time, and is expressed in depth units as in case of rainfall. When air is in contact with water, water is converted into water vapour to atmosphere. Evaporation needs some energy for water to change into vapour phase. This energy is usually supplied by the body from which the evaporation takes place. In the process, the evaporating body cools. For example, we suddenly feel cold when we come out of bathroom after bathing. Why? The vapour pressure of the water over the skin exceeds that of the air and heat from the body, and this supplies the energy to evaporate the water and we feel cold as the body temperature is less. Such cooling of the body becomes most important when air temperature exceeds body temperature. Heat conduction from the body then ceases and evaporation alone regulates the body temperature. It shows that evaporation has a powerful effect on temperature. We know that 585 calories of heat energy is required to evaporate one gram of water at 20°C. If one gram of water evaporates into 1 kg of air, the air will cool by 2.5°C. Hence, the water vapour in atmosphere depends on the availability of water at the earth's surface and its evaporation rate. Again, it depends mostly on the state of the evaporating surface, solar energy, temperature of the evaporating surface and the air just above it, vapour pressure deficit and wind speed just above the evaporating surface. The water vapour content in atmosphere is more if evaporation rate is high and vice versa. It is one of the important components in the hydrological cycle. The balance between rainfall and evaporation decides the total availability of water at a given location. A separate chapter is earmarked on evapotranspiration, keeping in mind its importance on phytoclimate and water management of crops.

5.1.1 Evaporation Rate Across India

Weather variables influence evaporation to a large extent if there is no dearth of water. The evaporation rate is less in the forenoon and more in the afternoon. The evaporation is measured using the open pan evaporimeter. If one examines closely, the pan evaporation at 0830 hr gives higher value when compared to the value recorded at 1430 hr irrespective of season and location. The reason is that the high value recorded in the morning was due to high temperature together with wind prevailing during the afternoon hours of the previous day. Though many weather variables influence the rate of evaporation, it is predominantly influenced by a few weather factors depending upon local weather conditions. For example, it is the surface air temperature and saturation vapour pressure deficit during summer which largely influence evaporation. It is more so in the arid and semiarid climates. On the other hand, dry wind influences the evaporation rate more than temperature during winter in the humid tropics, where wind speed is abnormally high. The evaporation rate is normally low during winter season across the country when compared to summer at a majority of the locations. The low evaporation rate is due to less temperature over evaporating surface during winter. The annual evaporation at the selected locations indicated that the evaporation is high (2637 mm) at Ahmedabad, followed by Bikaner (2583 mm), Hyderabad (2422 mm), Visakhapatnam (2304 mm), New Delhi (2300 mm) and Chennai (2003 mm). The evaporation rate is low along the West Coast, viz. Thiruvananthapuram (1531 mm), Mangalore (1600 mm) and Mumbai (1425 mm). The lowest annual pan evaporation is recorded at Srinagar (783 mm) followed by Kolkata (1101 mm). It also reveals that the pan evaporation rate varies between 4 and 7 millimetres per day on an average across the country except at Srinagar (2.1 mm/day). However, significant variations are noticed seasonally in pan evaporation from location to location, depending upon the local microclimatic conditions (see Table 5.1).

Table 5.1 Seasonal pan evaporation over selected stations across India

Station	*Pan evaporation* (mm)				
	Summer	*Monsoon*	*Post monsoon*	*Winter*	*Annual*
Ahmedabad	959(10.4)	824(6.8)	382(6.3)	472(5.2)	2637(7.2)
Bangalore	613(6.7)	460(3.8)	211(3.5)	407(4.5)	1691(4.6)
Bareilly	561(6.1)	532(4.4)	147(2.4)	157(1.7)	1397(3.8)
Bikaner	874(9.5)	1103(9.0)	324(5.3)	282(3.1)	2583(7.1)
Chennai	627(6.8)	771(6.3)	235(3.9)	370(4.1)	2003(5.5)
Hirakud	561(6.1)	450(3.7)	174(2.9)	214(2.4)	1399(3.8)
Hissar	782(8.5)	916(7.5)	242(4.0)	217(2.4)	2157(5.9)
Hyderabad	809(8.8)	768(6.3)	333(5.5)	512(5.7)	2422(6.6)
Indore	938(10.2)	620(5.1)	260(4.3)	336(3.7)	2154(5.9)
Kolkata	417(4.5)	366(3.0)	146(2.4)	172(1.9)	1101(3.0)
Mangalore	558(6.1)	362(3.0)	225(3.7)	45(5.1)	1600(4.4)
Mumbai	485(5.3)	423(3.5)	223(3.7)	295(3.3)	1426(3.9)
New Delhi	836(9.1)	845(6.9)	309(5.1)	310(3.4)	2300(6.3)
Patna	643(7.0)	573(4.7)	184(3.0)	222(2.5)	1622(4.4)
Srinagar	202(2.2)	430(3.5)	80(1.3)	71(0.8)	783(2.1)
Thiruvananthapuram	454(4.9)	457(3.7)	214(3.5)	406(4.5)	1531(4.2)
Vishakhapatnam	773(8.4)	730(6.0)	311(5.1)	490(5.4)	2304(6.3)

(Figures in parenthesis indicate pan evaporation in mm/day.)

In central Kerala (Palakkad and Thrissur districts), where the wind speed is high during winter due to Palakkad Gap, the evaporation rate on some days is very high and comparable to summer in the arid climates. The evaporation rate shoots up to 10 mm per day on some occasions across the Palakkad Gap during December and January. It is attributed to strong winds, which prevail during winter (*Mundakan* season locally), and high saturation vapour pressure deficit when compared to other places in Kerala. In the humid tropics where monsoon rainfall is very high, the evaporation is low during monsoon, post-monsoon and winter. It is high during summer as seen at Pilicode (see Fig. 5.2). The least is

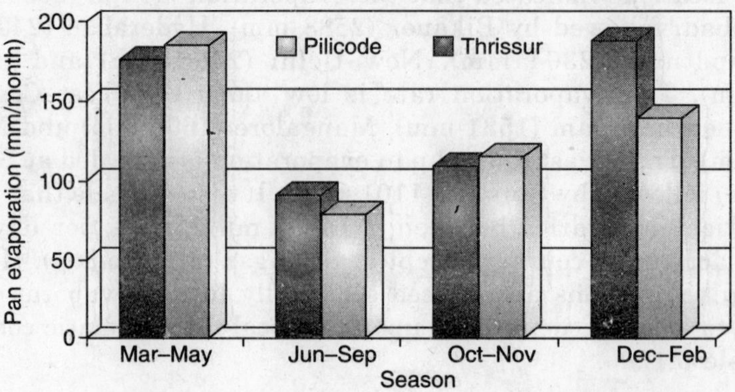

Fig. 5.2 Seasonal pan evaporation (mm) at Vellanikkara and Pilicode.

recorded during the monsoon (less than 100 mm per month, followed by post-monsoon (just above 100 mm per month). Unlike at Pilicode, the monthly pan evaporation at Vellanikkara is more than 180 mm in winter, when compared to that of summer (150 mm per month). It is an unique feature over the central part of Kerala due to high wind.

5.1.2 Daily Pan Evaporation in Humid Tropics

As seen in previous section, evaporation in the humid tropics is controlled by monsoon winds during the monsoon season (see Fig. 5.3). Due to low saturation vapour pressure deficit and cloudy weather, the evaporation rate is the least, varying between 1 and 4 millimetres per day. It is a unique feature in the humid tropics where heavy monsoon is noticed. Even in the central part of Kerala (Palakkad and Thrissur Disricts), the daily pan evaporation never exceeds 4 millimetres per day during monsoon. In the lull monsoon year 2002 (all-India drought year), a maximum of 4 millimetres per day only was noticed while just above 1 millimetre was the minimum. In contrast, the pan evaporation is very high during winter (not in summer), varied between 4 and 10 mm per day. In summer, it shoots up to 8 mm per day. More than temperature, it is the wind energy and vapour pressure deficit which control evaporation in the humid tropics.

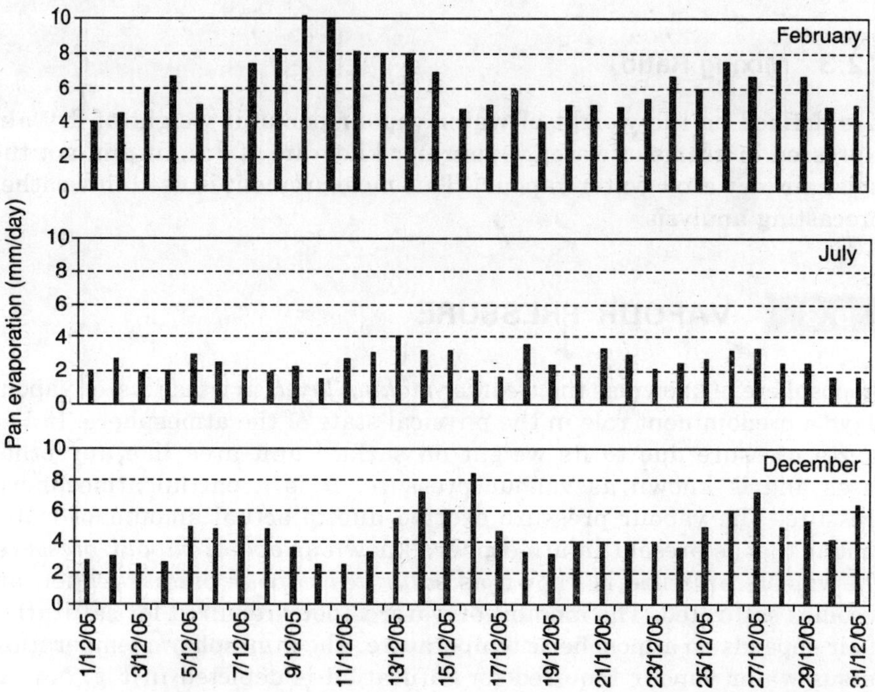

Fig. 5.3 Daily pan evaporation (mm) at Vellanikkara in selected months during 2005.

5.2 MEASUREMENT OF HUMIDITY

The actual moisture content of atmosphere or humidity is expressed in different forms, viz. absolute humidity, specific humidity, mixing ratio, vapour pressure, relative humidity and dew point temperature.

5.2.1 Absolute Humidity

The weight of water vapour per unit volume of air is known as absolute humidity and is expressed as grams of water vapour per cubic metre of air (g/m^3). It is not generally referred to, because absolute humidity varies with the expansion or contraction of air, though the amount of water vapour is constant.

5.2.2 Specific Humidity

It is the weight of water vapour per weight of a given mass of air, including the water vapour, and is expressed as grams of water vapour per kilogram of air (g/kg). It is commonly used because it represents the quantity of water vapour. Specific humidity of air changes only as the quantity of water vapour undergoes a change, which is dependent on temperature.

5.2.3 Mixing Ratio

It is defined as the weight of water vapour per unit weight of dry air, expressed in grams of water vapour per kilogram of dry air and not the moist air (dry air + water vapour). This measurement is used in weather forecasting analysis.

5.3 VAPOUR PRESSURE

Troposphere is known as the *weather-making layer*, in which water vapour plays a predominent role in the physical state of the atmosphere. It also exerts pressure due to its weight on surface unit area like any other gases and is known as vapour pressure. It is a partial atmospheric pressure. The vapour pressure exerted due to actual amount of water vapour that is present in atmosphere known as *actual vapour pressure*. The vapour pressure is known as *saturated vapour pressure* when air becomes saturated. The amount of water vapour required for saturation of air depends on atmospheric temperature. The atmospheric temperature versus water vapour required for saturation is depicted in Fig. 5.4. At lower temperatures, it requires very little amount of water vapour to saturate air and vice versa. For example, the amount of water vapour required is 10 g/kg of air at 15°C 20 g/kg at 25°C Celsius and 35 g/kg at 35°C. The saturation vapour pressure is a function of temperature. There is a logarithmic relationship between them. With temperature increase, the saturation vapour pressure increases exponentially.

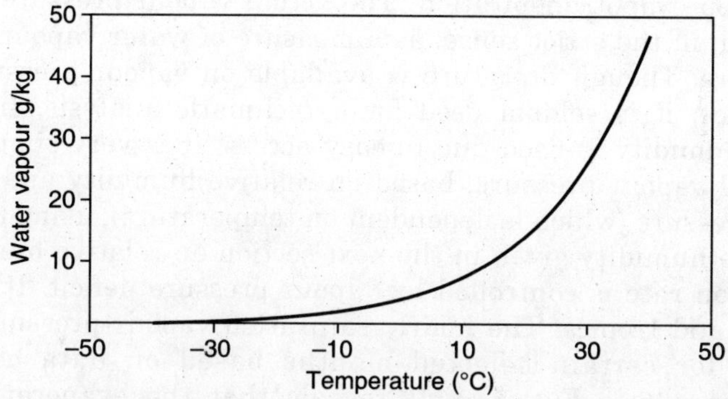

Fig. 5.4 Variation of water vapour with temperature.

Log 10 e_s = 0.02604T + 0.82488, where e_s is the saturation vapour pressure and T is temperature in °C. For water below 0°C, the surface for reference must be ice or super-cooled water rather than a liquid and then a different expression applies.

At higher atmospheric temperature, the saturation vapour pressure is high and vice versa. The saturation vapour pressure is 6.5 mm of Hg at 5°C, 9.2 mm of Hg at 10°C, 12.8 mm of Hg at 15°C, 17.5 mm of Hg at 20°C, 23.8 mm of Hg at 25°C, 31.8 mm of Hg at 30°C and 42.2 mm of Hg at 35°C (see Fig. 5.5). If the atmosphere is said to be unsaturated, it means that it can hold moisture further till it is saturated, which is a function of air temperature.

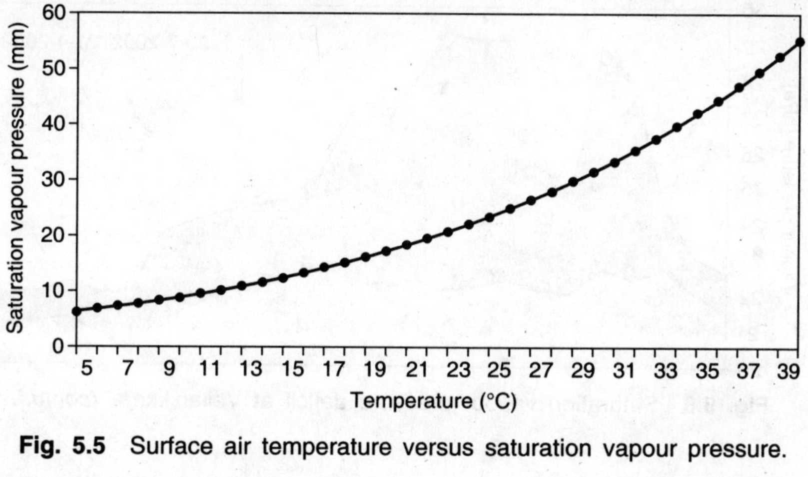

Fig. 5.5 Surface air temperature versus saturation vapour pressure.

5.3.1 Vapour Pressure Deficit

The difference between the saturation vapour pressure (the moisture holding capacity of air) and its actual water vapour pressure is termed *saturation vapour pressure deficit*. It is the one which plays a significant role in crop evapotranspiration. The actual vapour pressure is the one to be used, in the strict sense, as a measure of water vapour content in atmosphere. Though literature is available on vapour pressure and its significance, it is seldom used for agroclimatic analysis. Instead, the relative humidity is used due to easy access. It is very simple to work out actual vapour pressure, based on relative humidity and saturation vapour pressure (which is dependent on temperature), using the formula of relative humidity given in the next section on relative humidity. The evaporation rate is controlled by vapour pressure deficit. It is more so in the humid tropics. The hourly saturation vapour pressure deficit is depicted for certain selected months based on data collected at Vellanikkara (see Fig. 5.6). It reveals that the evaporation rate is influenced by the saturation vapour pressure deficit.

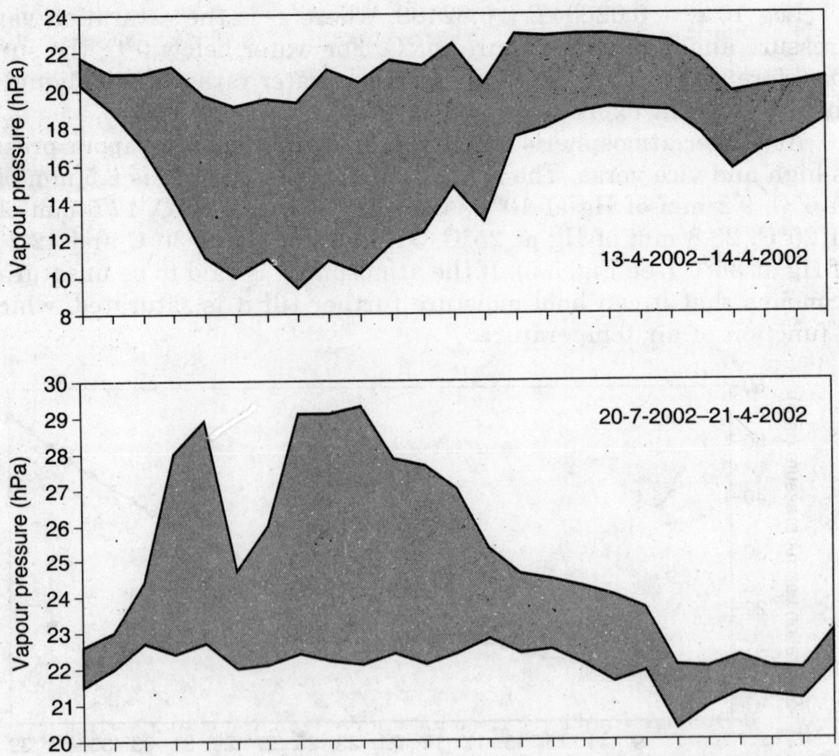

Fig. 5.6 Saturation vapour pressure deficit at Vellanikkara *(contd.)*...

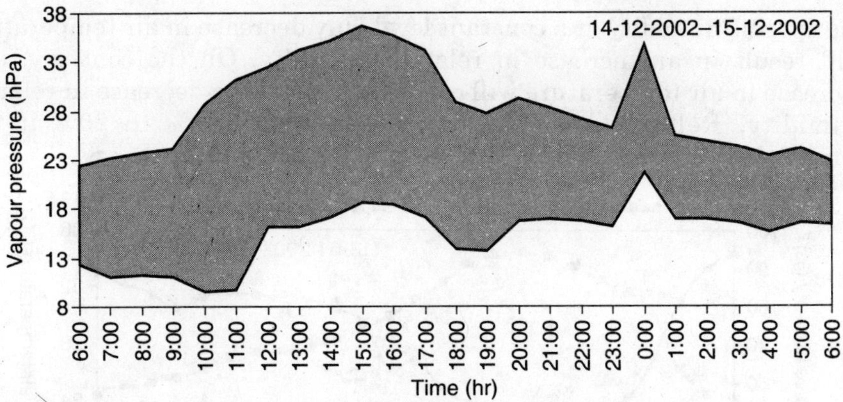

Fig. 5.6 Saturation vapour pressure deficit at Vellanikkara.

5.4 RELATIVE HUMIDITY

Relative humidity is the most commonly used parameter to express humidity of air, and all the students and researchers are familiar about it. It is defined as the ratio of the amount of water vapour that is actually present in air (actual vapour pressure) to the amount of water vapour that air can hold at its maximum capacity (saturation vapour pressure) at a given temperature and expressed as percentage. The relative humidity is 100% if air is saturated; this is otherwise known as *unsaturated air*. The relative humidity can be expressed as follows:

$$\text{Relative humidity} = \frac{\text{Vapour pressure } (e)}{\text{Saturation vapour pressure } (e_s)} \times 100$$

Many correlation studies with meteorological variables find that if any dependent variable has positive correlation with relative humidity, it has negative correlation with maximum temperature and vice versa. It is due to multi-collinearity effect, as the relative humidity is low if the maximum temperature is high and vice versa. One should be careful in analysing the data to understand the effect of relative humidity or maximum temperature on dependent variable.

5.4.1 Temperature and Relative Humidity

Relative humidity is high in early morning till 0830 hrs, when surface air temperature is low. A gradual decrease is noticed in relative humidity during the day. Again, it starts increasing in the afternoon, and is high in the nights and early morning hours due to low surface air temperature and its less specific humidity (air capacity to hold less water vapour at low temperature). It clearly shows that the relative humidity is a function of temperature and its water vapour holding capacity (see Fig. 5.7). With the specific humidity at a constant level, any decrease in air temperature will result in an increase in relative humidity. On the contrary, any increase in air temperature will cause corresponding decrease in relative humidity. Relative humidity decreases from 100% to 20% if the temperature increases from 5°C to 35°C (see Fig. 5.8).

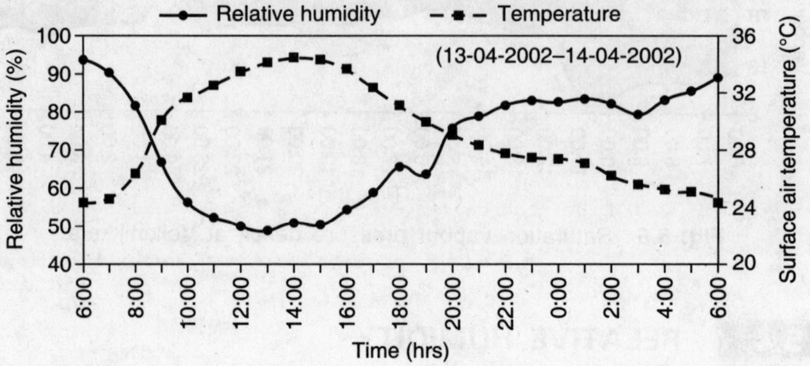

Fig. 5.7 Hourly relative humidity and surface air temperature at Vellanikkara.

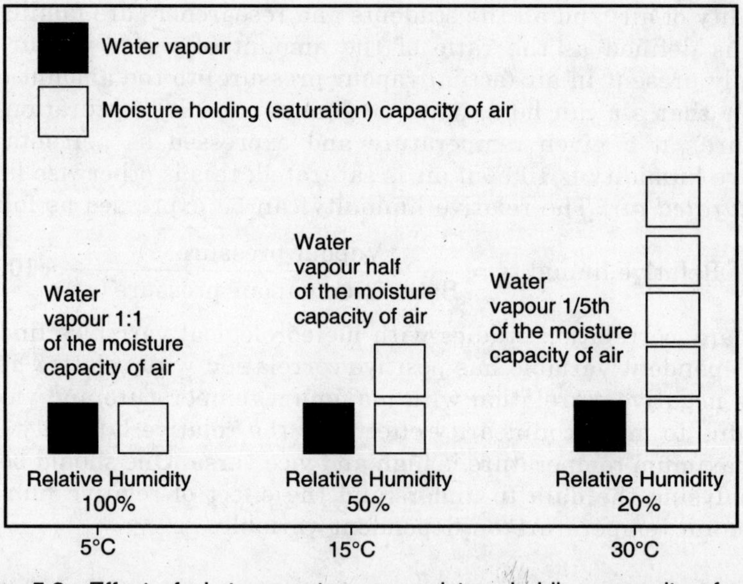

Fig. 5.8 Effect of air temperature on moisture holding capacity of air.

5.4.2 Monthly Distribution of Relative Humidity

A gradual increase in mean relative humidity is noticed from January (57%) to July (87%) and thereafters it decreases from 86% in August to 62% in December at Vellanikkara. Similar is the trend at Pilicode too, relatively with high relative humidity. It indicates that the relative humidity is low during winter and high during monsoon, post-monsoon and summer (see Fig. 5.9). It is the characteristic feature of the humid tropics as considerable atmospheric moisture is carried by monsoon wind during monsoon, which results in high relative humidity. Otherwise, the relative humidity is low during winter and high in summer in dry, semi-arid and arid climates.

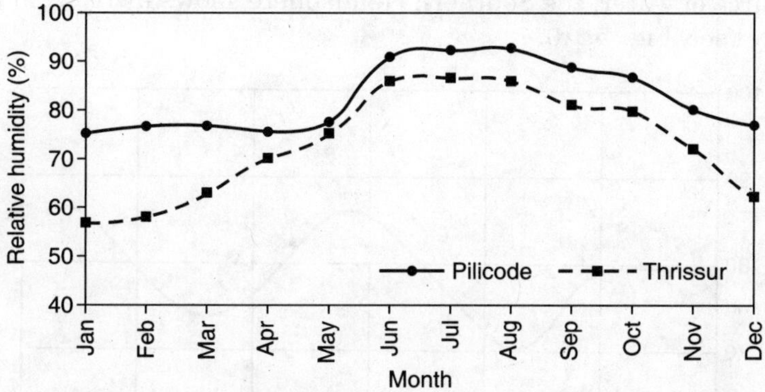

Fig. 5.9 Mean monthly relative humidity (%) at RARS, Pilicode and Vellanikkara.

5.4.3 Relative Humidity Over Coastal and Dry Climates

Evaporation is the main physical process which converts water from land and water surfaces to water vapour into atmosphere. Since oceans are the main source of water, water vapour content in atmosphere is more along the coast when compared to inland stations, where the influence of huge water mass like oceans is relatively less. Though Kerala, Tamil Nadu and Karnataka are very closely located geographically, it is the coastal climate which makes Kerala humid, while Tamil Nadu and Karnataka are dry in terms of relative humidity. Within Kerala too, there is variation in relative humidity. At Pilicode, the relative humidity is always higher (75–95%) while it is 57–87% at Vellanikkara, which could be explained due to the influence of nearness to sea. The relative humidity is still less at Palakkad and then at Coimbatore, though they are located within 100 km of surface distance. It reveals that the relative humidity is high along the coast and diurnal range is also less, as water vapour regulates the temperature due to sea breeze during the day and land breeze during the night, when compared to inlands where the relative humidity is less and diurnal range is also high.

5.4.4 Zonal Distribution of Relative Humidity

The maximum relative humidity is recorded at the equator, as equatorial air can hold more moisture due to high surface air temperature. A gradual decrease is seen towards the subtropical high-pressure belt. Again, a gradual increase is seen in relative humidity due to lower temperatures from the latitudes of 40° onwards to poles. The relative humidity is low over the Northern Hemisphere when compared to that of the Southern Hemisphere because of difference in distribution of land and ocean. The area under oceans is more in the Southern Hemisphere. Since this is the main source of water, the Southern Hemisphere shows increase in relative humidity (see Fig. 5.10).

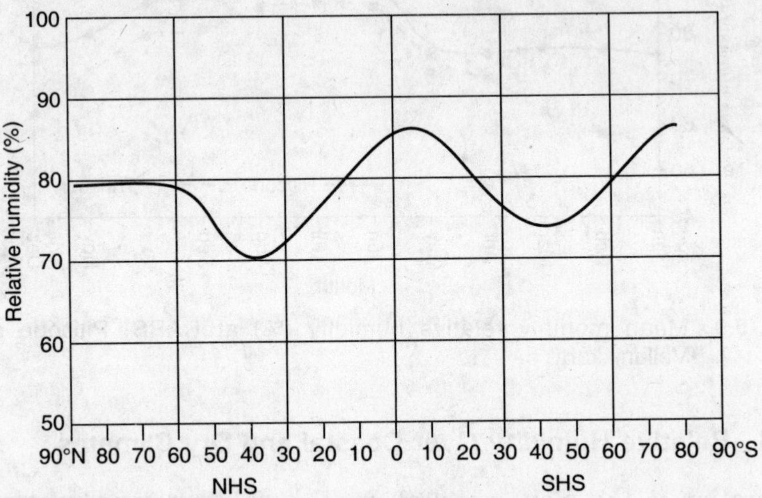

Fig. 5.10 The zonal distribution of relative humidity across the globe.

5.5 PSYCHROMETRY

The most convenient method of determining the humidity of the air for meteorological purposes is by the use of wet and dry bulb thermometers. This is known as *psychrometry*.

August's modification of Regnault's formula for deriving the vapour pressure, which has been in use in the India Meteorological Department since 1876 (with temperatures in °C instead of °F), is as follows:
For temperatures of the wet bulb below 0°C

$$x = \frac{f - 0.480(T - T')}{671 - T'} \times p$$

For temperatures of wet bulb above 0°C

$$x = \frac{f - 0.480(T - T')}{610 - T'} \times p$$

where

x = pressure vapour present in the air

f = saturation vapour pressure at temperature T' °C of the wet bulb
T = temperature of the dry bulb in °C
T' = temperature of the wet bulb in °C
p = pressure of air

It is well-known that the readings of the wet bulb thermometer depend not only on the humidity but also on the strength of the wind, which flows past the bulb, so that at a particular humidity of air, the depression of wet bulb below the dry bulb is less in still air than when the air is in movement. For temperatures of wet bulb higher than 0°C, Pernter modified Regnault's equation is in use so as to correspond to (i) calm wind (ii) light wind and (iii) strong wind. Under the conditions of exposure obtained in a Stevenson screen, the velocity of air past the wet bulb generally comes under 'light wind'. Pernter's modification of Regnault's formula for 'light wind' conditions almost exactly agrees with August's formula. August's equation has therefore been used in the India Meteorological Department for a long time for calculating the hygrometric tables. It must be remembered that these tables are strictly applicable only when the thermometers are in a Stevenson screen, or are exposed to a gentle breeze at the time of observation. They are not to be used with any form of artificially ventilated psychrometers, for which separate slide rules are provided to work out relative humidity based on dry and wet bulb thermometers.

5.6 DEW POINT TEMPERATURE

It is another measure of humidity. It is the temperature to which the air would become saturated if cooled at constant pressure without addition or removal of water vapour. In other words, the temperature at which the saturation vapour pressure is equal to actual vapour pressure is known as *dew point temperature*. If the dew point temperature is very close to the air temperature, the air becomes saturated. The dew point temperature could be worked out using hygrometric tables, based on dry and wet bulb thermometers readings at a given location. The diurnal variation in dew point temperature on 13.04.2002 and 14.04.02 indicated that the dew point temperature never coincided with actual air temperature at any point of time, though both the temperatures were close in early morning hours, indicating that the atmosphere is relatively in saturated condition during nights when compared to daytime (see Fig. 5.11). However the dry, wet and dew point temperatures were close during July, which is a heavy monsoon month. If the actual temperature falls below the dew point temperature, dew is deposited. It happens during winter months. It is a good water source, and favourable microclimate is produced in arid and semi-arid climates.

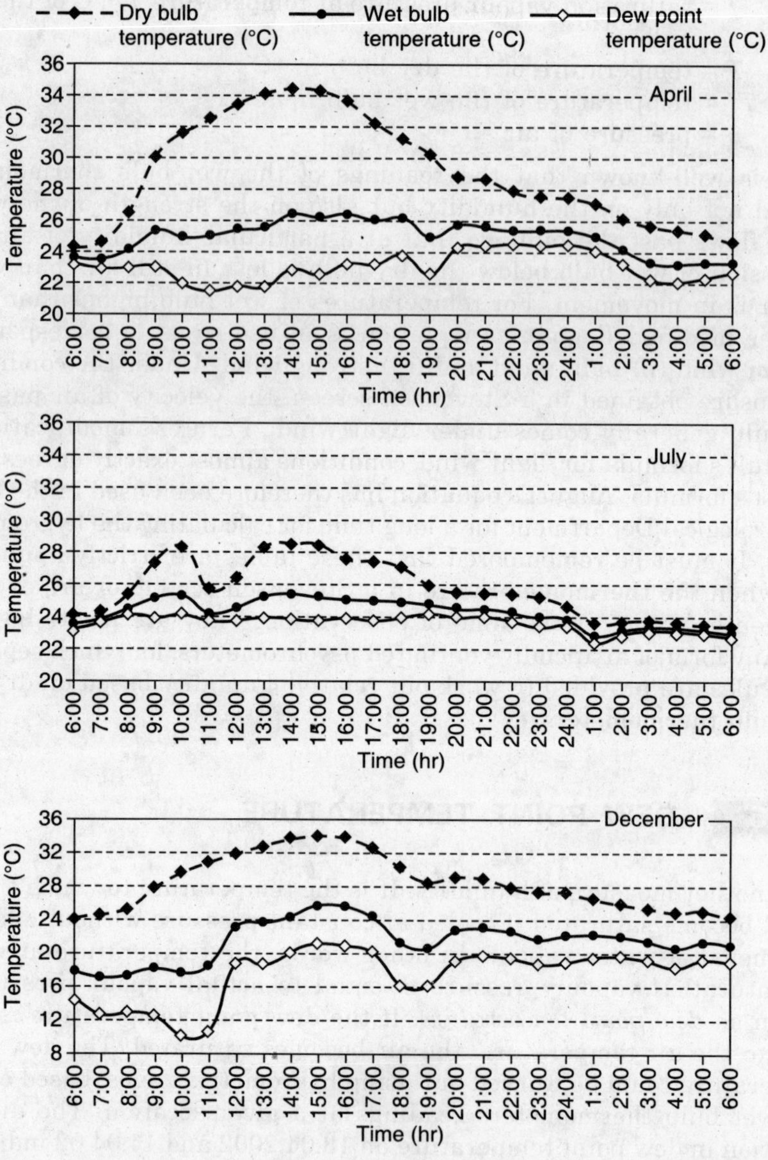

Fig. 5.11 Dry bulb, wet bulb and dew point temperatures at Vellanikkara.

5.6.1 Dewfall

Dew amount is another form of atmospheric moisture which refers to water condensed on the surface of soil, plant and other objects from atmospheric water vapour, due to radiation cooling during the night, when the temperature falls below dewpoint but above freezing. The dew deposits on leaf surfaces reduce the transpiration and water requirement of crops and wetting of leaves increases photosynthetic rate during early hours in the morning. Dewfall forms one of the sources of moisture during winter in the tropics towards north of the equator, starting from October to April in different parts of the country. In the humid tropics of high monsoon rainfall zones, whenever rainfall is not seen, dewfall is common during August and September due to high water vapour content with relatively low temperatures. The dew amount in case of fruit crops like mango and cashew play an important role in producing bisexual flowers. Moderate dewfall with more number of dewnights is conducive for better flowering in cashew. In the humid tropics like Kerala, the dew amount varies from 4 to 14 mm between October and March, the least being in Palakkad and Thrissur districts due to high wind from November to January due to Palakkad Gap. The dew amount also appears to be more in multi-tier cropping systems and mid and high lands due to low temperature when compared to the coastal belt, as temperature is relatively higher along the coast. Though the dewfall has advantage for crops in terms of evapotranspiration, soil moisture and flowering in some fruit crops, it has some disadvantages since dew is conducive for incidence and spread of crop diseases. The dewfall is measured using simple Duvdevani dew gauge. It varies with height, depending upon temperature profiles based on re-radiation cooling. Normally, the dewfall is high at 50 cm height (see Fig. 5.12). In case of Thrissur and Pattambi, the dewfall is high at the height of 100 cm from the ground. The bare soil cools more slowly than when it is covered with crops or vegetation. It is due to upward flow of heat from subsurface soils in case of bare soils. Dew forms more on vegetation than on bare soil. Calm and clear

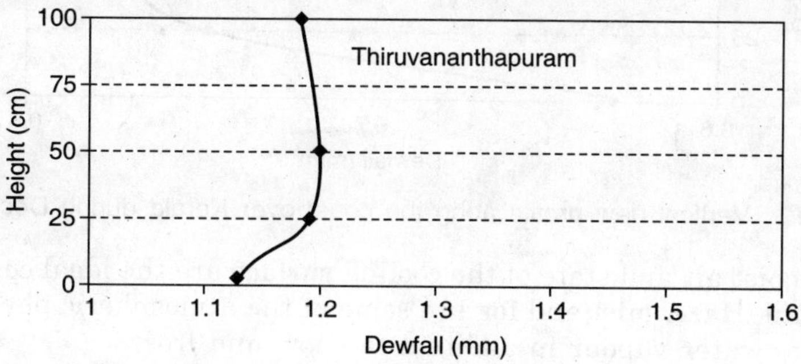

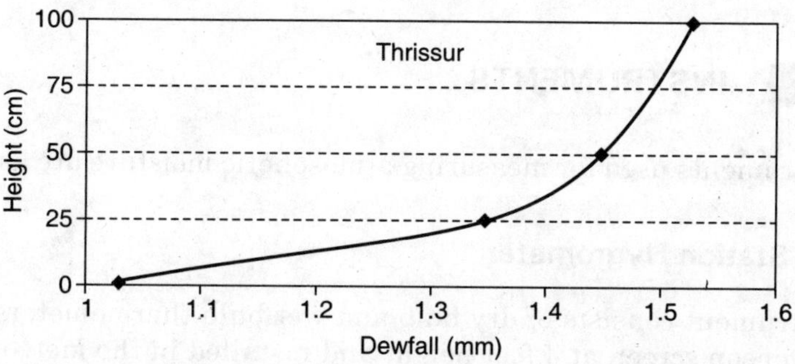

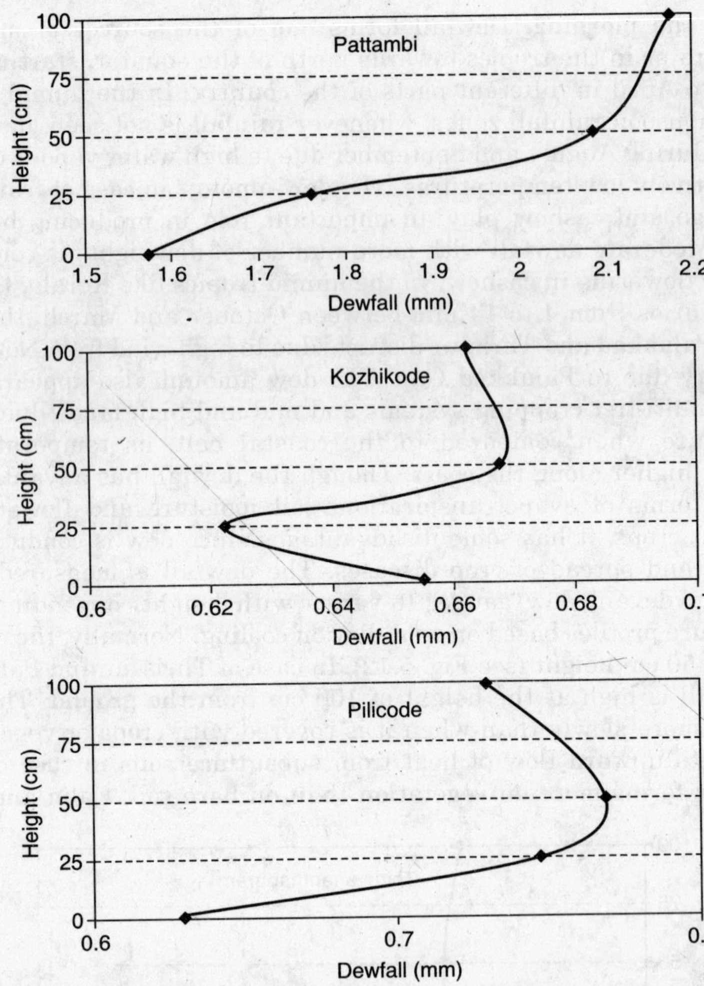

Fig. 5.12 Vertical dew profile aong the coast over Kerala during December.

nights, moist air and state of the cooling surface are the ideal conditions for dewfall. Haze, mist and fog are some of the atmospheric phenomena related to water vapour in addition to snow and frost.

5.7 INSTRUMENTS

The instruments used for measuring atmospheric moisture are as follows:

5.7.1 Station Hygrometer

The instrument consists of dry bulb and wet bulb thermometers kept in the Stevenson screen at 4 feet height and installed in the meteorological observatories to record dry bulb and wet bulb temperatures respectively (see Fig. 5.13). Using the above readings, the relative humidity is obtained from hygrometric tables. Since it is stationary and not movable, it is known as station hygrometer, and is the best to measure relative humidity of the surface atmosphere. However, it cannot be used for micrometeorological studies to measure relative humidity. It is measured in percentage.

Fig. 5.13 Station hygrometer.

5.7.2 Hygrometer

It is a simple device used in controlled environment. It is seen used mostly in greenhouses, mist chambers and incubators, and for household purposes to know relative humidity. Though it is not an accurate one in the strict sense, it gives idea on variations in relative humidity. In floriculture and nurseries, it is commonly used. The sensor in it is human hair, which expands and contracts depending upon variations in relative humidity. If the relative humidity is more, the needle goes up (human hair expands) and vice versa (see Fig. 5.14).

5.7.3 Hygrograph

It is a self-recording instrument in which the relative humidity of the air is continuously traced on a chart fixed to a rotating drum diameter whose diameter is 93.3 mm (see Fig. 5.15). The hygrograph depends upon the contraction and expansion of human hair with variation in relative humidity. A pen transfers these changes on the chart. It is not an accurate instrument but is useful for recording the sudden changes in amount of moisture in the atmosphere. The length of hair increases by about 2–2.5% as the relative humidity changes from 0 to 100%. The

Fig. 5.14 Hygrometer.

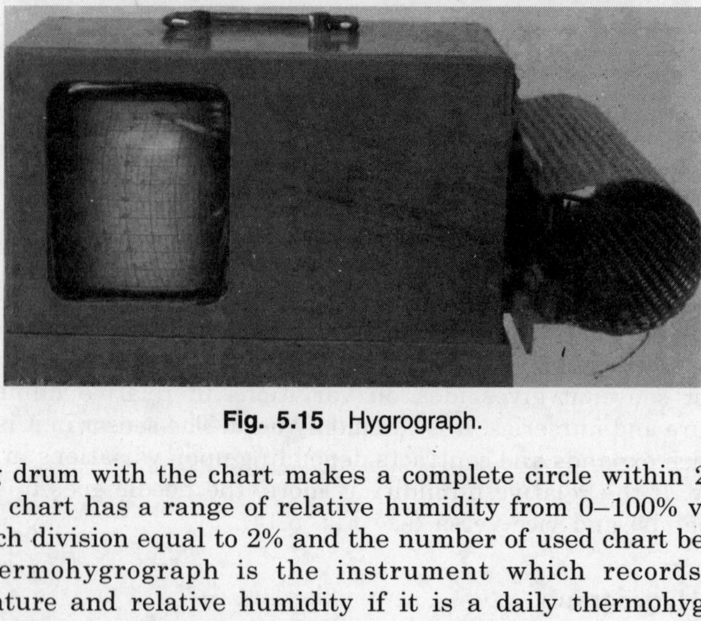

Fig. 5.15 Hygrograph.

rotating drum with the chart makes a complete circle within 24 hours and the chart has a range of relative humidity from 0–100% vertically with each division equal to 2% and the number of used chart being 398. The thermohygrograph is the instrument which records hourly temperature and relative humidity if it is a daily thermohygrograph while weekly record is obtained if it is a weekly one.

5.7.4 Leaf Wetness Recorder

It is a self-recording device used to measure continuously air temperature, relative humidity and leaf wetness. Like in a thermohygrograph, bimetal and human hair are used as sensors to measure temperature and relative humidity respectively in the leaf wetness recorder. The third important weather element used in plant pathological investigations is leaf wetness.

It is the leaf wetness and its duration are more important rather than temperature and relative humidity (see Fig. 5.16). Based on that, new devices like leaf wetness recorder and leaf wetness counters have been developed. The fine thread or strip of sunhemp fibre is the sensor used in leaf wetness recorder for measuring leaf wetness and duration. It is measured between the integers 0 and 5. The leaf surface is dry if it records only 0 and fully wet if it is 5. The dryness to relative wetness varies between 0 and 5.

Fig. 5.16 Leaf wetness recorder.

5.7.5 Leaf Wetness Counter

It is an electronic device, popularised by the ICRISAT to monitor leaf wetness and forewarn on late leafspot of groundnut based on the leaf wetness counter (see Fig. 5.17). If the count exceeds 450, it displays the

Fig. 5.17 Leaf wetness counter.

word 'SPRAY'. It is an indication to farmers for spraying against the late leafspot, which is a menace in groundnut belt of Andhra Pradesh. Such studies are not seen in other crops.

5.7.6 Open Pan Evaporimeter

The evaporimeter having a diameter of 122 cm and 25.5 cm depth is placed on a well-levelled wooden board having a height of 12.5 cm on the hard soil (see Fig. 5.18). One still well having 27.5 cm height and 8.5 cm diameter is kept in the evaporimeter to make the water free from any movement and inside it there is a pointed rod fixed. The purpose of the fixed rod inside the still well is to maintain the water level always constant at the time of observation. It acts as a reference point for constant water level. The quantity of water added to bring the water level to the fixed point inside the still well is nothing but evaporation since last observation. It is measured in millimetres.

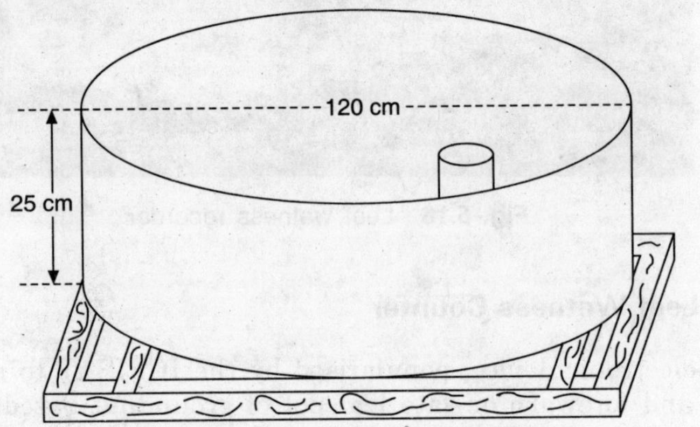

Fig. 5.18 Open pan evaporimeter.

5.7.7 Duvdevani Dewgauge

This instrument consists of rectangular wooden plates of $32 \times 5 \times 2.5$ cm^3 in dimension, coated with a special red paint (see Fig. 5.19). The size, form and distribution of dew drops on the surface of the plates are compared with a series of standard full-scale photographs (dew album), and a dew scale number is given for the amount of dew present on the surface of the plates in the morning. The reading of dew gauge should be taken before the sunrise. The wooden plates are kept at the standard heights of 5, 25, 50 and 100 cm from the ground after sunset and the dew amount is recorded for the above heights before sunrise. The dew conversion scale is as shown in Table 5.2.

The above dew scale is used to convert the dew scale into dew amount. This is a very important observation carried out at the agricultural research stations, where meteorological data are recorded.

Table 5.2 Dew scale and amount

Dew scale	*Amount of dew* (mm)
0	No dew
1	0.02
2a, 2b	0.045
3a, 3b	0.075
4a, 4b	0.11
5a, 5b	0.16
6a, 6b	0.21
7a, 7b	0.27
8	0.35
9	No dew but rain

Fig. 5.19 Duvdevani dewgauge.

REFERENCES/SUGGESTED READINGS

Chritchfield, H.J., *General Climatology*, 4th ed., Prentice-Hall of India, New Delhi, 1994, p. 453.

ICRISAT, *Proceedings of International Workshop on the Agrometeorological Research Needs of the Semi-Arid Tropics*, International Crops Research Institute for Semi-Arid Tropics, Hyderabad, 1978, p. 232.

Menon, P.A., *Our Weather*, National Book Trust, New Delhi, 1989, p. 349.

Molga, M., *Agricultural Meteorology*, Warsaw, 1962, p. 392.

Sacheti, A.K., *Agricultural Meteorology: Instructional-cum-practical Manual*, National Council of Educational Research and Training, New Delhi, 1985, p. 53.

Ventskevich, G.Z., *Agrometeorology*, National Science Foundation, Washington, 1961.

Vitkevich, V.I., *Agricultural Meteorology*, Israel Programme for Scientific Translators, Jerusalem, 1963, p. 320.

Wilsie, C.P., *Crop Adaptation and Distribution*, Eurasia Publishing, New Delhi, 1962, p. 448.

Sellers, W.D., *Physical Climatology*, University of Chicago Press, Chicago, 1965, p. 271.

QUESTIONS

1. Define the following:
   (a) Evaporation rate
   (b) Relative humidity
   (c) Dew point temperature
   (d) Vapour pressure
   (e) Saturation vapour pressure deficit
2. Distinguish between:
   (a) Evaporation and humidity
   (b) Vapour pressure and saturation vapour pressure
   (c) Absolute humidity and specific humidity
3. Write short notes on:
   (a) Evaporation rate across India
   (b) Psychrometry
   (c) Temperature and Relative humidity
   (d) Vapour pressure
   (e) Dewfall
   (f) Duvdevani dewgauge
   (g) Open pan evaporimeter

4. Fill in the blanks:
   (a) Water vapour content with height from the surface ———.
   (b) ——— per cent of water vapour is noticed near the earth's surface.
   (c) Water vapour in the atmosphere is ——— in the arid climates while ——— in the humid climates.
   (d) High humidity is favourable for incidence of ———.
   (e) The pan evaporation rate per day varies between ——— and ——— mm on an average across the country.
   (f) ——— influences evaporation rate during winter over the Palakkad and Thrissur regions.
   (g) Evaporation in the humid tropics is controlled by ——— winds during the monsoon season.
   (h) The saturation vapour pressure is a function of ———.
   (i) ———statistical relationship exists between the saturation vapour pressure and temperature.
   (j) ———, ———, ——— and ——— are the standard heights at which dewfall is measured.
5. State true or false:
   (a) Relative humidity is less at high temperature.
   (b) Relative humidity is less over Southern Hemisphere.
   (c) Relative humidity is high over the equatorial region as temperature is high.
   (d) Saturation vapour pressure increases with increase in temperature.
   (e) Dewfall is more over vegetation when compared to that of bare soil.
   (f) Dewfall is measured at three standard heights.
   (g) Dewfall is generally high over the ground surface when compared to the height of 50 cm from the ground.
   (h) Dewfall is high on a day when wind speed is more.
   (i) Clear skies during nights are not favourable for dew formation.
   (j) Evaporation rate is high when saturation vapour pressure deficit is less.
6. Elucidate the following:
   (a) Measurement of humidity
   (b) Instruments to measure relative humidity
   (c) Distribution of relative humidity

6 CLOUDS

Clouds are nothing but suspensions of minute water droplets or ice crystals, or combinations of both, in the lower atmosphere. One can watch them from the earth surface. They move along with the wind and remain stationary in calm winds. They grow vertically and horizontally, depending upon temperature of the cloud and its ambient temperature. They may dissipate or precipitate. If the ambient temperature of the clouds is more, they dissipate otherwise develop further. Clouds also play a major role in regulating earth's atmospheric temperature. The temporal satellite cloud pictures are useful in weather forecasting. Clouds are formed as a result of condensation of atmospheric water vapour. When low-lying atmosphere is in contact with ocean surface or large water bodies, it causes water to evaporate and rises into higher heights, becomes cooler and condense as clouds. Under further favourable atmospheric conditions, they burst and fall as rain. In the above process, they release large amounts of latent heat. This heat is the fundamental energy source during cyclones which intensifies rain and wind, causing heavy damage to crops and property. Nevertheless, clouds are the primary agents that provide rain and are thus of importance to plant life.

6.1 CLOUDS AND RADIATION BALANCE

Clouds may play a dual role in cooling and warming of the earth. If the sky is totally covered with clouds, clouds become the active surface instead of the earth surface. The incoming solar radiation is reflected back to a large extent from the top of clouds and thus low solar radiation reaches to the earth surface. This may help to keep the planet cool. On the other hand, they absorb longwave radiation and emit back in all directions and transmit little longwave radiation to space. In the process, the atmosphere traps huge amounts of heat emitted from the earth's surface and clouds, leading to the greenhouse effect. In cloudy nights, the effect of cloud is more and keeps atmosphere warmer. Hence, the normal radiation balance between the earth, atmosphere and the space is disturbed due to clouds significantly (see Fig. 6.1).

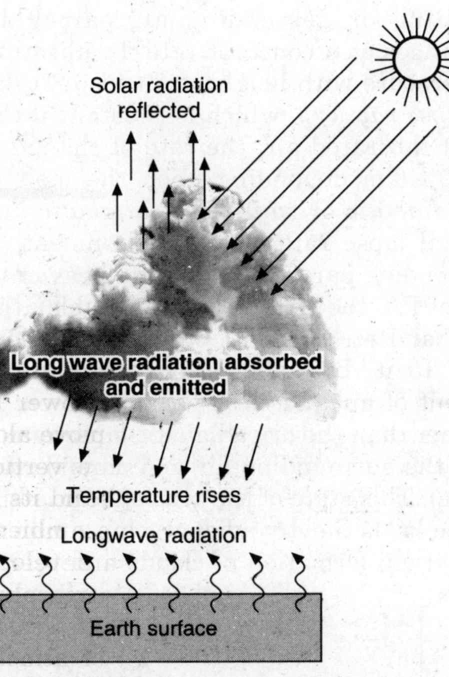

Fig. 6.1 Effect of clouds on radiation.

Clouds in the tropics are known as warm clouds, while they are called cold clouds in the temperate zones. Clouds over the oceans differ in character and heating effects to that of clouds over the continents. Their nature and effects also vary with the season and time of day. Clear skies are seen in winter and heavy clouds during summer. Tropical clouds trap three times as much heat as what clouds trap on a global average, but also reflect so much sunlight that the heating effect is cancelled out. Since the tropics account for 20% of the earth's surface and the effects of the tropical cumulonimbus clouds are disproportionate, the delicate heat balance they maintain is critical. A warmer ocean would likely to cause more clouds to form, increasing both their heating and cooling effect.

6.2 STABILITY AND INSTABILITY

The stability of air is determined by the distribution of air temperature in the atmosphere at various heights. In the free atmosphere, air temperature decreases with height in the troposphere. The decrease in air temperature with height is known as the *normal/environmental lapse rate* (γ). It is at the rate of 6.5°C/km. The lapse rate always varies with time and place. The rate of change of temperature in an ascending or descending air mass through adiabatic process is called the *adiabatic lapse rate.* In the adiabatic process, no heat is added to or subtracted from an ascending or descending air parcel. In this process, its temperature changes at a constant rate. In unsaturated air, the rate of change in temperature with height is about 9.8°C/km. This is known as *dry adiabatic lapse rate* (γ_d), which is constant if the air parcel is lifted adiabatically. In saturated air, the rate of change in temperature with height is 5 to 9°C/km depending upon the moisture content. This is known as *wet adiabatic* or *saturation adiabatic lapse rate* (γ_w).

At the normal lapse rate of the environment, the air is said to be stable as the ascending parcel is colder and heavier than the surrounding air (In the case of PA, the stratification is stable); thus the ascending air sinks downward and settles in its original position unless some outside force is applied to it. In stable atmosphere, the weather is fair. No vertical movement of air and clear skies are seen. If the environmental lapse rate is higher than the dry adiabatic lapse rate, the ascending parcel is warmer than the surrounding air and ascends further (In the case of PB, it is unstable). This state of the atmosphere is known as *unstable* (see Fig. 6.2). The line PC is the dry adiabatic lapse rate. This leads to vertical motion, condensation, formation of clouds and precipitation.

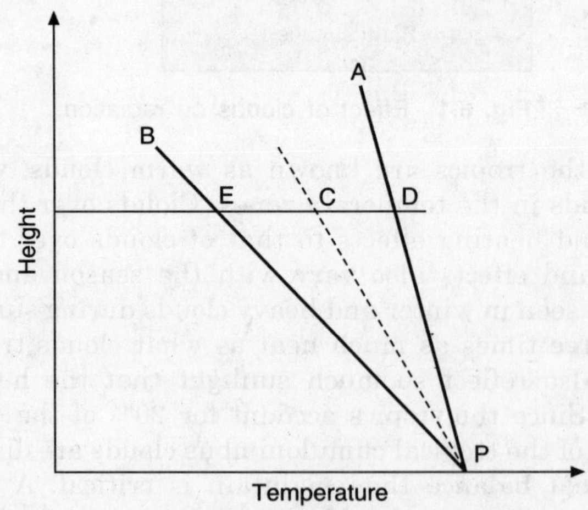

Fig. 6.2 Stability and instability based on hypothetical soundings of temperature.

If $\gamma < \gamma_d$, the atmosphere is said to be *stable.* In stable atmospheric conditions, weather is fair and calm winds are noticed. This weather phenomenon is common during winter. If $\gamma > \gamma_d$, the atmosphere is said to be *unstable.* This phenomena is common during summer, which leads to formation of convective clouds like cumulonimbus.

Atmospheric stability is defined as that condition in the atmosphere in which vertical motions are absent or definitely restricted. Atmospheric instability is the state wherein vertical movement of air is predominant.

6.3 CONDENSATION

The process in which water vapour is converted into the liquid form is called condensation. Water vapour can also be converted directly into solid form; this is called *sublimation.* The latent heat expended in evaporation is liberated in the process of condensation. Falling air temperature, rising water vapour content or the two processes acting together may lead to saturation of the given volume of air. Any further reduction in temperature or rise in water vapour content will thus lead to the process of condensation of water vapour. Therefore for condensation to take place, it is necessary that the air be saturated with water vapour and further cooling of air leads the excess water vapour into liquid. However, 'nuclei of condensation' must be present in the atmosphere before condensation. Condensation may take place due to re-radiation cooling, mixing of different air masses and adiabatic cooling of air masses.

6.3.1 Atmospheric Nuclei

The microscopic particles such as fine sand, crushed soil, crystals of salt, bacteria, viruses, plant pollen and other microscopic organisms which are present in the atmosphere are known either as atmospheric nuclei, condensation nuclei or hygroscopic nuclei. The microscopic particles, which absorb or attract water vapour are called hygroscopic nuclei.The hygroscopic particles, namely sea salts which include sodium chloride and magnesium chloride, sulphur dioxide (SO_2) which oxidises to sulphur trioxide (SO_3) and becomes sulphuric acid when mixed with water, and oxides of nitrogen (specially nitric oxides) act as the atmospheric nuclei. Active volcanoes, ocean spray, fires, and meteors which burn out in the atmosphere, are the sources of a large number of such hygroscopic nuclei. Salt particles carried over the continents from the oceans by wind are supposed to be the most powerful condensation nuclei. The salt nuclei may vary in size from 0.1 to 1 micron with a few giant nuclei as large as 5 to 6 micron. The number of salt nuclei in the air is enormous and varies from 10 to 1000 per cm^3. The combustion nuclei are generally smaller and their number varies considerably with industrial activity. The size and the name of different nuclei are given in Table 6.1.

Table 6.1 Size and the name of different nuclei

Name of the nuclei	*Size of the particle*
Aitken nuclei	< 0.1 micron
Large nuclei	0.1 to 1 micron
Giant nuclei	> 1.0 micron

1 micron $= 10^{-6}$ m

Aitken, the famous meteorologist from USA, discovered the condensation nuclei particles to explain the phenomena of condensation in 1879. Hence, the name *Aitken nuclei* is used for microscopic particles having less than 0.1 micron.

6.3.2 Different Forms of Condensation

Dew, frost, fog and clouds are the different forms of condensation. Of these, the first three are given below, while the discussion on clouds is presented in Section 6.4 onwards.

Dew

Dew consists of tiny droplets of water produced by condensation on surface objects rather than on nuclei in air above the surface. Clear skies, calm air and appreciable quantity of moisture in the air are the necessary conditions for dew formation. Dewfall is seen in different parts of India from September to April. However, dewfall at nights and dew amount are more during December, January and February depending upon the location. It is also discussed in the previous chapter under *Atmospheric moisture.*

Frost

Frost occurs when the dew point temperature of air falls below the freezing point (0°C). When condensation starts with temperature below 0°C, the water vapour in air passes directly from the gaseous to solid state. The northernmost states in India are susceptible to frost during winter.

Fog

Fog is defined as almost microscopically small drops of water condensed from and suspended in air near the surface of the earth in sufficient number to reduce the horizontal visibility to less than one kilometre. Fog is also defined as a cloud with its base at or very near the ground level. In case of a dense fog, visibility is reduced to a few metres. Accidents are common in the early morning, during winter in high ranges in the tropics and at higher latitudes, due to poor visibility.

6.4 CLOUDS

The different forms of clouds and their appearance are depicted in Fig. 6.3. High clouds are also known as *Cirrus clouds*. They are thin and pure white, having thin ice sheets or flakes only. They do not produce rain. They produce red colour during the dawn and the dusk. They dissipate with the progress of sunshine during the day, if they are noticed in the early morning. They are also good indicators for formation of middle clouds in the night if they form during the evening hours. They appear mostly during winter, which is the season for clear skies. The middle clouds are relatively thicker, having ice crystals and minute water droplets. The middle clouds are known as *Altus*. *Stratus* is the lowest horizontal spreading cloud. The low clouds are still thicker, having more water droplets than ice crystals. They grow vertically with uniform base. The low clouds with vertical development are due to high atmospheric instability. They are known as Cumulus and Cumulonimbus clouds. They form due to high convective activity during summer. They carry huge amounts of water droplets. The cumulonimbus clouds give plenty of rain, causing surface run-off and soil erosion. They create *thunderstorms*. A similar activity is known as *duststorm* if it is formed in deserts. They are common in western Rajasthan.

Fig. 6.3 Different forms of clouds with height.

6.4.1 Cloud Genera

All the cloud types have been categorised the world over into four groups, depending on the height at which the base of cloud forms in the atmosphere. They are further subdivided into ten families. It is known as the cloud genera. The cloud genera in different regions of the globe is schematically presented in Fig. 6.4.

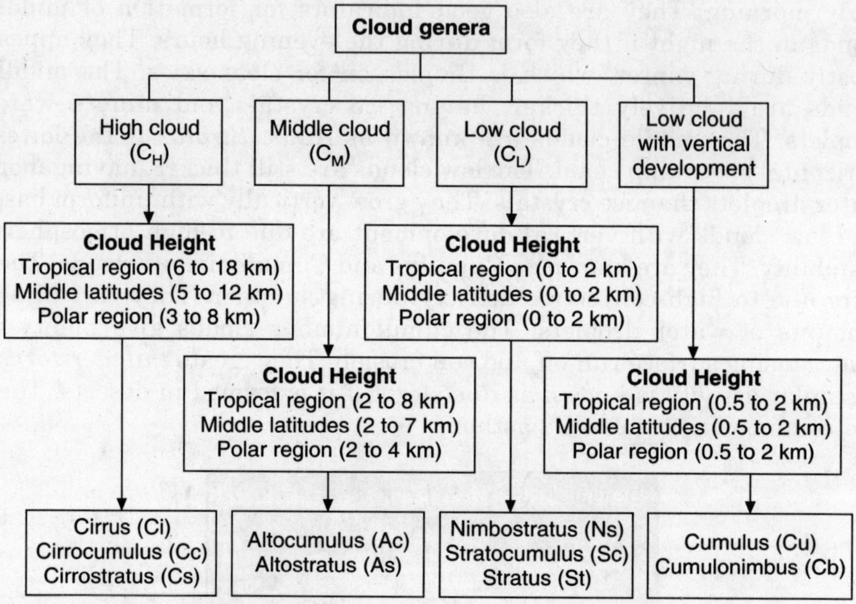

Fig. 6.4 Schematic presentation of different types of clouds.

6.4.2 Name of the Clouds

Cloud names have originated from Latin. The meaning of each type of cloud is given in Table 6.2.

Table 6.2 Cloud names and their meaning

Name of the cloud	*Meaning in Latin*
Cirrus	A lock of hair, a tuft of horse hair, a bird's tuft
Cumulus	An accumulation, a heap, a pile
Stratus	To extend, to spread out, to flatten out, to cover with a layer
Altus	Height
Nimbus	Rainy cloud
Cirrocumulus	*Cirrus* and *cumulus*

(Contd.)...

Table 6.2 Cloud names and their meaning *(Contd.)*

Name of the cloud	*Meaning in Latin*
Cirrostratus	*Cirrus* and *stratus*
Altocumulus	*Altum* and *cumulus*
Altostratus	*Altum* and *stratus*
Stratocumulus	*Stratus* and *cumulus*
Nimbostratus	*Nimbus* and *stratus*
Cumulonimbus	*Cumulus* and *nimbus*

Source: International Cloud Atlas, W.M.O., 1956.

6.4.3 Description of Clouds

The cloud description of all ten families is given along with cloud plates under each height so as to understand better and identify them easily (see Figs. 6.5(a) to (j)).

High clouds (above 6 km)

Cirrus (Ci): The height of formation of cirrus cloud is relatively near tropopause. It is the highest cloud. It is also the thinnest cloud. It appears white and full of ice crystals or flakes due to reflection of sunlight (see Fig. 6.5(a)). The temperature at the height of the cloud is the lowest. It has silky cotton appearance. The moon or sun is seen through them. Cirrus group of clouds does not give rain.

Fig. 6.5(a) Cirrus clouds.

Cirrocumulus (Cc): It occurs in small globular masses or white flakes with or without shadows or with very slight shadows, arranged in groups and often in lines (see Fig. 6.5(b)). They form below the height of cirrus. As a whole, cirrocumulus clouds are very thin with ice flakes. They do not produce shade and look as if they are arranged in sequence. They do not give rain.

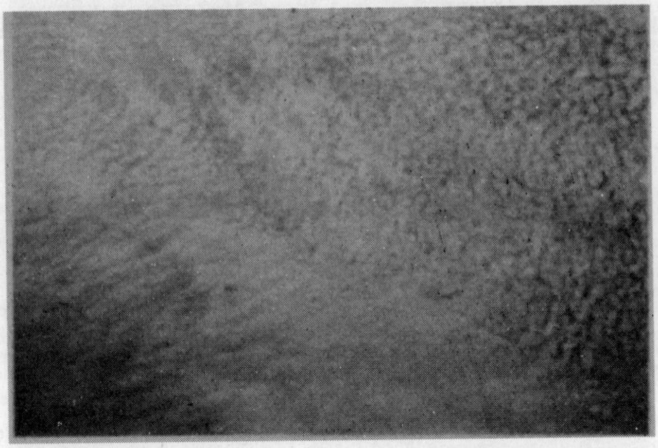

Fig. 6.5(b) Cirrocumulus clouds.

Cirrostratus (Cs): Cirrostratus appears in patches as a thin whitish sheet, covering the sky with a whitish veil, and produces halos around sun and moon (see Fig. 6.5(c)). It is similar to the cirrocumulus (spreading in layers) but is lower. It is composed of ice crystals. It produces different colours at dawn and dusk.

Fig. 6.5(c) Cirrostratus clouds.

Middle clouds (2.5 to 6 km)

Altocumulus (Ac): They are rather large ball-like masses (see Fig. 6.5(d)). They are white or greyish and partially shadded. They are arranged in groups or lines and closely packed, so that their edges appear to touch. They are constantly blown around by wind and often moulded into rolls. They don't produce rain even when they overcast the sky. If they appear continuously, it is an indicator of a low pressure system. They develop into depression.

Fig. 6.5(d) Altocumulus clouds.

Altostratus (As): It is just like cirrostratus, but a thick cloud without halo phenomena (see Fig. 6.5(e)). It falls under middle clouds. The height of this cloud is below the height of cirrocumulus. Its horizontal spread is more and looks like a uniform sheet of milky white. If it becomes thicker and thicker, it gives continuous moderate to heavy rain. It may be seen for two or three days at a stretch.

Fig. 6.5(e) Altostratus clouds.

Low clouds (from earth's surface to 2.5 km)

They commonly cover the whole sky causing grey overcast (mean upper level 6500 feet). Stratocumulus, Stratus and Nimbostratus fall under low clouds.

Nimbostratus (Ns): This is very dense stratiform cloud. This makes a bright day as dark as late evening. This is a low, amorphous layer, usually nearly uniform and of a dark grey colour. It seems feebly illuminated seemingly from the inside. It is thick enough throughout to blot out the sun (see Fig. 6.5(f)). It is the ordinary rain cloud, though not the type which gives the heaviest rainfall. The duration of the cloud depends upon its size and development. A very brief moderate downpour normally occurs for ten or fifteen minutes and stops.

Fig. 6.5(f) Nimbostratus clouds.

Stratocumulus (Sc): The cumulus cloud becomes stratocumulus when top of cumulus get flattened due to stable upper air layer. It is dark in colour and becomes larger when two or three clouds with considerable vertical extent grow combined (see Fig. 6.5(g)). It sometimes produces rains. It starts with big drops of rain, which later become rain with smaller drops and finally a drizzle, before the cloud completely melts.

Fig. 6.5(g) Stratocumulus clouds.

Stratus (St): This is a low, uniform layer of cloud resembling fog, but not resting on the ground. Generally it is a grey cloud layer with a fairly uniform base, which gives drizzle, ice prisms or snow grains (see Fig. 6.5(h)). These clouds are common in States like Uttaranchal, Himachal Pradesh and Jammu and Kashmir. We can observe these clouds during winter in the above regions. Snow grains are common in January and February due to stratus clouds.

Fig. 6.5(h) Stratus clouds.

Clouds with vertical development

They create a broken sky with clear areas between individual clouds.

Cumulus (Cu): Cumulus is a thick cloud with vertical development. The base is nearly horizontal. The upper surface is dome-shaped like cauliflower or cotton bales. The top of the clouds contains ice crystals and the base is with full of water droplets (see Fig. 6.5(i)). If they grow vertically with thick mass, they are known as cumulonimbus clouds.

Fig. 6.5(i) Cumulus clouds.

Cumulonimbus (Cb): This is a heavy mass of cloud with great vertical development. The upper part rises like a mountain, often spreading out in the form of an anvil (see Fig. 6.5(j)). They generally produce heavy showers of rain. This cloud is responsible for thunder and lightning. The anvil shape is composed chiefly of ice crystals. This cloud forms due to convection, mostly seen in the equatorial belt and the tropics. Landslides, floods and surface run-off are the common features while it rains. The lifespan of this type of clouds varies between 20 and 45 minutes in the majority of the cases.

Fig. 6.5(j) Cumulonimbus clouds.

6.5 MEASUREMENT OF CLOUDINESS

When weather observations are made, the extent of the cloudiness of the sky is estimated visually and recorded. The standard unit for reporting the cloud amount is *Okta* or *eighth of the sky*. The total amount of cloud is recorded as 0 when sky is free from clouds and is recorded as 8 when the sky is completely overcast. The symbols used for reporting cloud amount in weather maps according to the number measured in Okta is shown in Table 6.3.

Table 6.3 Symbols for reporting cloud amount

OKTA	0	1	2	3	4	5	6	7	8	9
SYMBOL	○	⦶	◔	◔	◑	◕	◕	◕	●	⊗

It can be converted into percentages very easily. If the measured Okta is 2 the cloud amount is 2/8 which means 25% of sky is covered with clouds. If the sky is obscured, it is reported with a cross in an open circle. The code is 9 for obscured sky. Cloudiness is a very important measure in weather forecasting. The number of bright sunshine hours per day is also another measure for cloud amount.

The following is the cloud amount reported in weather forecast as clear sky, partly cloudy, cloudy and overcast sky:

○ Clear ◔ Partly cloudy ◕ Cloudy ● Overcast

6.6 CLOUD SEEDING

It is discovered that the lack of rain from clouds is due to non-availability of sufficient quantities of hygroscopic/freezing nuclei for condensation or sublimation to induce the growth of raindrops. In 1930, Bergeron demonstrated that raindrops start forming in a cloud when ice crystals appear at a temperature below –15°C. These ice crystals grow rapidly at the cost of super-cooled droplets and fall out of the clouds as rain, hail or snow. Thus, rainfall can be made to occur by introducing suitable hygroscopic or freezing nuclei to accelerate the process of formation of ice crystals. After the end of World War II, experiments were conducted and it was demonstrated that ice nuclei in the air could be increased by introducing into the clouds micron-size particles of silver iodide and solid carbon dioxide. It was also discovered that comparatively warm clouds could be seeded for rainfall with sodium chloride particles. With the discovery of these agents, cloud seeding experiments were initiated and successfully conducted in the USA, the USSR, Australia and several African and Asian countries.

The Indian Institute of Tropical Meteorology has been using common salt (NaCl) for rain forming in the monsoon clouds. The technique involves seeding of clouds with massive doses of common salt. This results to the condensation of water vapour on the salt particles. They induce more moisture into the clouds and their explosive growth is expected. It may lead to rains on further development. Very recently, the governments of Maharastra and Andhra Pradesh (in India) took initiative in cloud seeding as they experience frequent droughts and crop failures have become common. The Government of Andhra Pradesh claims the cloud seeding since 2005 and 2006 yielded positive results as rainfall increased over the cloud-seeded areas. However, the benefits of cloud seeding is yet to be fully realized as it is in the initial stage on operational mode. It is felt that the exploitation of rain in one region through cloud seeding may upset the hydrological balance over the other regions of the globe. It may lead to frequent natural disasters, which may be detrimental to humankind, if the radiation and hydrologic cycles are disturbed through human interventions.

6.7 CLOUD OBSERVATION

The description of clouds can be understood much better by watching them regularly. Observation of all types of clouds during winter is remote because mostly clear sky or Cirrus type only could be seen. The students are also taught that Cumulus clouds are seen in the summer while clear sky and Cirrus in winter. It may not be true always. To my surprise, Cumulus clouds appeared in winter on (1-2-2003 at around 1000 hr and the sky was full of Cumulus by 1300 hr) over Vellanikkara. The cumulus clouds all around started growing vertically as if they were Cumulonimbus and by 1500 hr they became stationary. Being a weatherman, I was expecting that a few of them may grow further into cumulonimbus and expected downpour when I saw the way they were growing. However, all the clouds appeared milky white except a few in grey colour. All the cumulus clouds dissipated by 1700 hr and clear sky was noticed unexpectedly. At about 1900 hr, Cirrocumulus and thick Cirrus were noticed. It was a cloudy night. It rained an amount of 23.6 mm in the night with thunders between 0200 and 0300 hr. Naturally, rain must be from Cumulonimbus clouds. On the next day (2-2-2003) at 0600 hr, the sky was full of Altocumulus with overcast condition and by 0700 hrs, the sky was full of Stratus and Cumulus clouds along the horizon. At the zenith, clear sky was noticed. By 0900 hr, clear skies appeared all around. Again by 1100 hr, the sky was full of Cumulus with overcast condition. It was a cloudy day without rain.

The immediate second event was on 8-2-2003. Relatively big rain drops fell at 0600 hrs for a while, followed by drizzle up to 0730 hr. Drizzle of this nature is very uncommon in this part of Kerala, that too in February during which convective activity initiates. Thereafter Stratocumulus remained till 1100 hr without rain and dissipated. Clear sky was observed throughout the day. The feature of Stratocumulus is that it initiates with big rain drops, followed by drizzle, remains for two or three hours and dissipates (see Fig. 6.5(g)). It reveals that the weather systems are independent and unpredictable. However, regular notes on visual cloud monitoring will be a useful tool to the weathermen in nowcasting or short range weather forecasting.

REFERENCES/SUGGESTED READINGS

Agee, E.M. and Asai, T., *Cloud Dynamics*, D. Reidel Publishing, Dordecht, 1982, p. 423.

Bureau of Meteorology, *Observing the Weather*, Australian Government Publishing Service, Canberra, 1979, p. 39.

Hobbs, P.V. and Deepak, A. (Eds.), *Clouds—Their Formation, Optical Properties and Effects*, Academic Press, London, 1981, p. 497.

IMD, *Users Requirements for Agrometeorological Services*, India Meteorological Department, 1998, p. 336.

Lal, D.S., *Climatology*, Chaitanya Publishing, Allahabad, 1899, p. 420.

WMO, *International Cloud Atlas*, World Meteorological Organization, Geneva, 1956, p. 137.

QUESTIONS

1. Define the following:
   (a) Stability and Instability
   (b) Fog
   (c) Cloud
2. Distinguish between:
   (a) Environmental lapse rate and dry adiabatic lapse rate
   (b) Dew and frost
   (c) Frost and fog
3. Write short notes on:
   (a) Clouds and radiation balance
   (b) Atmospheric nuclei
   (c) Forms of condensation
   (d) Cloud seeding
   (e) Cloud genera
4. Fill in the blanks:
   (a) Dry adiabatic lapse rate of the atmosphere is ———°C/km.
   (b) Environmental lapse rate of the atmosphere is ———°C/km.
   (c) During condensation process ——— heat is liberated.
   (d) The diameter of the Aitken nuclei is ——— μ.
   (e) ——— and ———are the favourable conditions for the formation of dew.
   (f) Cumulonimbus cloud is a ——— cloud.
   (g) Fog is another form of ——— cloud.
   (h) ——— is the unit for measurement of the clouds.
   (i) ——— is the alternative device to measure cloud amount.
   (j) Cloud observation will be a useful tool in ———.
5. Match the following:

(a) Giant nuclei	Stratus
(b) Large nuclei	Altus
(c) Aitken nuclei	$< 0.1\ \mu$
(d) High cloud	Convection
(e) Low cloud	$>1.0\ \mu$
(f) Middle cloud	Cirrus
(g) Cumulonimbus	$0.1–1.0\ \mu$

6. Elucidate the different types of clouds.

7 PRECIPITATION

Precipitation is the deposition of atmospheric moisture on the ground in the form of rainfall, drizzle, snowfall and hail. Precipitation is also defined as particles of liquid water or ice formed within a cloud and falling towards the ground. The precipitated moisture falling on the ground depends on several factors viz. the temperature at which condensation takes place, the type of clouds and their height, and the precipitation generating processes. In the temperate climates, snowfall is common due to low temperature, and rainfall is common in the tropical climates due to warm temperature. Hailstorms are noticed very often during summer due to high convective cumulonimbus clouds.

Precipitation is the principal source of water to all living beings over the earth surface. All the water sources at the earth surface viz. the ground water, the river water, the canal water, the surface wells, the lakes and the ponds depend on precipitation only. Rainfall supports the plant growth throughout its life cycle. The crop growing season depends on the length of the rainy season. The amount and distribution of rainfall decide the final crop output if other environmental factors are not limiting. Among all the weather variables, rainfall and temperature play directly a major role in crop growth, development and yield. Though the selection of crops is based on thermal regime like the temperate and tropical crops, it is the rainfall distribution which finally decides crop production. Paddy can be grown during the monsoon in the humid tropics under rainfed conditions, where rainfall and number of rainy days are high. In contrast, *kharif* pulses and millets are only grown under rainfed conditions in the semi-arid and arid climates due to less number of rainy days and poor rainfall. Water-intensive crops like rice and sugarcane could also be grown well in the arid and semi-arid climates if irrigation is assured through the ground or canal water. It reveals that the cropping pattern could be changed with irrigation facilities in the tropics where the thermal regime is not a constraint unlike moisture regime.

Rainfed crops are given top priority in India as they occupy major share in terms of cultivable area. The rainfed crops are highly vulnerable to weather aberrations. A good monsoon across the country gives a better crop and vice versa, as the rainfed crops are sown with the onset of monsoon, and the crop growth and development depend on the monsoon behaviour, showing the relevance of monsoon rainfall in foodgrains production.

7.1 PRECIPITATION PROCESSES

It may be noted that the only process capable of producing condensation and the resultant precipitation from extensive air masses is the adiabatic cooling. It is also equally true that not all condensation is followed immediately by precipitation. Even though all clouds contain water, some produce precipitation while others do not. In certain cases, precipitated moisture does fall from the clouds, but it gets evaporated in the atmosphere before actually reaching the earth's surface. Only when the cloud droplets, ice pellets or ice crystals grow to such a large size so as to overcome the normal buoyancy and updrafts in the atmosphere, precipitation occurs. It means that a sort of rainfall mechanism must operate in a cloud to reach rainfall at the ground.

7.1.1 Precipitable Water

The total amount of water vapour in the atmospheric column above a point on the earth's surface or in a specific atmospheric layer in kg per unit area is known as the *precipitable water*. In other words, it is nothing but the amount of water that is available to fall as precipitation. It also indicates the amount of moisture available in the cloud-free atmosphere to interact with various radiation streams and thus influences atmospheric heating rates.

7.1.2 Rainfall Mechanisms

The exact process or the processes which cause the cloud droplets to join together to form large raindrops, capable of falling to the earth as precipitation, are not fully understood. However, ice-crystal and coalescence theories have been postulated to explain the above processes.

Ice-crystal theory of Bergeron

Tor Bergeron, an eminent meteorologist from Norway, proposed the ice-crystal theory in 1933. This process is common in cold clouds in which water droplets do not freeze at 0°C. If ice nuclei is present at the top of the clouds, vapour diffuses rapidly from air to ice crystals due to difference in saturation vapour pressure. The saturation vapour pressure is lower over ice than water. As ice crystals grow rapidly at the expense of water droplets, it generates crystals large enough to fall. This results in precipitation.

Collision-coalescence theory

This takes place in warm clouds, which are common in the equatorial and tropical regions. The temperature in the uppermost part of these clouds seldom falls below 7°C. The collision-coalescence theory was proposed by E.F. Bowen from Australia in the 1940s. The occurrence of precipitation from warm clouds involves the coalescence of cloud droplets of different sizes. Since the rate of fall of these unequal particles is different, they collide with each other within the cloud and the larger drops grow at the expense of the smaller ones. In clouds with great vertical thickness and abundant moisture, cloud droplets are repeatedly carried upward and downward by ascending and descending air currents. The larger droplets grow in big size and fall on the ground on its own weight in the form of precipitation. Whatever process or processes are at work, a continuous supply of moisture to the atmosphere over a particular location is the most important condition for ensuing a prolonged precipitation.

7.2 TYPES OF PRECIPITATION

Precipitation is caused due to several mechanisms as explained above. Precipitation is possible if atmosphere is instable and adiabatic cooling takes place, for which the causal factors must be either convectional, orographic or cyclonic. If the precipitation is caused due to the above factors, it is known as *convectional*, *orographic* or *cyclonic* precipitation, respectively.

7.2.1 Convectional Precipitation

Solar radiation is the main source of heat to produce convection currents in the air. It happens mostly during summer. When the solar radiation is high, thermal low is created at the earth's surface due to high surface temperature. The thermal low follows adiabatic cooling results in ascending of air, and results in condensation. With enough continuous flow of atmospheric moisture, heavy rainfall is expected. The pre-monsoon showers are due to convectional precipitation (see Fig. 7.1). This type of precipitation is of a very short duration and consists of heavy showers. The cloud involved in this type of precipitation is cumuli-form. They are mostly cumulonimbus.

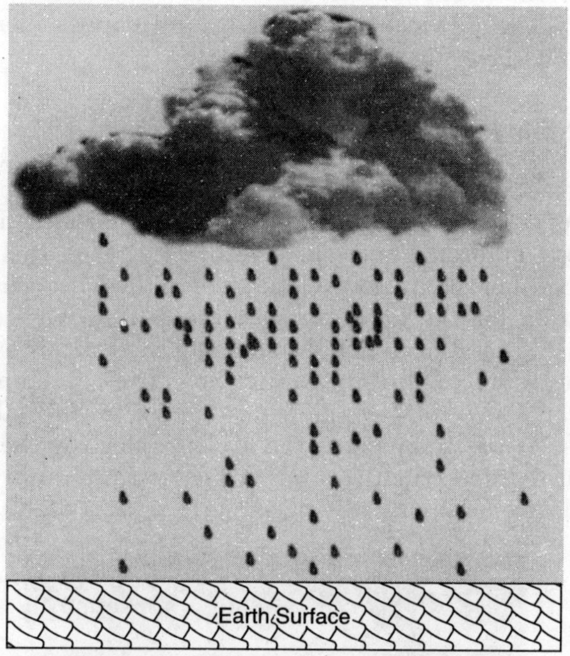

Fig. 7.1 Convectional precipitation.

7.2.2 Orographic Precipitation

In India, southwest monsoon gives copious rainfall on the windward slope of mountains, whereas on the leeward side there are extensive rainshadow regions. The Western Ghats, the Himalayas and the Aravalli

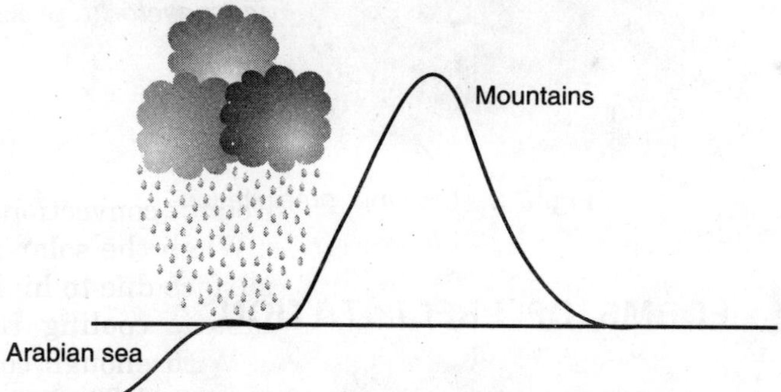

Fig. 7.2 Orographic precipitation.

mountain range are the best examples why rainfall is copious towards windward. When the mountain or high lands acting as barriers to the flow of air, it is forced to rise and air cools adiabatically, resulting in formation of clouds. Precipitation occurs due to the above process during the monsoon. The north-eastern states, east Rajasthan and Kerala are found along the windward side of mountain ranges, benefitting from orographic rains (see Fig. 7.2).

7.2.3 Cyclonic Precipitation

It occurs when deep and extensive air masses are made to converge and move upwards so that their adiabatic cooling results. It takes place when a cyclone is formed over the sea (see Fig. 7.3). In India, the east coast is vulnerable to frequent cyclones, while the west coast is relatively free. The reason for the above is that the sea surface temperature is more, and this is conducive for development of cyclones over the Bay of Bengal when compared to the Arabian Sea. The occurrence of cyclones is common from April to May and from October to November. They give copious rains. At the same time, the cyclones destroy the crops as well as the physical infrastructure due to heavy rains and strong winds, causing the loss of millions of rupees.

Fig. 7.3 Cyclonic precipitation.

7.3 FORMS OF PRECIPITATION

Rain, drizzle, snow, sleet and hail are the different forms of precipitation. The form of precipitation depends upon the instability of atmosphere and temperature. The type of precipitation during winter is snow and sometimes sleet in north India, while it is rain and drizzle during summer or monsoon seasons. In south, snowfall or sleet are not seen due to warm temperature, while rain or drizzle are common depending upon the type of cloud. Of course, hailstorms are not uncommon in India during summer due to high instability in the atmosphere.

7.3.1 Rain

It is the precipitation of liquid water particles either in the form of drops having diameter greater than 0.5 mm or in the form of smaller widely scattered drops. When the precipitation process is very active, the lower air is moist and the clouds are very deep, rainfall is in the form of heavy downpour. On occasions, falling raindrops completely evaporate before reaching the ground. Such streaks of rainfall are called *virgae*.

7.3.2 Drizzle

It is fairly uniform precipitation composed of fine drops of water having diameter less than 0.5 mm (see Fig. 7.4). When the drops of falling precipitation are very small and of uniform size, and seem to be floating in air, it is referred to as drizzle. If the drops in a drizzle completely evaporates before reaching the ground, the condition is referred to as *mist*.

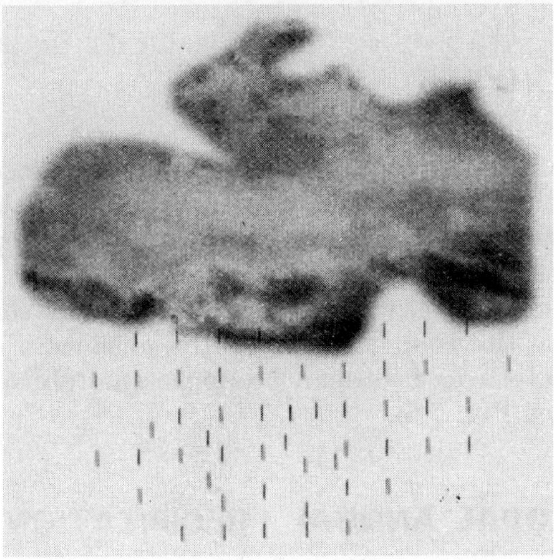

Fig. 7.4 Drizzle.

7.3.3 Snow

It is the precipitation of white and opaque grains of ice. Snow is the precipitation of solid water, mainly in the form of branched hexagonal crystals or stars. In winter, when temperatures are below freezing in the whole atmosphere, the ice crystals falling from the Altostratus do not melt and reach the ground as snow.

7.3.4 Sleet

It refers to precipitation in the form of a mixture of rain and snow. It consists of small pellets of transparent ice, 5 mm or less in diameter. It refers to a frozen rain that forms when rain falling to the earth passing through a layer of cold air and freezes. This happens when temperature is very low. It is not commonly seen in India except in high ranges, that too in winter, in extreme north and northwest India.

7.3.5 Hail

Precipitation of small pieces of ice, with diameter ranging from 5 to 50 mm or something more, is known as hail. Hailstorms are frequent in the tropics. In India, the period from March to May offers the ideal condition for hailstorms. It is the most dreaded and destructive form of precipitation produced during thunderstorms or from cumulonimbus clouds. The crops grown in Telangana region of Andhra Pradesh were badly affected due to hailstorms that occurred in April 1996 and 1997. The wheat crop was damaged in March 1998 due to hailstorm in some parts of Haryana and northern Uttar Pradesh.

7.4 ISOHYETS

Isohyets are the lines, on a map, connecting various locations having an equal amount of rainfall. They display the spatial distribution on rainfall maps over a region or area. Based on the respective isohyet maps, the spatial distribution of rainfall over the world, country, state, district or taluk could be understood well. Rainfall also can be compared between the regions using the isohyet maps. In the absence of actual data, the rainfall of any given location can be approximately worked out using isohyet maps (see Fig. 7.6).

7.5 GLOBAL ANNUAL PRECIPITATION

The average annual precipitation over the whole earth is 857 mm, distributed unevenly. The equatorial regions and monsoon areas of southeast Asia record the highest rainfall. The temperate regions, on the other hand, receive moderate amounts of precipitation. The dry regions of sub-tropical high pressure belt and polar regions receive little precipitation. In the hydrological cycle, if 857mm of rainfall is considered equal to 100 units, 77 units directly go to oceans and the remaining 23 units only received on the earth surface. Out of 23 units, 16 units go back to atmosphere in the form of evaporation and 7 units as surface run-off to oceans. A total of 84 (77 + 7) units from oceans go back as evaporation. That is how the global hydrological cycle is maintained.

7.6 RAINFALL OVER INDIA

The mean annual rainfall over India is 1087 mm, 78% of which is received during the monsoon season (June–September) while 10% and 9% each is received during the post-monsoon season (October to November), and summer season from March to May, respectively (see Fig. 7.5). The remaining 3% is seen during winter (December to February). This is the trend in all the States except over Tamil Nadu and Jammu & Kashmir, where the monthly rainfall is influenced by northeast monsoon. Tamil Nadu receives 49% of annual rainfall from October to December while Kerala receives only 18% of annual rainfall during this time.

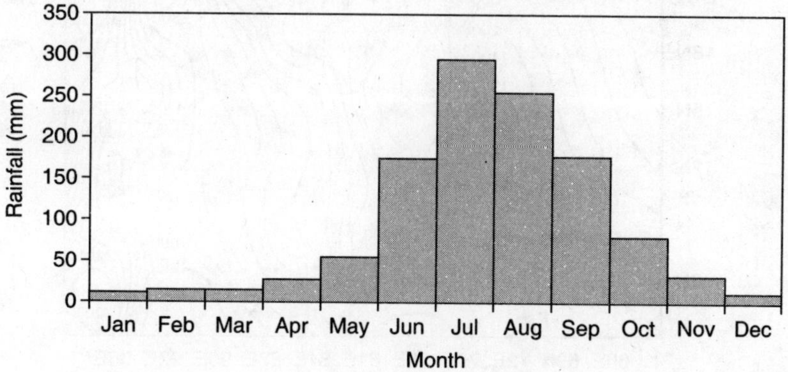

Fig. 7.5 Mean monthly rainfall (in mm) of India.

As per the data compiled by the weather bodies, the spatial distribution of annual rainfall over India (see Fig. 7.6) indicates that the mean annual rainfall increases from the extreme west in Rajasthan (200 mm) to northeastern States (2800 mm) in the north. Similar trend is noticed from extreme north to south of the country. The west of the Western Ghats, the northeast of the Himalayas and the east of the Aravallis received relatively high rainfall due to the influence of orography as they are located along the windward of the monsoon. The west coast and the northeastern States receive the highest rainfall in the country. Mawsynram in Meghalaya records the highest (11410 mm) mean annual rainfall while Cherrapunji, also in Meghalaya, records the mean annual rainfall of 10870 mm. The world's highest rainfall is received in the above region. The distribution of isohyets along the mountains is very narrow, indicating high spatial variability in rainfall between the windward and leeward sides. Kerala, Karnataka and Maharashtra are classic examples of regions towards west which receive high rainfall when compared to the east. It is more pronounced in Karnataka and Maharashtra when compared to Kerala. If the isohyet analysis is carried out at micro-level with more number of data points the actual rainfall scenario may be different, though the trend is similar to that of micro-level.

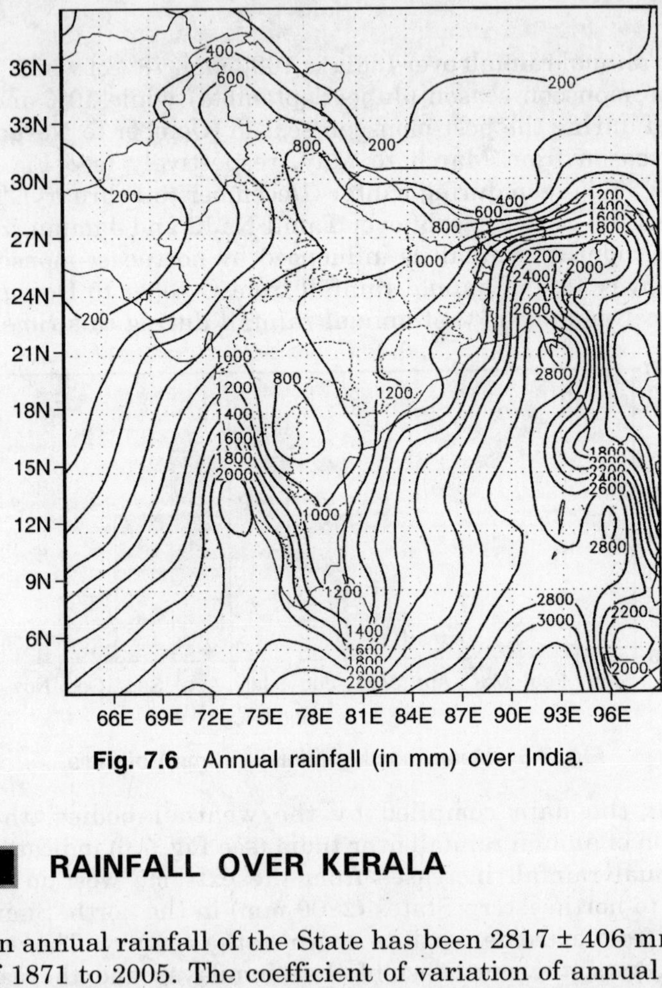

Fig. 7.6 Annual rainfall (in mm) over India.

7.7 RAINFALL OVER KERALA

The mean annual rainfall of the State has been 2817 ± 406 mm over the period of 1871 to 2005. The coefficient of variation of annual rainfall is only 14.4%, indicating that it is highly stable and dependable, though the monthly coefficient of variation varied from 28.4% in June to 146.1% in January. However, the monthly rainfall is also relatively stable in June (CV–28.4%) and July (33.2%) during the southwest monsoon and October (CV–37.5%) due to northeast monsoon as the coefficient of variation is relatively low (see Table 7.1). June and July are the rainiest months while January and February receive the least rainfall.

Table 7.1 Mean monthly normal rainfall (1871–2005) over Kerala

Month	*Rainfall* (mm)		
	Normal	*Standard deviation*	*CV* (%)
Jan	11	17	146.1
Feb	17	19	115.4
Mar	36	28	78.5
Apr	112	52	46.5
May	246	159	64.6
Jun	684	194	28.4
Jul	632	209	33.2
Aug	373	157	42.0
Sep	224	122	54.7
Oct	288	108	37.5
Nov	156	85	54.4
Dec	38	39	102.1
Annual (mm)	2817	406	14.4

7.7.1 Seasonal Rainfall over Kerala

The season-wise rainfall contribution over Kerala indicated that 68% of annual rainfall is received during the monsoon, followed by post-monsoon (16%). The least is seen in winter (see Table 7.2). The seasonal rainfall during monsoon is dependable as the CV is 19.6%. At the same time, the rainfall during winter is not dependable as the CV is very high (71.5%), followed by summer (41.6%).

Table 7.2 Season-wise normal rainfall (1871–2005) and its percentage contribution to the annual

Season	*Normal rainfall (1871–2005) in* mm	*Percentage contribution to annual rainfall*	*Standard deviation* (mm)	*CV* (%)
Monsoon (June–Sept)	1914	68	378	19.7
Post-monsoon (Oct–Nov)	444	16	138	31.2
Winter (Dec–Feb)	65	2	47	91.8
Summer (Mar–May)	394	14	164	41.5
Annual	2817	100	406	14.4

The highest rainfall (5883.8 mm) is recorded at Neriamangalam (Ernakulam district) and the lowest (651.3 mm) at Chinnar (Idukki district). The annual rainfall increases from 1479 mm at Parassala in the south to 3562 mm at Hosdurg in the north of Kerala. The annual rainfall increases from the coast to the foot hills and then decreases on the hill tops. This trend is partially disrupted in the Palakkad region (see Fig. 7.7). Though the annual rainfall in the northern region is more, the effective rainfall is relatively high in the southern region due to even distribution of rainfall. July is the rainiest month in the northern districts, while June is the rainiest month in the southern districts. The southern parts extending from Ponnani to Thiruvananthapuram (except Devikulam) show bimodal monthly rainfall due to the influence of both the monsoons viz. southwest monsoon and northeast monsoon. The northern districts, especially Kasaragod and Kannur, show only unimodal monthly rainfall (southwest monsoon only) and experience a prolonged dry spell if the pre-monsoon showers fail.

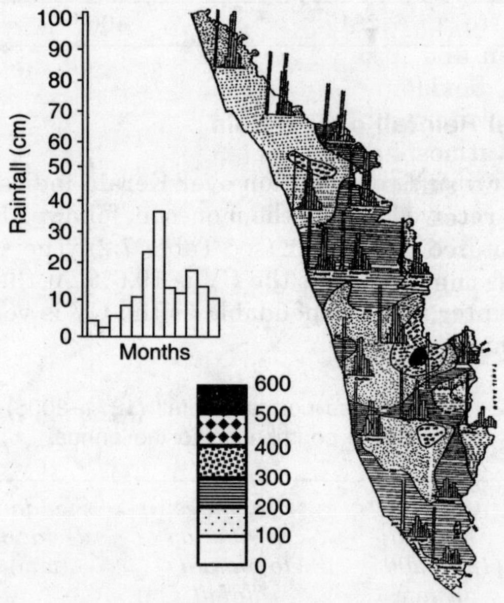

Fig. 7.7 Mean annual rainfall (in mm) over Kerala.

The monthly rainfall distribution and the high amount of rainfall are favourable for paddy cultivation in Kerala under rainfed conditions, and no irrigation is required at any stage of the crop under wetland situations. More than 90% of paddy growing area is rainfed during *Virippu (kharif)* while it could be grown only under assured irrigated conditions during *Mundakan (rabi)*. Many of the plantation crops also require irrigation, starting from December to May. The above situation is more prevalent in northern districts of Kerala where predominantly unimodal rainfall is noticed.

7.8 THUNDERSTORMS

Thunderstorms have been defined as storms, produced by cumulonimbus clouds and always accompanied by lightning and thunder. They are accompanied by strong wind and heavy rain. It consists of a series of cumulonimbus clouds or cells, each cell going through a sequential development, often called a life cycle, in about half an hour. Its prerequisites are such that the atmosphere must be unstable, usually highly unstable, and it should be highly humid. They are usually less vigorous in subsequent development than the initial ones, simply because each cloud serves to redistribute the vertical temperature and thus decreases the instability.

7.8.1 Benefits of Thunderstorm

Thunderstorms fix the atmospheric nitrogen. The molecular atmospheric nitrogen is converted into reactive nitrogen in the presence of high temperatures (up to 10,000°C) during lightning, which mixes with atmospheric oxygen and forms oxides of nitrogen such as nitric oxide (NO) and nitrogen dioxide (NO_2). They are water-soluble and reach the soil through rainy water which in turn is available to plants. Although 78% of the earth's atmosphere possesses nitrogen, nitrogen compounds are rarely present in sufficient quantities in the soil to allow plants to grow at maximum rates. Hence, ammonium and nitrogen fertilizers are artificially manufactured and fed into the soil to increase crop production. About 30–50% of the nitrogen compounds are produced due to thunderstorm lightning.

7.8.2 Hazards of Thunderstorm

Lightning, hails, high winds, duststorms, heavy rains, flash flooding and landslides are the common hazards during thunderstorm activity. Loss of human lives and property are the concern during the thunderstorm activity.

7.9 HYDROLOGICAL CYCLE

The circulation of water within the earth-atmosphere system is known as the *hydrological cycle* or *water cycle*. Evaporation and transpiration from the earth's surface add water vapour to the atmosphere. Precipitation may occur through condensation and cloud formation under favourable conditions. As precipitation falls, some of it may evaporates, the rest reaching the ground. The ground reaches it may be a water surface or a vegetation surface. Of the precipitation reaching vegetation, some is held (intercepted) by the canopy and eventually evaporates again. Some will run down vegetation stems or drip from the canopy and reach the ground. Sometimes, a portion of it may be taken off by winds.

Of the water reaching the ground, some will infiltrate into the surface (see Fig. 7.8). The infiltration rate is extremely variable, depending upon factors such as soil type, condition and moisture content. In addition, the influence on infiltration of man-made surfaces such as roads, paths and buildings is considerable. If rainfall intensity is greater than the infiltration rate, then some water accumulates in surface depressions such as ponds and lakes. When the depressions are full, water begins to move over the surface, usually towards channels. Slope and vegetation cover affect the rate of water movement over the surface and hence the time available for infiltration. Some of the water stored in depressions will evaporate and some will eventually be able to infiltrate into the soil.

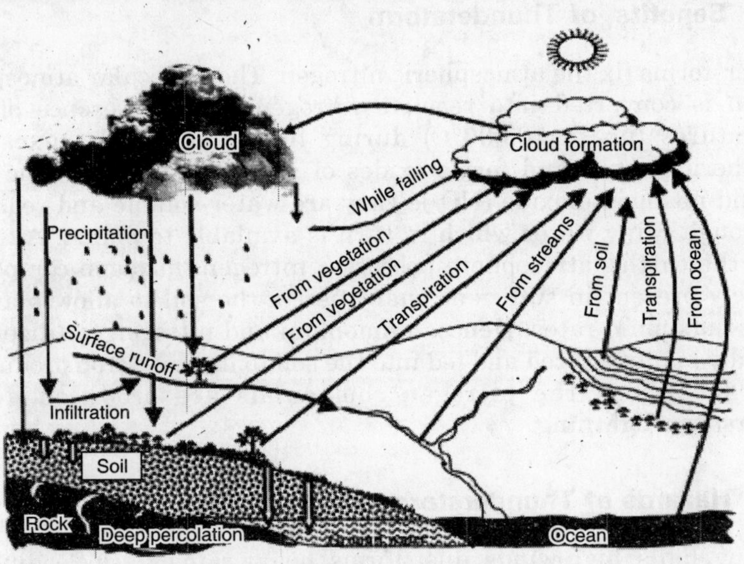

Fig. 7.8 Hydrological cycle.

Some of the water entering the soil may evaporate back into the atmosphere, and some will be taken up by plants and transpired. In the upper layers of the soil, sub-surface lateral movement of water (through flow) is considerable and may reach streams or lakes. Some water may percolate to deeper layers to become ground water, where it may be held for very long periods. Ground water can move up over a limited depth by capillary action where it may evaporate or be taken up by plants and transpired. Some ground water, by lateral movement, may eventually seep into streams, lakes, seas or oceans.

Of the water reaching open-water surfaces, either directly as precipitation or indirectly as surface, sub-surface or ground water, some will evaporate and some will seep into the ground.

The above processes are nothing but the hydrological cycle which is a complex system. However, human interference at one stage or other has repercussions throughout the whole cycle. Changing the land use over a river catchment, for example, may have a great impact upon the path followed by water falling as precipitation, perhaps leading to problem of flooding, soil erosion, increased dry season, reduced river flow, or conversely, having reciprocal beneficial effects.

7.10 INSTRUMENTS

The instruments used for measuring precpitation are described in the following sections.

7.10.1 Ordinary Raingauge

It consists of a base and outer drum with a funnel (see Fig. 7.9). The funnel is placed on the base. There is a plastic jar inside the base in which rainwater is collected through the funnel. It is collected using a measuring jar graduated in 20 millimetres. This measuring jar is meant for the standard area of the funnel, having 200 cm^2 surface area of the funnel. One should be very careful about the measuring jar, as they vary in terms of the area of funnel through which rainfall is received. The standard time of rainwater measured is at 0830 hrs daily. The height of the instrument is 12 inches from the ground. It gives the total amount of rainfall during the last 24 hours in millimetres. It is better to maintain a wooden platform very near the raingauge to place the measuring jar. This avoids error in taking readings. A special type of snowgauge is used to measure snowfall.

Fig. 7.9 Ordinary rainguage.

7.10.2 Self-recording Raingauge

It is an automatic device (see Fig. 7.10). A continuous rainfall record for 24 hours is obtained on a daily chart if rain is seen. Otherwise, it simply gives a horizontal line if there is no rain. The rainfall events viz. onset and cessation of rainfall, duration of rainfall, intensity of rainfall and the total amount of rainfall in millimetres can be obtained from the rainfall chart used in the device. Unless the total amount of rainfall recorded is same as the rainfall amount of the ordinary raingauge, it is not dependable for measurement of other rainfall events. That is why ordinary raingauge is a must, where self-recording raingauge is available, to ensure that the data recorded by self-recording raingauge is correct. The working principle of this raingauge is based on floating chamber. When rain water enters the floating chamber, the water level rises and at the same time the float above the water surface also rises. When it rises, the pen arm attached to the floating stem will mark on the chart. When it reaches to its maximum, the rain water will be siphoned out. Then the pen arm shows zero value. That is how rainfall is recorded using the self-recording raingauge. The data on intensity of rainfall is used in soil erosion and river flow measurement studies.

Fig. 7.10 Self-recording raingauge.

7.10.3 Automatic Weather Station

It is an electronic device wherein various weather sensors are connected to the system which automatically records and stores the data (see Fig. 7.11). The system requires hardware as well as software to log in, store, display, transfer and analyse the data. The data can be transferred to the computer as it is a computer compatible system. The time interval for recording weather data can be either a minute, 30 minutes or one hour, which can be programmed. If there is no specific purpose for monitoring weather data, one-hour interval is recommended to avoid enormous data storage without use. Print-outs should be taken at regular intervals to maintain as hard copy so that the data are not lost. The standard weather parameters recorded through this electronic device are air temperature, radiation, soil temperature, relative humidity, wind speed and direction, leaf wetness and rainfall. The sensor used in the device is dependent upon the type of weather element to be recorded. For example, tipping bucket mechanism in the case of rainfall and thermisters in the case of air temperature are the sensors used in the device of automatic weather station. The whole system is battery-operated and rechargeable through a solar panel which is part of the system. It is mounted on a 2-metre mast. Otherwise, a separate mast should be erected on the basis of the height of the crop if it is meant for micro-meteorological studies in plantation crops. Since it is an automatic device, it can be installed in remote places too. However, it should be closely monitored for better maintenance. Now, satellite based automatic weather stations are being installed by IMD and ISRO across the country with the indigeneous technology.

Fig. 7.11 Automatic weather station.

REFERENCES/SUGGESTED READINGS

Chorley, J.R., *Introduction to Physical Hydrology*, Methuen, United Kingdom, 1977, p. 211.

Cosgrove, B., *Weather*, Dorling Kindersley, London, 1991, p. 64.

Das, P.K., *The Monsoons*, National Book Trust, New Delhi, 1986, p. 162.

Kumar, K.R., *Climate Maps: CPC Merged Analysis of Precipitation, 1979–2000*, Personal Communication, Indian Institute of Tropical Meteorology, Pune, 2003, p.12.

Lal, D.S., *Climatology*, Chaitanya Publishing, Allahabad, 1986, p. 420.

Lutgens, F.K. and Tarbuck, E.J., *The Atmosphere—An Introduction to Meteorology*, Prentice Hall, New Jersey, 1979, p. 413.

Menon, P.A. and Rajan, C.K., *Climate of Kerala*, Classic Publishing, Kochi, 1989.

Partridge, J.I., (Ed.), *Will it rain?*, The State of Queensland, Deptartment of Primary Industries, Brisbane, 2002.

Ranganathan, C., *Cyclone Havoc and Other Weather Phenomena*, Tamil Nadu State Council for Science and Technology, Chennai, 2000, p. 115.

Sacheti, A.K., *Agricultural Meteorology: Instructional-cum-Practical Manual*, National Council of Educational Research and Training, New Delhi, 1985, p. 53.

Sellers, W.D., *Physical Climatology*, University of Chicago Press, Chicago, 1965, p. 271.

Tarakanov, G.G., *Tropical Meteorology*, Mir Publishers, Moscow, 1982, p. 206.

USDA, *The Yearbook of Agriculture*, Indian Edition, Oxford & IBH, New Delhi, 1969, p. 751.

QUESTIONS

1. Define the following:
   (a) Precipitation
   (b) Precipitable water
2. Distinguish between:
   (a) Ordinary raingauge and self-recording raingauge
   (b) Convectional and orographic precipitation
   (c) Cyclonic and convectional precipitation
   (d) Rainfall and drizzle
   (e) Snow and sleet
   (f) Thunderstorms and duststorms
   (g) Hails and virgae
   (h) Mist and drizzle

3. Write short notes on:
   (a) Precipitation mechanisms
   (b) Types of precipitation
   (c) Forms of precipitation
   (d) Rainfall over India
   (e) Rainfall over Kerala
   (f) Thunderstorms
   (g) Hydrological cycle
4. Fill in the blanks:
   (a) ——————is the average annual global precipitation
   (b) ——————is the average annual precipitation over India
   (c) ——————is the average annual precipitation over Kerala
   (d) ——————is the average monsoon rainfall over Kerala
   (e) ——————units of precipitation directly goes to oceans
   (f) ——————is known as isohyets
   (g) ——————is available to plants during thunderstorms
   (h) ——————is known as the hydrological cycle
   (i) ——————is measured using Self-recording Raingauge
   (j) ——————is the standard time to measure daily rainfall
   (k) ——————is the season during which hailstorms are noticed
   (l) —— and —— are the major components in hydrological cycle
5. Match the following:

(a) E.F. Bowen	East coast
(b) Tor Bergeron	Lightning
(c) Orographic precipitation	Raingauge
(d) Cyclonic precipitation	Ice-crystal theory
(e) Thunderstorm	West Coast
(f) Amount of rainfall	Collision-coalescence

6. Elucidate the following:
   (a) Rainfall over India
   (b) Rainfall over Kerala

8 MONSOON

The word 'Mausam' in Hindi means *weather*, while 'Mausim' in Arabic means *season*. The literature on monsoon says that the word 'Monsoon' is originated from the Arabic word 'Mausim', which was used by seamen seven or eight centuries ago to describe a system of alternating winds in the Arabian Sea. Two seasons have been identified in India based on wind direction viz. southwest monsoon and northeast monsoon. Wind blows from the southwest for six months during the summer and from the northeast for another six months during the winter across the country. The southwest monsoon is noticed during summer and northeast monsoon during winter over India.

The beneficial effects of monsoon are enormous as agriculture in India depends largely on monsoon. The failure or break in monsoon could well produce a crisis of considerable magnitude. The intermittent breaks of monsoon may lead to no rainfall, thus indicating a dry spell. The loss in *kharif* foodgrains production was extended up to 20% due to the all-India drought that occurred during monsoon 2002. Similar was the case in 1987. Despite advanced technology, monsoon influences even today the Indian foodgrains production to a considerable extent. The *kharif* foodgrains production is adversely affected due to monsoon break or failure as cultivable area become less under rainfed conditions. The crops already sown with the advent of monsoon are also adversely affected due to dry spells which result in soil moisture stress. The recurrence of such dry spells is noticed very often across the country, which affects the seasonal crops adversely. Heavy and prolonged rain from June to September, followed by a dry spell, is the characteristic feature of the humid tropics. The behaviour of monsoon across Kerala decides the total output of paddy during Virippu (*kharif* season), beginning from June to September, as it is grown under rainfed conditions. Similar is the case in almost all the States except Tamil Nadu and Jammu & Kashmir, where the crops are grown under the rainfed conditions during the northeast monsoon.

8.1 MONSOON COUNTRIES

The monsoon rains are seen not only in India but also in several countries. All the monsoon countries are interested in understanding the behaviour of the Indian Asiatic monsoon. The monsoon regions viz. India, Pakistan, Bangladesh, Nepal, Bhutan, Burma, Sri Lanka, Malaysia, northern tip of Indonesia, China, Japan, northwest Australia, East and West Africa, southeast USA and Brazil have international collaboration to understand the Asiatic monsoon, as it influences the economy to a considerable extent.

8.2 SUMMER OR SOUTHWEST MONSOON

In the Northern Hemispheric summer, the Asian land mass gets heated while the surrounding oceans remain relatively cool. A well-marked heat low develops over Pakistan by the end of May and during June a trough (which is an elongated low pressure of 994 hPa) develops, extending the heat low to the northwest Bay of Bengal across the plains of north India, at the mean sea level. This trough or low pressure over north India is known as the *monsoon trough.* Along the monsoon trough, the Inter Tropical Convergence Zone (ITCZ) also shifts towards India from the equator by the end of June or July. At the same time, the surface atmospheric pressure is high over the southern Indian Ocean. The difference in atmospheric pressure over land and ocean provides the driving force for the generation and development of the Asiatic monsoon circulation. The southeast trade winds move from the Southern Hemisphere to the Northern Hemisphere due to difference in atmospheric pressure. The winds blowing from southeast in the Southern Hemisphere get deflected on crossing the equator due to the effect of the earth's rotation, advance northwards and become southwesterlies. This southwesterly wind strikes the Kerala coast towards the end of May or beginning of June and constitutes the southwest monsoon. The summer or southwest monsoon in the Indian subcontinent is observed from June to September. It has two branches viz. the Bay of Bengal current and the Arabian Sea current. The Bay of Bengal current covers north-eastern States while the south peninsular India is covered by the Arabian Sea current. Both the sea currents merge somewhere in north India by the end of June and travel together to northwest India by mid-July, covering the whole country. The shift in the monsoon trough towards north or south from the normal position may lead to monsoon breaks or failure.

8.2.1 Onset and Withdrawal of Southwest Monsoon

The monsoon arrival is very often a gradual beginning with a short period of transition from extreme heat to a more humid atmosphere with light rain. The moisture laden winds are pleasant with moderate speed one or two days ahead of the onset of monsoon. The normal date of onset of monsoon can be worked out at each location based on normal rainfall computed for five-day periods, which are commonly known as *pentads*. A sudden increase in normal rainfall of a particular pentad over one or two preceding pentads in May, June or July, depending upon location, is considered the monsoon pentad for defining onset of monsoon. A mid day of that particular pentad is taken as the normal date of onset of monsoon. The India Meteorological Department has produced maps indicating normal dates of onset of monsoon for the country as a whole (see Fig. 8.1). The India Meteorological Department also declares the actual onset of monsoon every year for the country as a whole. Since the monsoon enters from Kerala (it is popularly known as the Gateway of Monsoon), the normal onset of monsoon over Kerala is the normal onset

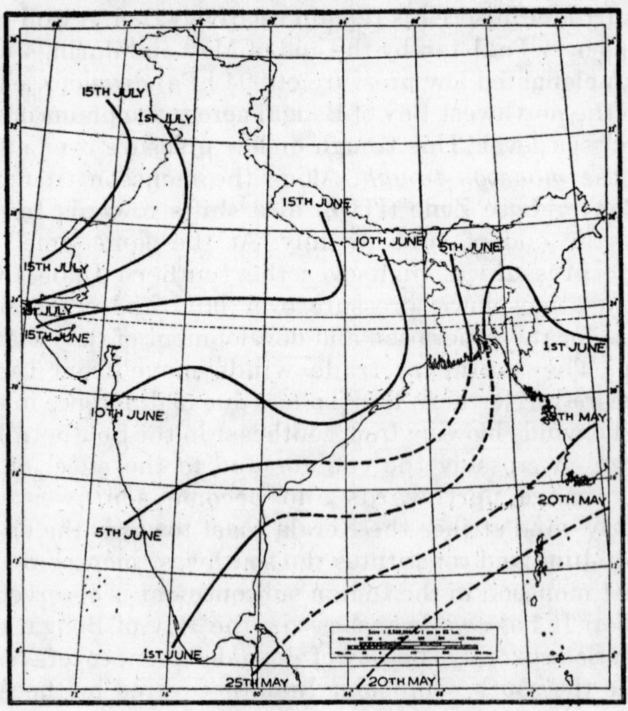

Fig. 8.1 Onset of monsoon over India.

of monsoon over India. However, it varies from one region to another across the country. The monsoon reaches the extreme south of the Indian peninsula by 1st June and reaches extreme northwest of India by 15th July, taking 45 days to cover the whole of India. The normal onset of monsoon over Mumbai is 10th June and 7th June over Kolkata and Visakhapatnam. It appears around 3rd July over New Delhi. The onset of monsoon over Rajasthan, where monsoon is very slow and rainfall is low, is depicted in Fig. 8.2. It is on 15th June over extreme southeast Rajasthan and 15th July over west Rajasthan. In contrast to the late onset of monsoon over Rajasthan, the monsoon begins to withdraw early from Rajasthan by the 1st of September. The withdrawal of the monsoon is a far more gradual process than its onset. The monsoon usually withdraws from the whole country, except the southern peninsula, by 15th October (see Fig. 8.3). The onset and withdrawal of monsoon influence the crop growing season and selection of crops to a large extent.

For example, the duration of monsoon over western Rajasthan is hardly 60 days while it is 165 days over Kerala (both the monsoons together). It reveals that short-duration crops like pulses and millets only could be taken up in Rajasthan, where the monsoon duration is short and rainfall is less and highly unpredictable. In contrast, water-intensive crops with long duration can be taken up in Kerala, where the duration of monsoon is long and rainfall is very high. It is one of the reasons why natural vegetation (tropical forests) is supported by heavy monsoon over Kerala and deciduous forests in Rajasthan. The natural vegetation is a good indicator to know whether a place is wet or dry.

8.2.2 Onset of Monsoon Over Kerala

The normal date of onset of monsoon over Kerala is on 1st June with a standard deviation of seven days, varying between 25th May and 8th June. The earliest onset of monsoon was on 11th May in 1918 and belated monsoon on 18th June in 1972. The march of onset of monsoon, since last one and a half decades, showed that it was after 1st June on many occasions (see Fig. 8.4(a)). The trend in normal onset of monsoon also tends to be towards the first week of June, revolving around 1st June. It is always stable, ranging between 25th May and 8th June as only 27.7% of the years fell below or above one standard deviation of the normal onset of monsoon. The occurrence of maximum frequency (32.1% of years) of the onset of monsoon was between 30th May and 3rd June, followed by June 4th–8th (21.2% of the years) and May 25th–29th (19.0% of the years). The chances of early monsoon before 25th May are relatively less (1.5–5.8% of the years). The same is true in the case of occurrence of late monsoon beyond June 8th, as only 3.6–11.7% of the years fell under the late category. 72.3% of the years fell in the range of 25th May and 8th June (see Fig. 8.4(b)) indicating that there is no change in normal date of onset of monsoon since the last 137 years if the period

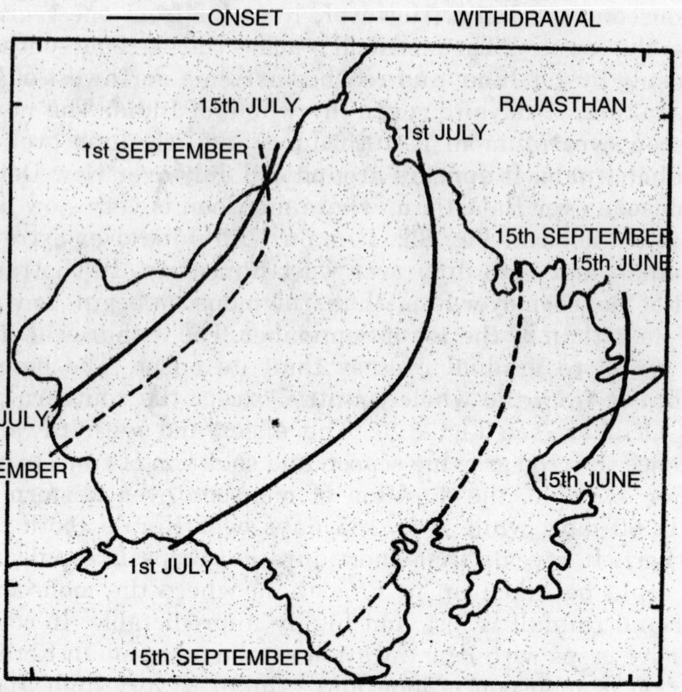

Fig. 8.2 Onset and withdrawal of monsoon over Rajasthan.

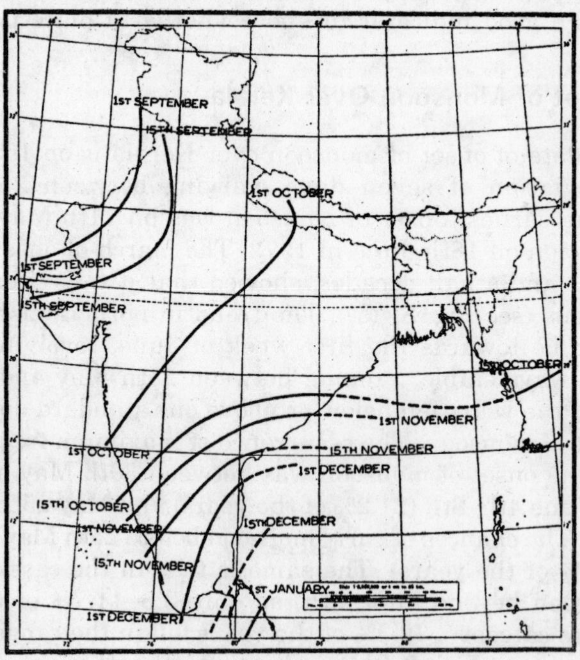

Fig. 8.3 Withdrawal of monsoon over India.

from 1870 to 2006 is considered as a whole, though the inter-annual variation was noticed within the limits of ±1 standard deviation.

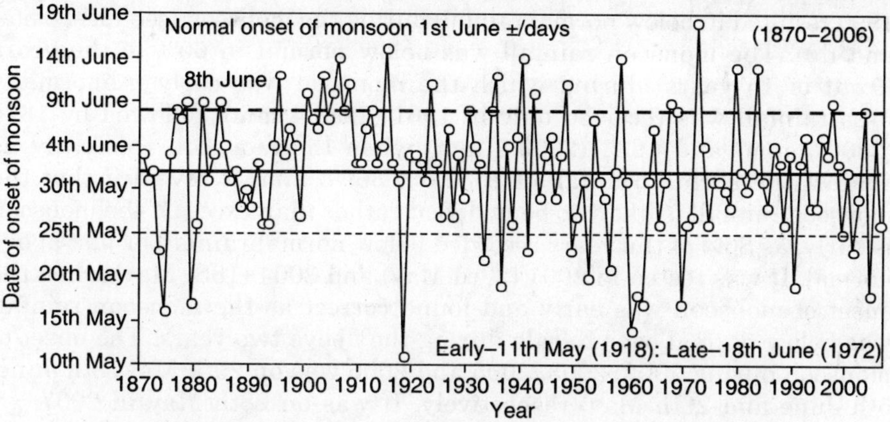

Fig. 8.4(a) Onset of monsoon over Kerala from 1870–2006.

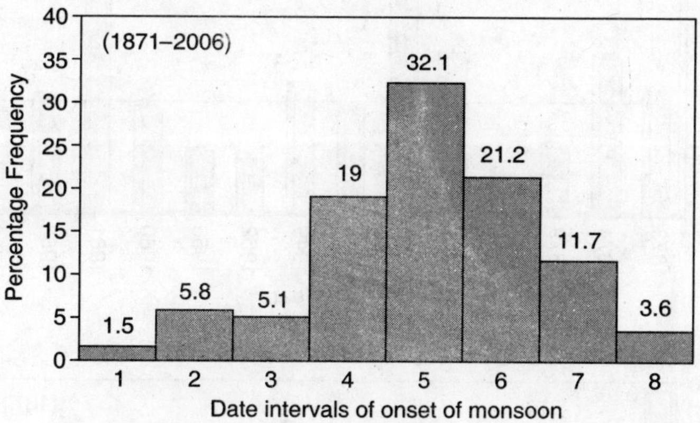

1	2	3	4	5	6	7	8
May 10–14	May 15–19	May 20–24	May 25–29	May 30–June 3	June 4–8	June 9–13	June 14–18

Fig. 8.4(b) Percentage frequency distribution of onset of monsoon over Kerala (1870–2006).

8.2.3 Onset of Southwest Monsoon versus Monsoon Rainfall over Kerala

The onset of monsoon was early (before 25th May) on 15 occasions (1873, 1874, 1879, 1918, 1933, 1936, 1941, 1949, 1952, 1956, 1960, 1961, 1962, 1969 and 1990) out of 131 years, accounting for 11.5% of the years under early monsoon. Interestingly, the early onset of monsoon was seen only in 1990 (19th May) after a lapse of 20 years (17th May in 1969) in the recent past. Incidentally, the early onset of monsoon during 1969 and 1990 resulted in below normal rainfall during the monsoon (see Figs. 8.5(a) and (b)). The monsoon rainfall was below normal in 60% of the years (9 out of 15 years) during which the monsoon was early. Abnormally high rainfall was received only in 1961 (2943.4 mm), followed by 1874 (2300.7 mm) and 1933 (2303.8 mm) when the monsoon was early as against the normal rainfall of 1932.6 ± 369.3 mm. It revealed that the monsoon rainfall is likely to be in deficit rather than excess if the monsoon is early, as 80% of the years recorded below normal rainfall or just above normal. It was tested in 2001 (23rd May) and 2004 (18th May) when the onset of monsoon was early and found correct as the monsoon rainfall was below normal over Kerala during the above two years. The onset of monsoon during 2002, 2003, 2005 and 2006 was on 29th May, 8th June, 5th June and 26th May, respectively. It was on 28th May in 2007.

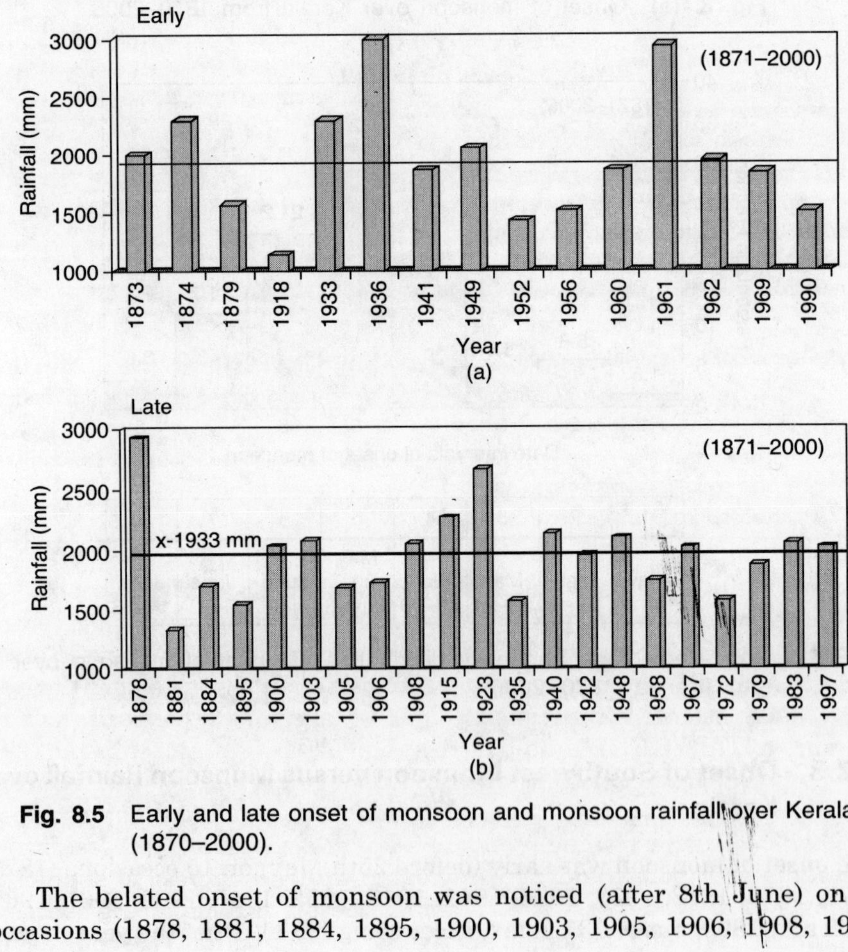

Fig. 8.5 Early and late onset of monsoon and monsoon rainfall over Kerala (1870–2000).

The belated onset of monsoon was noticed (after 8th June) on 21 occasions (1878, 1881, 1884, 1895, 1900, 1903, 1905, 1906, 1908, 1915,

1923, 1935, 1940, 1942, 1948, 1958, 1967, 1972, 1979, 1983 and 1997) out of 131 years, accounting for 16% of the years under late monsoon. Out of 21 late-onset-of-monsoon years, 42.8% of the years (9 out of 21 years) only recorded below normal rainfall. However, the year 1881 was the only one that could be classified as *meteorological drought year*. The meteorological drought is defined when the annual rainfall was below 25% of the normal annual rainfall. A significant rainfall deficit was noticed in 1895 (–20.3%) and 1972 (–18.6%) in addition to 1881 (–30.7%). Abnormally excess rainfall (51.4% above normal) was noticed in 1878 when the monsoon was late, followed by 1923 (37.9% excess of normal). 57% of the years (12 out of 21 years) noticed excess rainfall when compared to normal if the monsoon was late.

It is likely that the chances of the occurrence of deficit rainfall when compared to that of normal during the monsoon season is relatively high if the onset of monsoon is early (before 25th May) and no such trend is seen when the monsoon is late.

8.2.4 Onset of Effective Monsoon

Advance knowledge of the probable dates of onset of effective rains is of paramount importance to start agricultural operations such as land preparation and sowing. It is essential to identify the onset of effective monsoon as distinct from pre-monsoon showers. The onset of monsoon is defined on the basis of a sudden increase in pentad rainfall while the onset of effective monsoon is defined on the basis of rainfall and evapotranspiratiom demand, which is relevant in case of crops (Raman and Murthy, 1971).

(a) The first days's rain in a period of 7 days should be at least 5 mm. Any rainfall below this could only be adequate to meet the evapotranspiration loss after a long dry spell. It is assumed that the 5 mm rain at most will be equal to the daily average evapotranspiration.

(b) The total rainfall during a 7-day period should be 35 mm or more. At least 4 out of these 7 days should be'rainy' days (a day is a rainy day if the rainfall of that day is at least 2.5 mm). Even though the daily evapotranspiration is assumed to be 5 mm, during rainy days it is likely to be much less. Thus in a period of 7 days, out of which 4 are rainy days, the evapotranspiration can safely be assumed to be at the most 25 mm. The second criterion assumes that at least 10 mm of rain will be available to provide the soil profile with adequate moisture for initiating agricultural operations.

The above criteria can be adopted to work out the effective monsoon or commencement of effective rains. Similar concepts also can be used for defining wet and dry spells depending upon the purpose.

8.2.5 Onset of Wet and Dry Spells

Several techniques are in use to work out wet and dry spells. The initial and conditional probabilities as per the first order Markov chain model is widely used world over to understand the crop-growing seasons based on dry and wet spells. Instead of mean rainfall, assured rainfall is also defined at 75% probability level. It is computed on the basis of rainfall distribution, which is assumed as incomplete gamma distribution. It holds good wherever rainfall distribution is influenced only by southwest monsoon and follows unimodal. In India, except for a few pockets in the south, where bimodal rainfall is noticed, all the locations experience unimodal rainfall and thus incomplete gamma distribution holds good for computing assured rainfall. The dependable rainfall or assured rainfall levels at different probability levels also can be worked out using the percentile rank method. However, a simple definition is given here for defining wet spell and dry spell based on weekly rainfall.

Wet spell

The interval between the onset of effective rains and the commencement of the first dry spell is called the wet spell. The effective rain may be a rainfall spell of more than 35 mm spread over a period of 7 days with at least 4 rainy days to begin with. The subsequent wet spell may be based on weekly rainfall of 35 mm spread over a week without considering the number of rainy days. Thus the wet spells are spells of prolonged rain without dry intervals. The concept need not be universal. It can be quantified using the first order Markov chain model. Again, weekly rainfall of 35 mm is an arbitrary one. One can choose the amount of rainfall based on weekly crop water requirement in different agroclimatic zones.

Dry spell

Initial dry spell may be defined as a week during which the weekly rainfall is less than 35 mm with less than 4 rainy days. Of course, if the initial requirement for a wet spell is not met, this means it is a dry spell. 2.5 mm or more rainfall per day leads to a rainy day as per the IMD definition. No rains continuously for a week may be critical for upland paddy in the humid tropics and several other crops grown under rainfed.

The initial and conditional probabilities for 30 mm of weekly rainfall is worked out at Vellanikkara as an example. It reveals that the initial wet spell with 75% probability of having more than 30 mm rainfall commences on 28th May and ceases on 9th September. The conditional probability commences one week ahead. Hence, the crop season may start on 21st May and cease on 9th September. The optimum time of sowing is the last week of May under normal climatic conditions at the assigned probability level of 75% (see Fig. 8.6). This type of rainfall probability analysis can be examined before taking up crops for cultivation.

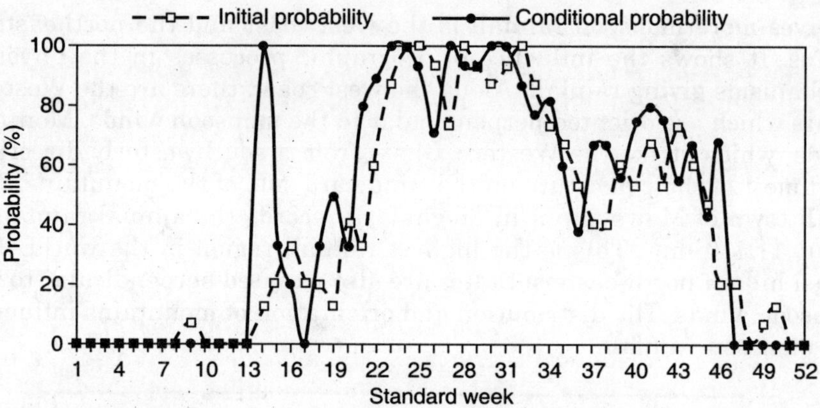

Fig. 8.6 Initial and conditional probability of weekly rainfall at Vellanikkara.

8.2.6 Monsoon Rainfall of India

The mean all-India rainfall is 848 mm during the southwest monsoon season out of 1087 mm received as total annual rainfall. 78% of the annual rainfall over India is recorded during the southwest monsoon between June and September. Rainfall received during the remaining period is relatively less, indicating the relevance of southwest monsoon for the country as a whole (see Table 8.1).

Table 8.1 Mean seasonal rainfall over India (1871–2003)

Season	*Rainfall* (mm)	*CV* (%)	% *of Annual rainfall*
Southwest monsoon (June–September)	848	10	78
Post-monsoon (October–November)	108	31	10
Winter (December–February)	36	42	3
Summer (March–May)	95	22	9
Annual	1087	9.5	100.0

The rainfall trend during the monsoon is similar to that of annual rainfall as 60–90% of it is seen during the monsoon period depending upon the region. The isohyetal map compiled by the weather bodies (see Fig. 8.7) for the monsoon reveals that rainfall increases from north to northeast and west of the Western Ghats. Interestingly, the southern region receives between 400 and 800 mm (decreases from west as well as from east), where the influence of northeast monsoon is also noticed. Its distribution is highly variable in space and time. The region which receives more monsoon rainfall is the west coast and the northeastern States. It shows the influence of orographic processes in the physical mechanisms giving rainfall. Along the west coast, there are the Western Ghats which are oriented perpendicular to the monsoon winds. Monsoon winds, which strike the Western Ghats from a southwesterly direction, shed most of their moisture on the windward side of the mountains. The small town of Mawsynram in Meghalaya records the annual rainfall of about 11,410 mm. This is the highest rainfall region in the world. The Khasi hills of north-eastern States are also situated perpendicular to the monsoon winds. The distribution and orientation of mountains influence the monsoon rainfall.

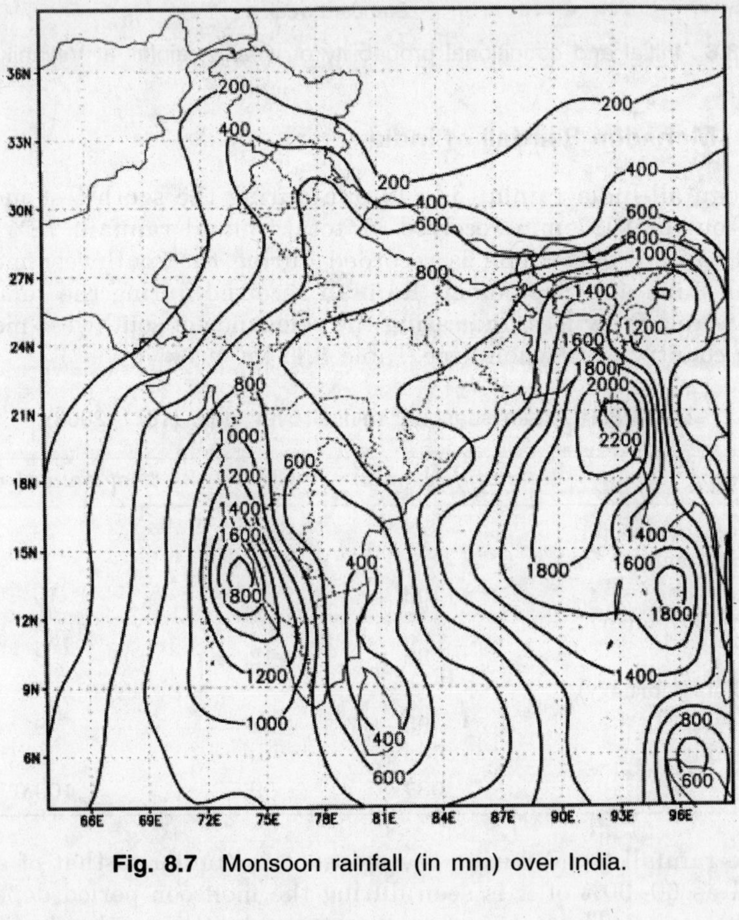

Fig. 8.7 Monsoon rainfall (in mm) over India.

8.2.7 Monsoon Rainfall Over Kerala

The normal rainfall during southwest monsoon over Kerala from 1871 to 2005 is 1914 ± 378 mm. The monsoon rainfall is dependable and the inter-annual rainfall variability is relatively less, as the coefficient of variation is only 19.7% (see Table 8.2). The seasonal rainfall during the post-monsoon and summer seasons over Kerala is 444 mm and 394 mm respectively. The least (65 mm) is received during winter (see also Table 7.2). The dependable rainfall during monsoon at 75% level is 1647 mm.

Table 8.2 Mean seasonal rainfall over Kerala (1871–2005)

Season	*Rainfall* (mm)	*CV* (%)	% *of Annual rainfall*
Southwest monsoon (June–September)	1914	19.7	68
Post monsoon (October–November)	444	31.2	16
Winter (December–February)	65	91.8	2
Summer (March–May)	394	41.5	14
Annual	2817	14.4	100

The occurrence of excess and deficit rainfall when compared to that of normal was seen in more or less equal number of years. As a whole, 32.6% (44 years out of 135) of the years recorded either excess or deficit rainfall. The rainfall variability was high during excess rainfall years when compared to that of deficit rainfall years. It is interesting to notice that in the last two decades there was no excess rainfall year except 1981. At the same time, the occurrence of deficit rainfall was high in the recent past. Overall, there was a decrease of 232.6 mm in monsoon rainfall, indicating a decline of 12.2% over the normal rainfall of 1914 mm

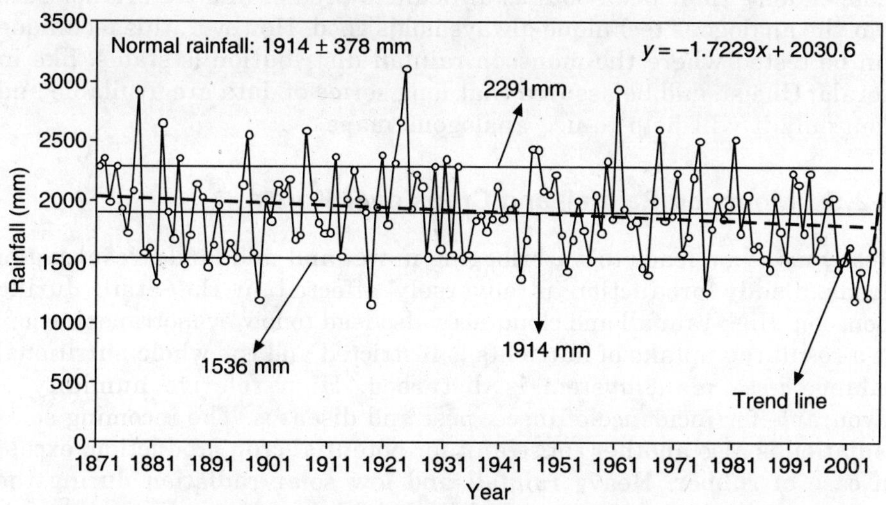

Fig. 8.8 Yearly trend of monsoon rainfall over Kerala from 1871–2006.

(see Fig. 8.8). The declining trend was more evident in the rainiest months of June and July. Reports also indicate that the decline in rainfall is insignificant towards north of Kerala when compared to south. However, acute shortage of water is seen during the summer months in recent years.

8.2.8 Analogous Technique

The distribution of monsoon rainfall could be predicted at different locations if the daily rainfall data are available for more number of years using a simple analogous technique. This technique simply is as follows:

(a) Collect daily monsoon rainfall over a period of 30–50 years.
(b) Prepare daily monsoon rainfall maps for individual years.
(c) Compare daily rainfall of June and July of the current year with that of the corresponding months of previous years.
(d) Pick up the year in which the daily rainfall of June and July is same to that of current year.
(e) Forecast the rainfall distribution for August and September of the current year based on the rainfall distribution of June and July, presuming that the same daily rainfall trend will follow as in the case of selected year in which the daily rainfall of June and July is same as to that of current year.

The above technique can be used as one of the tools for proper crop management practices if found satisfactory consecutively for three to five years. The vast database that is available on daily monsoon rainfall over a long period of time at different locations could be used for developing 'analogous monsoon maps'. This can be used as a forecasting tool under LRF when the monsoon is advanced. As weather systems are independent, their behaviour is difficult to predict and we are not sure that the analogous technique always holds good. However, this technique can be tested where the monsoon rainfall distribution is stable, like in Kerala. One should be assured that long series of data are available and then only it will help to use analogous maps.

8.2.9 Monsoon Rainfall and Crops over Kerala

Heavy monsoon leads to waterlogging in wet and mid lands, resulting in floods. Paddy production is adversely affected in the State during monsoon. High rainfall and cloudiness also lead to low evapotranspiration, as a result the uptake of nutrients is restricted and the whole nutritional balance in a plant system is disturbed. High relative humidity is favourable for incidence of insect pest and diseases. The incoming solar radiation is also another constraint for potential crop production, except in case of rubber. Heavy rainfall and low solar radiation during the monsoon and insignificant rainfall and high solar radiation during summer are the major climatic constraints in the humid tropics (see Fig. 8.9). The broad negative and positive effects of heavy monsoon are indicated on the basis of the author's long experience in the field of agrometeorology of plantation crops.

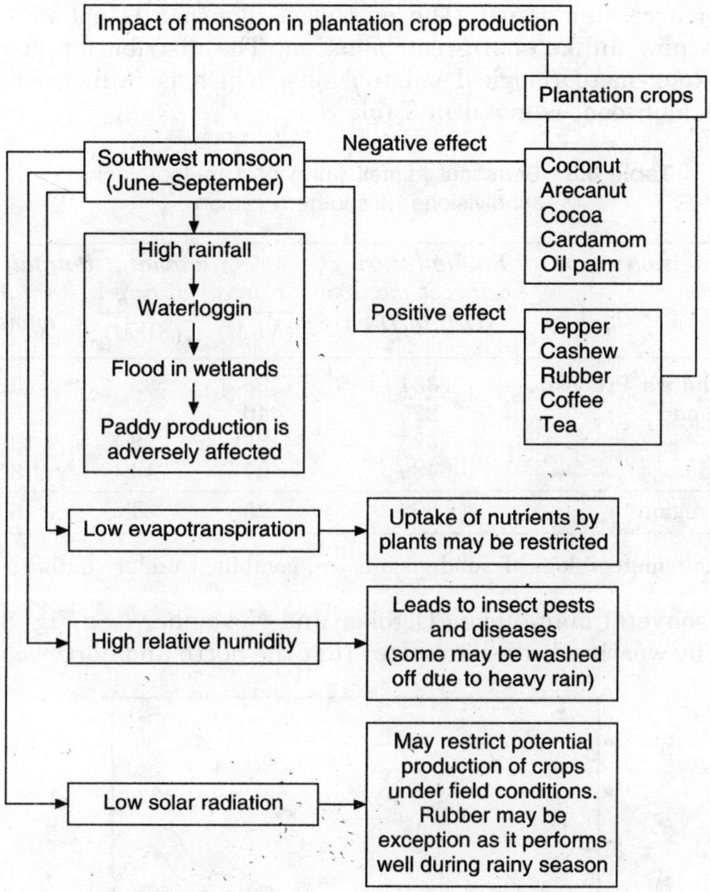

Fig. 8.9 Impact of monsoon on plantation crop production.

8.3 WINTER/NORTHEAST MONSOON

The onset of northeast monsoon normally over south peninsula (Tamil Nadu, south coastal Andhra Pradesh, Rayalaseema, interior south Karnataka and southern districts of Kerala) is on 15th October and the withdrawal is complete by 1st December. The northeast monsoon season is from October to December. During this season, the pressure pattern reverses its distribution. Winds blowing from north India are cold and dry during winter. They are known as *northeasterlies*. During winter, the winds pick up moisture from the Bay of Bengal and produce rain along the east coast. The influence of northeast monsoon is seen towards the southern districts of Kerala only. Its influence is not significant over the northern districts of Kerala. The main crop season of the State of Tamil Nadu is during the northeast monsoon. The occurrence of cyclones during the northeast monsoon only will decide whether the monsoon is normal or excess or deficit. The northeast monsoon is not understood well even now unlike southwest monsoon. The distribution of rainfall over the four meteorological subdivisions, which is influenced by the northeast monsoon, is given in Table 8.3.

Table 8.3 Seasonal rainfall (mm) of 4 meteorological subdivisions of southern region

Subdivision	*Rainfall during northeast monsoon (Oct. to Dec.)*	*% of annual rainfall during*		*Rainfall during southwest monsoon*
		(NEM)	*(SWM)*	
Coastal Andhra Pradesh	331	38	32	582
Rayalaseema	225	40	20	396
Kerala	525	28	18	1906
Tamil Nadu	489	27	49	337
Southern region*	392	25	34	601

*All the four meteorological subdivisions are combined under southern region.

The isohyetal map during October and November (see Fig. 8.10) as compiled by weather bodies indicates that the north and northwest India

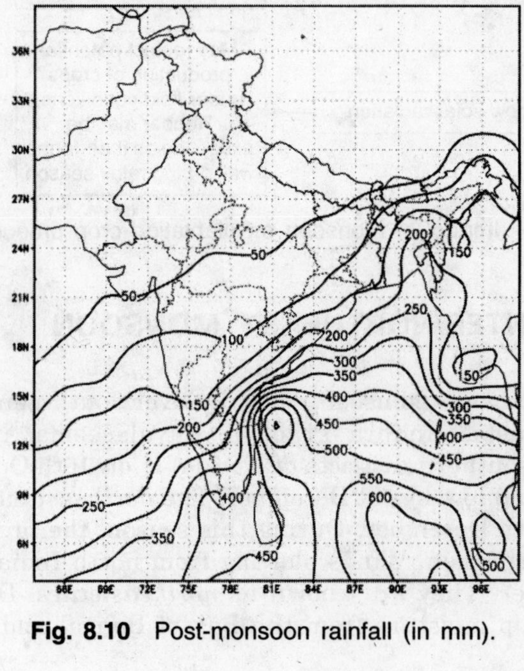

Fig. 8.10 Post-monsoon rainfall (in mm).

receive no rainfall and relatively less (50–100 mm) over central India and increases towards Tamil Nadu and Kerala (300–400 mm). In winter also, a similar trend is noticed. The only difference is that the northern States viz. Jammu & Kashmir, Punjab, Haryana, Himachal Pradesh, Uttaranchal and Delhi also receive rains during northeast monsoon (see Fig. 8.11). Of course, some pre-monsoon showers are also noticed from March to May across the country (see Fig. 8.12).

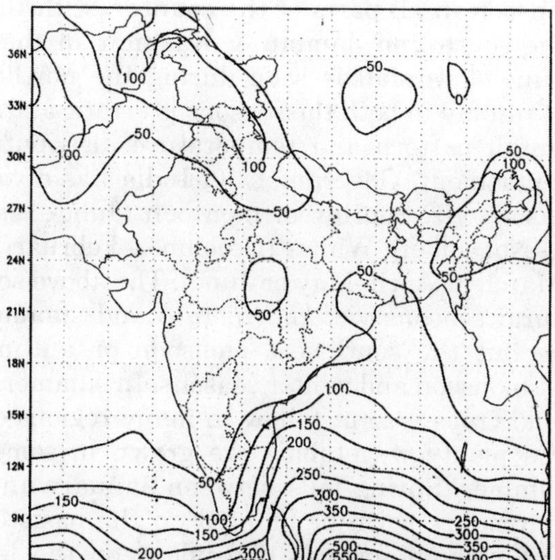

Fig. 8.11 Winter rainfall (in mm).

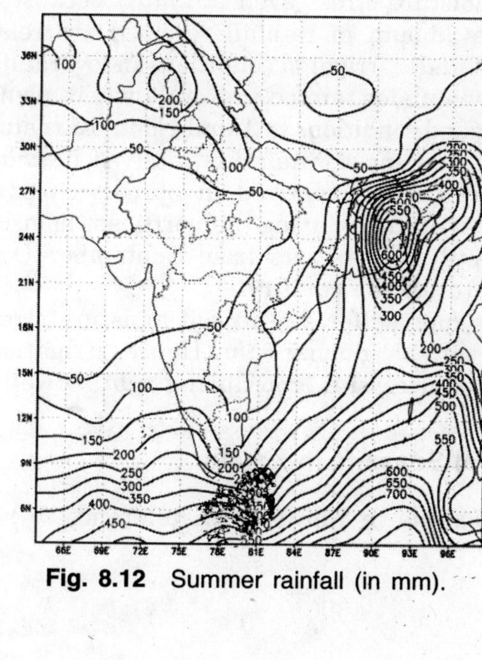

Fig. 8.12 Summer rainfall (in mm).

8.4 AGRICULTURAL SEASONS

Crops require soil moisture during the growth and development phase. The only source for soil moisture is rainfall under rainfed conditions. Since crops depend on rainfall, they must be adjusted to a particular period during which rainfall is assured. Such periods over India are southwest and northeast monsoons. Plenty of rainfall is noticed during the southwest monsoon in all parts of the country, with the exception of Tamil Nadu in the south and Jammu & Kashmir in the north, where considerable amount of rainfall is seen during the northeast monsoon. The distribution of rainfall in both the seasons is erratic and unpredictable. Based on the rainfall pattern and temperature distribution, the India Meteorological Department, Government of India has divided the whole year into four seasons viz. Southwest monsoon (June–September), Post monsoon (October–November), Winter (December–February) and Summer or pre-monsoon (March or April–May or June). The above seasons coincide with the agricultural seasons viz. *kharif, rabi* and summer. The *kharif* season is nothing but the southwest monsoon or autumn while *rabi* coincides with post monsoon and winter seasons. In summer, the cultivable land under seasonal crops is kept fallow in many regions of the country. Wherever water is plenty, vegetables are grown in some parts of the country during summer. Spring falls between January and March.

All the field crops are sown or planted on the basis of the onset of monsoon. The type of crop sown is dependent on the length of rainy season. The crops are never in shortage of water during the first crop season in major parts of India if monsoon is normal except in northwestern parts of the country. This is more so in case of Rajasthan where crops are grown under soil moisture stress even in *kharif* because the number of rainy days and the amount of rainfall received are relatively less. Of course, rice is grown under irrigated conditions in low rainfall regions. For example, rice is grown under rainfed in Kerala where plenty of rainfall is seen and under irrigated conditions in Tamil Nadu, Karnataka and Andhra Pradesh, where low amount of rainfall is received in *kharif* and *rabi.*

In *rabi,* some rainfall is received in northern parts of India and Tamil Nadu gets 49% of rains during the northeast monsoon. In fact, the crop season in Tamil Nadu starts from September–October onwards unlike in other States of the country.

Almost all the crops suffer due to soil moisture stress if irrigation facilities are not available during *rabi.* Hence, irrigation facilities are prerequisite to raise successful crop during *rabi* as well as summer.

8.4.1 Agricultural Seasons in Kerala

The agricultural seasons in Kerala with reference to rice crop are as follows:

Virippu/Autumn/*Kharif*/First crop season:
April–May to September–October
Mundakan/Winter/*Rabi*/Second crop season:
September–October to December–January
Puncha/Summer/Third crop season:
December–January to March–April

The optimum time of planting is the first fortnight of August for second crop in Onattukara. The optimum season in Kuttanad area is October–January for the *Puncha* crop. The time of planting plays a major role in deciding crop duration and grain yield in photosensitive varieties or cultivars of rice. In case of Allikkannan, it is observed that the first week of June is the optimum time of transplanting for obtaining maximum grain and straw yields in the wetlands of Kannur and Kasaragod districts of Kerala. It appears that all the photoinsensitive varieties perform well if transplanted during the third week of October in *rabi* in the northern districts of Kerala.

All the field or seasonal crops are raised in Kerala without irrigation during *virippu* season while crops are under severe moisture stress in *mundakan* and *puncha* seasons if irrigation is not provided. This is more so in case of the northern districts of Kerala. To some extent, the second crop of rice can be raised even under rainfed conditions in southernmost districts of Kerala due to rains received because of the influence of northeast monsoon. As the plantation crops are predominant in Kerala and grown round the year, irrigation is a must from December to May for successful cultivation of many plantation crops.

8.5 MONSOON MANAGEMENT

As mentioned in earlier sections, the monsoon plays a major role in Indian economy. A good monsoon makes the country go forward in terms of foodgrains production as was realized during the all-India drought year 2002. Its negative impact is directly (a whopping 3% negative growth was noticed during 2002–03) and indirectly enormous on the Indian economy, showing the relevance of Indian summer monsoon. Realising its importance, Dr. M.S. Swaminathan stated that the 'Monsoon Management Boards' at the Centre and State levels can only change the saying "Indian agriculture is a gamble of the monsoon" into "India's strength lies in its ability to manage monsoons". His concept is put in a schematic presentation (see Fig. 8.13). The technological innovations and awareness on weather should be brought under the umbrella of Monsoon Management Boards, wherein climate managers and crop weather watch groups with compensatory crop production programmes play a crucial role in pro-active monsoon management, so as to sustain Indian agricultural production.

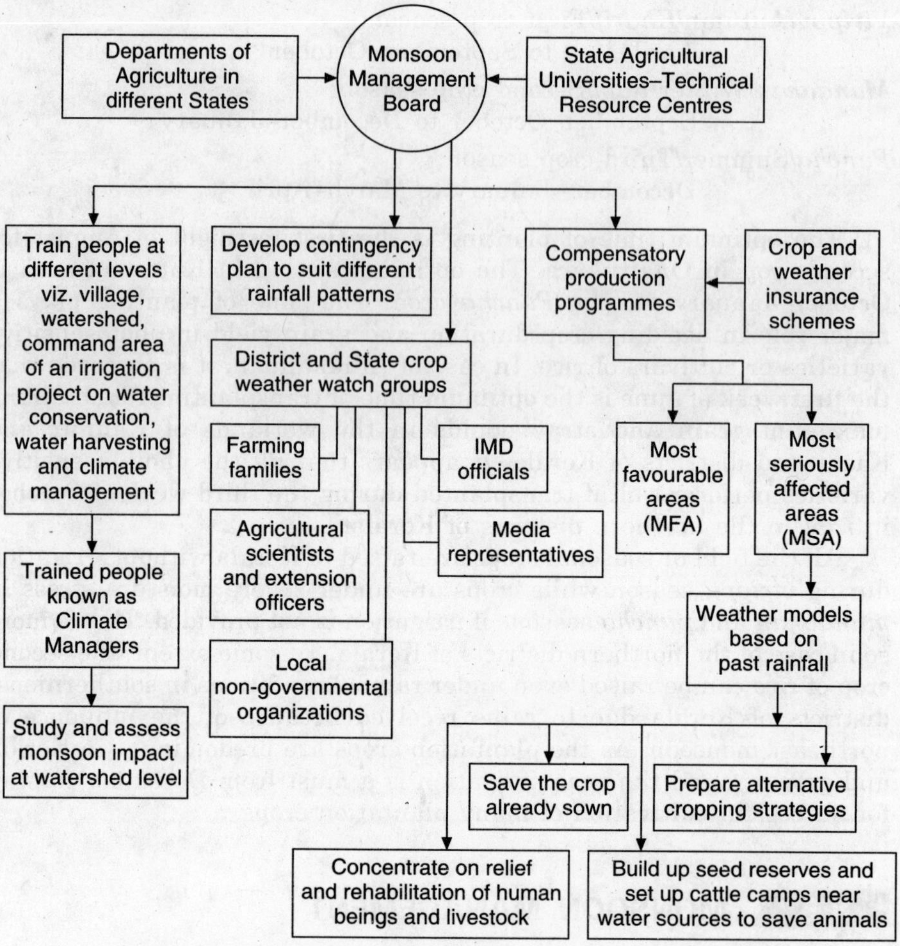

Fig. 8.13 Schematic presentation of monsoon management.

REFERENCES/SUGGESTED READINGS

Ananthakrishnan, R. and Soman, M.K., 'The onset of the southwest monsoon over Kerala: 1901–1980', *Int. J. Climatology*, 8, 283–296, 1988.

Ananthakrishnan, R. and Soman, M.K., Onset dates of the Southwest monsoon over Kerala for the period 1870–1900, *Int. J. Climatology*, 9, 321–322, 1989.

Das, P.K., *The Monsoons*, National Book Trust, New Delhi, 1968, p.162.

Kulshrestha, S.M., *Be Weatherwise*, National Institute of Science Communication, Council for Scientific and Industrial Research, New Delhi, 1998, p.106.

Kumar, K.R., *Climate maps: CPC Merged Analysis of Precipitation, 1979–2000*, Personal Communication, Indian Institute of Tropical Meteorology, Pune, 2003, p.12.

Parthasarathy, B., Kumar, K.R. and Munot, A.A., 'Forecast of rainy season foodgrains production based on monsoon rainfall', *Indian J. Agric. Sci.*, **62**(1), 1–8, 1992.

Raj, P.C.A., *Onset of effective monsoon and critical dry spells, A Computer based forecasting technique*, Water Technology Centre, Indian Agricultural Research Institute, New Delhi, 1979.

Raman, C.R.V. and Murthy, B.S., *Water availability periods for crop planning*, Sci. Rep. 173, India Meteorological Department, Pune, India, 1971.

Swaminathan, M.S., 'Pro-active monsoon management plan is needed' (Interview), *Frontline*, **17**, 77–81, 2000.

TNAU, 'Capturing the benefits of seasonal climate forecasts in agricultural management', *Proceedings of short course,* June 20–29, 2001, Tamil Nadu Agricultural University, Coimbatore, p.196, 2001.

QUESTIONS

1. What are the monsoon countries?
2. What are the agricultural seasons over India?
3. What is the concept of monsoon management?
4. Distinguish between:
   (a) Southwest monsoon and northeast monsoon
   (b) Onset and withdrawal of monsoon
   (c) Onset of monsoon and effective onset of monsoon
   (d) Dry spell and wet spell
   (e) Kharif and Rabi
   (f) Virippu and Mundakan
5. Write short notes on:
   (a) Southwest monsoon
   (b) Onset and withdrawal of southwest monsoon
   (c) Onset of monsoon over Kerala
   (d) Onset of monsoon versus monsoon rainfall over Kerala
   (e) Onset of effective monsoon
   (f) Dry and wet spells
   (g) Monsoon rainfall over India
   (h) Monsoon rainfall over Kerala
   (i) Analogous technique
   (j) Northeast monsoon
   (k) Monsoon rainfall and crops over Kerala

6. Fill in the blanks:
   (a) ——— number of seasons have been defined based on wind direction.
   (b) Monsoon is originated from the Arabic word ————.
   (c) Kerala is popularly known as ————.
   (d) ———— is the onset of southwest monsoon over India.
   (e) ———— is the onset of southwest monsoon over Kerala.
   (f) ———— and ———— are the States in India to receive less monsoon rainfall.
   (g) ——— and ——— are the States receive significant rains during norhtheast monsoon.
   (h) ———— is the main crop season in India.
   (i) ———— is the main crop season in Tamil Nadu.
   (j) ———, ——— and ——— are the main rice crop seasons in Kerala.
7. Match the following:

(a) Poor monsoon	Main crop season
(b) Heavy monsoon	Kharif
(c) Northeast monsoon	Rajasthan
(d) Southwest monsoon	Rabi
(e) Virippu	Tamil Nadu
(f) Mundakan	Summer
(g) Puncha	Kerala

8. Elucidate the following:
   (a) Agricultural seasons
   (b) The monsoons

9 WEATHER AND CROPS

Climate influences the distribution of crops over different regions of the world, while weather influences crop production and productivity. Of the various weather elements affecting crops, temperature and rainfall are perhaps the most important. It is the reason for the development of special flora and fauna in different climatic zones of the world as seen in the tropical, temperate and frigid zones. The tropical regions get abundant solar heat and the vegetation is high while it is low in the temperate regions. The vegetation is very little in the frigid zones. Similarly, natural vegetation is at its poor growth in the arid climates and better in the humid tropics. Xerophytes are confined to the arid climates where low rainfall is seen, while hydrophytes are confined to the tropics. Mesophytes are intermediate in their water requirement. Most of the crops are included in this group.

9.1 ROLE OF WEATHER VARIABLES IN ENVIRONMENTAL STRESS

The biotic and abiotic stresses are the two major components in environmental stress under which crops are grown. The influence of weather factors on biotic stress due to incidence of insect pest and diseases, epiphytes, weeds, rodents and grazing animals is indirect rather than direct. Of course, the incidence of some of the pests and diseases is directly influenced by weather elements. Under abiotic stress, crop management practices play an important role in deciding crop output in addition to the soil environment (see Fig. 9.1). The crop-weather

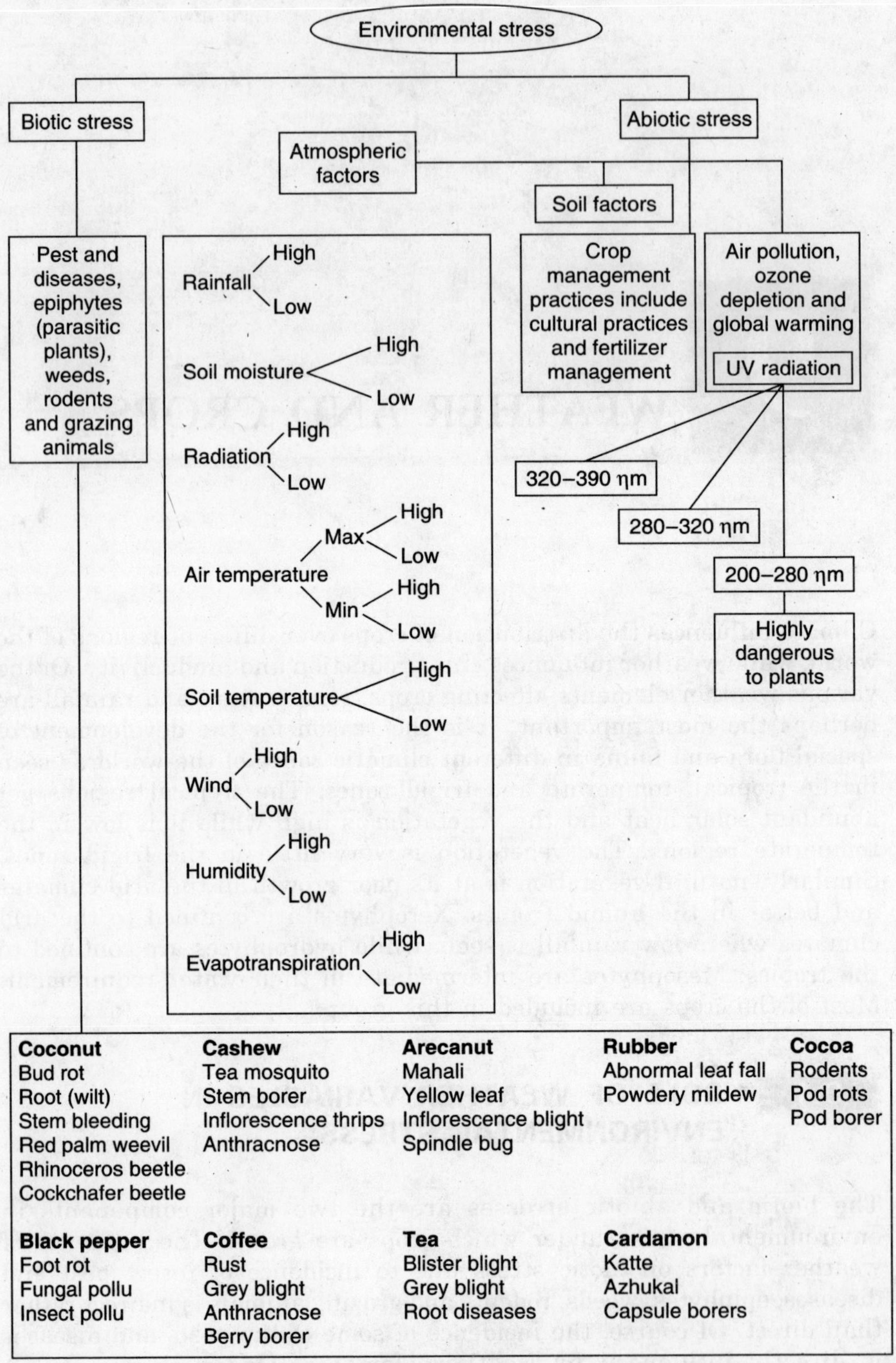

Fig. 9.1 Effect of biotic and abiotic stresses on crops.

relationship could be better understood if the various aspects of environmental stress on crops are taken into account. The impact of UV-B radiation due to ozone depletion on crops is another area of interest which is yet to be understood. Unless the crops are protected through better crop management, it is difficult to quantify the effects of weather elements on crop growth, development and yield. It is one of the reasons why the concept of crop simulation modelling has been brought to highlight the yield potential constraint analysis under different crop growing environments of a given crop. Rainfall, air and soil temperatures, photosynthetically active radiation (PAR), soil moisture, wind, humidity, and evapotranspiration are the atmospheric variables which play a predominant role if the other environmental factors are not limiting. The weather abnormalities with respect to crop and location are also to be defined to understand the adverse or beneficial effects of the same on crop output. No doubt, many studies on crop-weather relationships are presented in relevant literature. They are, however, seldom used for operational purposes as they either lack sound conceptual basis or are mere statistical tools applied for interpretation to understand the crop-weather relationship.

9.2 CLIMATE TYPES AND VEGETATION

The interaction of rainfall with temperature has a very profound effect on vegetation and soil types. The temperature of a place is largely influenced by latitude, altitude and continental factors. Based on the above, the natural vegetation is classified according to the climatic types viz. tropical, temperate, taiga, tundra and polar (see Table 9.1).

Table 9.1 Climate and vegetation

Climate	*Vegetation*
Tropical	Rainforests, desert and grasslands
Temperate	Grasslands and deciduous forests
Taiga	Coniferous forests
Tundra	Low shrubby growth, mosses, lichens
Polar	Snow and ice

9.2.1 Temperature and Vegetation

The classification of vegetation and crops based on temperature is given in Table 9.2.

Table 9.2 Temperature and crops in different regions

Class	*Region*	*Temperature conditions*	*Type of vegetation*	*Common crops cultivated*
Mega-therms	Equatorial and Tropical	High temperature throughout the year	Tropical rain forest	Tropical crops like rubber, coffee, cassava, cocoa, banana, rice and sugarcane
Meso-therms	Tropical and Subtropical	High temperature alternating with low temperature in winter	Tropical deciduous forest	Subtropical crops like maize, sorghum and other millets
Micro-therms	Temperate and high altitudes (up to 1200 ft in the tropical and subtropical)	Low temperature	Mixed coniferous forest	Temperate crops like wheat, oats, potato and apple
Hekisto-therms	Arctic and Alpine regions*	Very low temperature	Alpine vegetation	Pines and spruce

*(> 1600 ft in the tropics and 1200 ft in the temperate regions).

9.3 INFLUENCE OF ALTITUDE ON CROP DISTRIBUTION

The altitudinal sequence of crops in Kerala, which is related to temperature distribution, is given in Table 9.3.

Table 9.3 Altitudinal sequence of crops in Kerala

Class	*Region*	*Temperature conditions*	*Altitude (AMSL)*	*Crops*
Mega-therms	Lowland	Moderate temperature throughout the year	0–7.5 m	Coconut, rice and cashew
Mega-therms	Midland	Moderate temperature throughout the year	7.5–75 m	Coconut,cocoa, arecanut, rubber, cashew and black pepper
Meso-therms	Highland	Moderate to low temperature throughout the year, winter temperature is low with altitude	75–700 m	Tea, coffee (robusta), rubber, black pepper and coconut
Micro-therms	High ranges	Low temperature throughout the year	700–1500 m	Cardamom, coffee (arabica) and tea

The above crop distribution is a classic example of the influence of temperature within the tropics. The distinct feature is that the crops which are grown in the high ranges are not seen in the lowlands and midlands. It is due to temperature distribution which is reflected by altitude.

9.4 INFLUENCE OF MONSOON IN THE HUMID TROPICS

Low and ill-distributed rainfall are common features of dryland farming, wherein crops like sorghum, pearlmillet, fingermillet and *kharif* pulses are grown. Rice is grown in wetlands along the coastal belts and inlands, if supported with irrigation. It is also grown in flat areas of the Western Ghats of India, while tea, coffee, cardamom, rubber, black pepper, cashew, coconut and arecanut are grown in slopes and valleys, where heavy rainfall is seen. The above crop distribution is seen in Kerala with altitude, which falls under the humid tropics as seen in Table 9.3.

High rainfall during the southwest monsoon (June to September) leads to waterlogging, less bright sunshine, low intensity of solar radiation and low evapotranspiration. Scanty and erratic rainfall from November to May lead to soil moisture stress or drought, and high soil and atmospheric temperatures. At the same time, the crop evapotranspiration demand is high during summer. The above situation is more predominant in the northern districts of Kerala viz. Malappuram, Kozhikode, Kannur and Kasaragod. In general, the two extreme climates adversely affect all the crops grown in the northern parts of Kerala in one way or other.

9.4.1 Weather and Rice

Rice is indigenous to the humid areas of tropical and subtropical regions. Rice can be grown as a transplanted or directly sown crop during the first crop season, depending on the availability of water, if other environmental conditions in the humid tropics are not limiting. During *rabi* and summer, rice cannot be grown successfully without irrigation as seasonal droughts, which adversely affect the crop, are common.

Rainfall

Rice is grown mostly under rainfed conditions in Kerala during the first crop, which receives more than 2000 mm during the crop season in some rice-growing areas of Kerala. Rainfall received during the first crop season is more than sufficient to meet crop water requirements. Hence, the crop is never under soil moisture stress in the first crop season where it is grown in wetlands. However, premonsoon showers during April and May are very important for the rice crop in Kerala as it is mostly a directly sown crop during *kharif*. If premonsoon showers are erratic or inadequate, germination and survival of seedlings become doubtful and crop suffers heavy loss initially. It means poor grain yield during the first crop season. Based on pre-monsoon and monsoon rainfall, 37% of yield variability can be explained during *kharif* in the case of paddy. The paddy crop is under waterlogged or flooded condition depending upon the distribution of rainfall in *kharif* under wetlands. The wet spells are detrimental from flowering to maturity in case of paddy. Rainfall is not adequate in the other two seasons (*Mundakan* and *Puncha*) to meet the water requirement of rice. Hence, irrigation should be assured for successful crop during *mundakan* (*rabi*) and *puncha* (*summer*) seasons. In other rice-growing States, it is mostly cultivated under assured irrigation conditions as the rainfall is inadequate to meet crop evapotranspiration demand. It is a transplanted crop even during the *kharif* in other rice-growing States unlike in Kerala and Orissa.

Temperature

Temperature influences the growth, development and yield of rice. Being a tropical or subtropical plant, rice generally requires high temperatures above 20°C but not above 35–40°C. The optimum appears to be near 30°C for the day maximum and near 20°C for the night minimum. The crop duration varies with the maximum temperature. It can be expressed in terms of heat units (degree-days). It indicates that if the maximum temperature is high, the number of days required for crop maturity is proportionately less (see Fig. 9.2). It reveals that the photo-insensitive cultivars take less time for crop maturity during *rabi* and summer when compared to that of *kharif*. Unlike in *kharif*, the crop is exposed to high temperature during the ripening phase in *rabi* and summer.

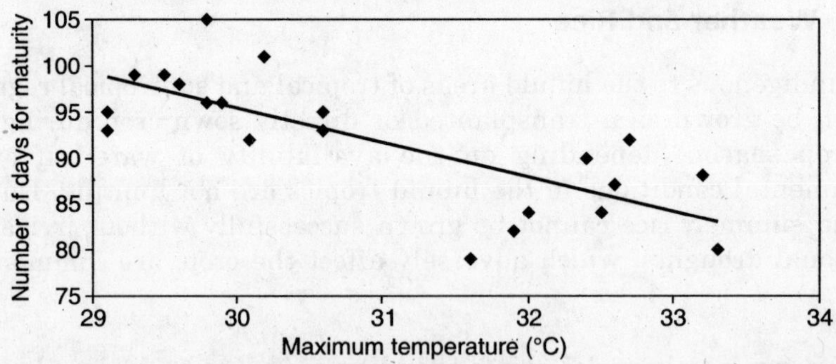

Fig. 9.2 Maximum temperature versus crop duration in Jaya at RARS, Pattambi (Kerala).

Solar radiation

The intensity of solar radiation has a profound influence on rice yield. Its role in photosynthesis, photoperiodism and photonasty are well-known. Light intensity fluctuates highly within a day, within a month and among the seasons. It is reported that the amount of solar radiation received during the period of 45 days before the harvest is the factor on which the final crop output depends. The grain weight per spikelet can be improved if the solar radiation is optimum during the maturity period if other factors are not limiting.

Atmospheric water vapour

Rice grown in the humid tropics is subject to unusual stresses due to continuous high water vapour content in the atmosphere during the first crop season. Such humid conditions are favourable for the development and spread of many fungal and bacterial diseases. It is also conducive for various crop insect pests.

Wind

Wind is normally an unimportant factor in rice production. A light wind is said to be beneficial as it stirs air and transports CO_2 to the leaf canopy. Very strong winds do occur during cyclones and lead to severe damage. During pollination, strong wind may induce sterility and increase the number of abortive endosperms. Hence, strong winds during flowering phase may not be conducive for rice crop. It is the case during the mundakan season across the Palakkad and Thrissur districts. High wind speed during flowering may cause pollen dehydration and consequent spikelet sterility. Identification of suitable windbreaks during the second crop season may be beneficial, wherever wind speed is high.

9.4.2 Weather and Wheat

Rice and wheat are the two main cereals in India. Unlike rice, wheat is grown in temperates and subtropics. The majority of the States in central and north India grow this crop during *rabi* (November to March). Among the cereals, wheat is very sensitive to surface air temperature. Rise in maximum temperature during December adversely affects the active tilling phase of the crop while the reproductive phase is affected if the rise in maximum temperature is noticed during February and March. The cold wave between December 2002 and January 2003 as well as the heat wave during March 2004 adversely affected the crop to a great extent. Wheat yield will be low if the minimum temperature goes below 6°C or if the maximum temperature goes above 30°C. The wheat yield is even low if low relative humidity coincides with maximum temperature of more than 32°C during the reproductive phase. That is the reason why heat and cold waves during the crop season adversely affect the wheat yield to a considerable extent. Due to aberrations in temperature for the last 4–5 years, the wheat production across the northern belt of the country was not to the levels of plan estimates. The climate projections also indicated that the rise in temperature over north India may result in reduction in wheat area as well as wheat production. All these indicate that wheat is highly sensitive to high maximum and low minimum temperatures among different weather variables. The response of the crop to temperature again varies depending upon rainfed and irrigated conditions. The effect of temperature on crop growth is more under rainfed conditions. Delay in sowing may lead to early maturity, as it is exposed to higher temperature, and thus low yields are expected. Of course, early rains during sowing time also will improve the crop yields under rainfed conditions.

Growing degree days for wheat crop

Wheat is highly sensitive to surface air temperature and thus to growing degree days under different crop environments. If the growing degree days are more the wheat yield is likely to be more (see Fig. 9.3(a) and (b)). There are reports that the growing degree days varies between 1700 and 1800 day degree Celsius from emergence to physiological maturity over Ludhiana region. However, different varieties will have varied growing degree days depending upon the duration of the variety, which influences wheat yield. In a variety of crops, the yield is more if the growing degree days are more, within a certain limit. The duration of various biotic events also affects growing degree days and it finally reflects on yield.

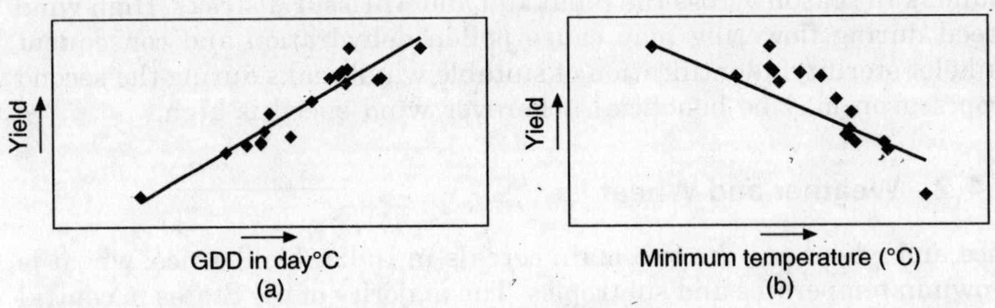

Fig. 9.3(a) and (b) Growing degree days and minimum temperature versus wheat yield.

9.4.3 Weather and Coconut

Coconut is relatively tolerant to weather aberrations, unlike seasonal crops. It tolerates high rainfall and temperature to some extent. Coconut production under rainfed conditions is influenced significantly by the length of dry spells at critical stages. The initiation, ovary development and button-size nut development in coconut are sensitive to soil moisture availability. The crop performs better under irrigated conditions during summer.

Rainfall

Rainfall appears to be the most important weather element which determines coconut growth and development as it is grown under rainfed conditions. Rainfall during summer months (March, April and May) augments coconut yield in the following year, while heavy rains lead to waterlogging in coconut gardens. High rainfall during the monsoon may adversely affect the coconut yield in the following year, as the yield is less than average whenever the annual rainfall exceeds 4275 mm (see Fig. 9.4). The button shedding appears to be high during the rainy season. Hence, summer rainfall has positive effect and high rainfall during monsoon may be having negative effect. There is a lag period between the influence of weather and crop yield as the initiation of primodium in coconut takes 44 months before the final harvest. However, the duration of prolonged dry spell from November to May adversely affects the coconut production in the following year under the field conditions (see Fig. 9.5(a) and (b)).

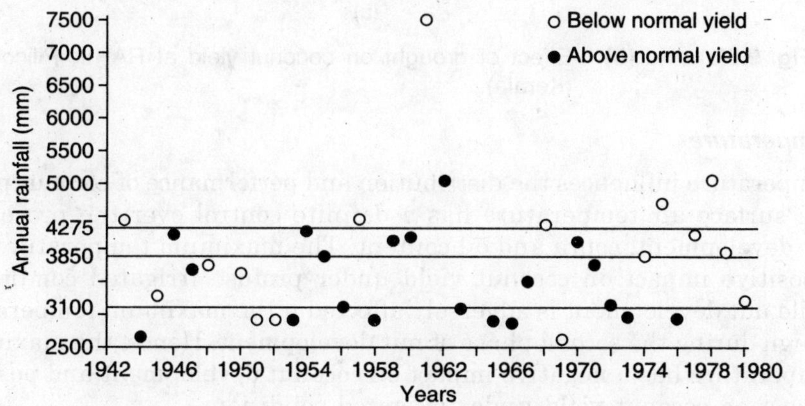

Fig. 9.4 Annual rainfall versus coconut yield in the following year at RARS, Pilicode (Kerala).

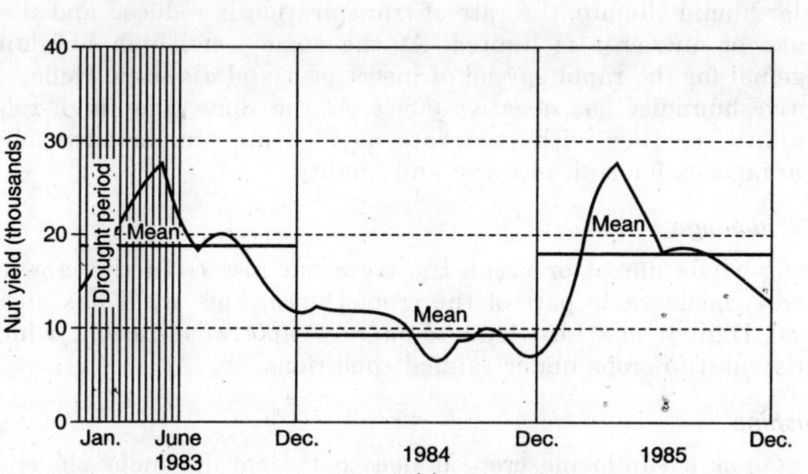

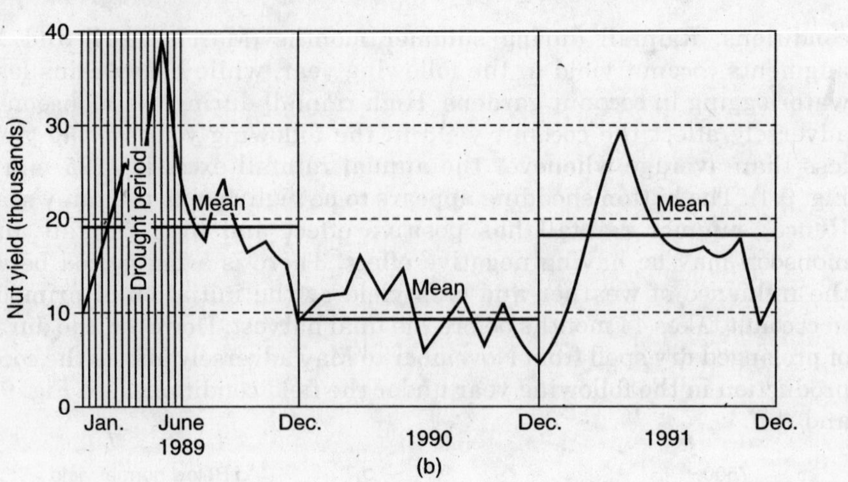

Fig. 9.5(a) and (b) Effect of drought on coconut yield at RARS, Pilicode (Kerala).

Temperature

Temperature influences the distribution and performance of coconut palm. The surface air temperature has a definite control over the process of nut development, copra and oil content. The maximum temperature has a positive impact on coconut yield under profuse irrigated conditions, while nut development is adversely affected if the maximum temperature is high during the second phase of nut development. Hence, the maximum temperature has a negative impact on coconut development and positive impact on coconut yield under irrigated conditions.

Humidity

Under humid climate, the rate of transpiration is reduced and thereby uptake of nutrients is limited. At the same time, high humidity is congenial for the rapid spread of insect pest and diseases. Hence, high relative humidity has negative effect. At the same time, high relative humidity together with moderate surface air temperature may be advantageous for both nut size and quality.

Wind damage

Strong winds uproot or break the trees but also twist the crown and destroy considerable part of the crop. Hence, high wind has negative effect. High winds also lead to more evapotranspiration, which is detrimental to crops under rainfed conditions.

Sunshine

Coconut is a sun-loving crop. It does not come up under shade. Low sunshine during the rainy season may be one of the inhibitant factors for low coconut yields in Kerala. The coconut palms along the northern rows in coconut gardens grow towards sunlight, which shows that the coconut palms require a good amount of sunlight.

9.4.4 Black Pepper

Pepper requires a warm and humid climate. An annual rainfall of 2500 mm is ideal for proper growth of the crop. Rainfall is believed to be the chief pollen vector in pepper. The flowering process in pepper is initiated by the application of water equivalent of 70 mm or more rainfall within the period of three weeks, followed by a dry spell. Rainfall influences the flower bud differentiation process. Spike elongation and berry development in pepper may be ceased if there is prolonged dry spell immediately after good summer showers. Hence, good summer showers followed by a dry spell may be detrimental to black pepper production. Moderate and continuous rainfall till berry initiation may be favourable to black pepper. It may not come up well in high ranges, where low temperature prevails throughout the year.

9.4.5 Rubber

Rubber can be grown from almost sea level up to an altitude of 500 m in areas receiving well-distributed annual rainfall of not less than 2000 mm with a warm, humid, equable climate (21 to 35°C). The rate of growth is maximum in rainy season when compared to summer. Hence, high rainfall during summer as well as during monsoon has positive effect on rubber.

9.4.6 Arecanut

The arecanut palm is capable of growing under a variety of climatic and soil conditions. It grows well from almost sea level up to an altitude of 1000 m in areas of abundant and well-distributed rainfall or under irrigated conditions. Arecanut is mostly grown in valleys under profuse irrigation in summer. Arecanut may not come up well under waterlogged conditions. The climatic requirements of arecanut appear to be similar to that of coconut, but it is relatively more sensitive to soil moisture stress.

9.4.7 Cashew

Continuous rain without critical dry spell may delay the bud break in cashew. A cashew plant undergoing a first dry spell may wait for the second dry spell to occur for bud break. It appears that both the dry spells, at the time of bud break and thirty days prior to bud break, with bright sunshine are critical for bud break in cashew. However, the influence of dry spell and bright sunshine on bud break of cashew is yet to be understood. Cashew flowering is highly erratic and depends on many environmental factors (see Fig. 9.6).

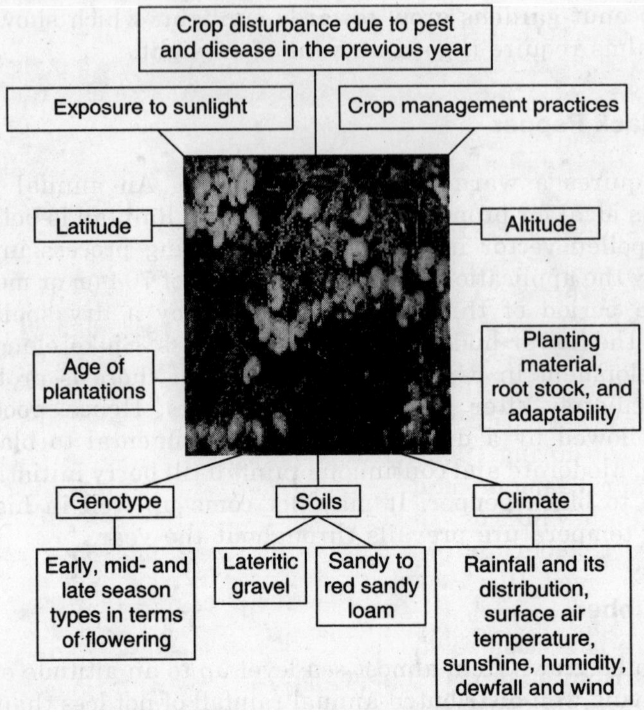

Fig. 9.6 Effect of abiotic and biotic factors on flowering of cashew.

The soil moisture stress has no relevance in bud break of cashew, as it commences much before soil moisture stress starts across the west coast. Unusual rain between November and December may lead to delay in the reproductive phase of cashew. The delay in crop duration could be explained with geographical coordinates, once the bud break in cashew is initiated immediately after the rain spell is over.

Relatively dry atmosphere with mild winter may be conducive for better flowering in cashew. Mild winter may be defined as "low minimum surface air temperature ranging between 15 and 20°C coupled with more dew nights having moderate dew". Better availability of soil moisture during flowering period (December and January) may not influence flowering in cashew.

Cashew responds to phototropism. Cashew is not only photosensitive but also thermosensitive, as the biotic events respond to surface air temperature. It seems that a temperature range between 15°C and 35°C may be optimum for better growth and production of cashew. The surface air temperature is conducive across the West Coast for better cashew yield, while it may not be so across the East Coast.

The kernel weight appears to be dependent on the availability of soil moisture. It is very important during nut setting and development. The nut and kernel weights may be better with altitude in early season types, while it may not be the case with late season types. Depending upon the genotype and its adaptability, the nut and kernel weights are superior in both the West and East Coasts of India. However, the shelling percentage is lower across the East Coast compared to the west coast. The nut quality is affected if a wet spell is noticed during the nut development of cashew.

9.4.8 Cardamom

The important physiological stages like panicle initiation and subsequent growth (forwarding) depend on receipt of showers during January to May. The failure of showers during this period result in poor growth and crop yield. The unprecedented drought in summer 1983 adversely affected the growth and yield of cardamom in Coorg district, and the same trend prevailed in other cardamom-growing tracts of south India. The drought that prevailed during 1982–83 resulted in as high as 50% mortality in some of the cardamom estates.

In the Palney hills of Tamil Nadu, both panicle and flower production are seen in October and November, during which northeast monsoon rains are received. It indicated that both panicle and flower production in cardamom are commensurate with the rainfall pattern of the cardamom tract. There is a close relationship between cardamom production and rainfall distribution during summer. Summer rains during March, 2008 have benefited the cardamom crop of 2008–2009 upto 20–30%.

9.4.9 Coffee

The major factors that affect production of the arabica and robusta varieties of coffee are given in Table 9.4. The blossom and backing showers

Table 9.4 Effect of climatic factors on arabica and robusta

Climatic factors	*Arabica*	*Robusta*
Lack of blossom showers	Absence of rains in March–April	Absence of rains in March
Rain on opening day of blossom	Partial to complete failure	Partial failure
Hailstorm	Injury to the floral and vegetation phase	Partial failure
Backing showers	Absent in May–poor crop set	Absent in April–poor crop set
Severe western exposures (with lowering of sub-soil moisture)	Partial to complete failure	Partial to complete failure
Excessive wetness and waterlogging	Partial to complete failure	Partial to complete failure
Wind	Both are sensitive to wind disturbance	

are very important in the case of coffee. Coffee yield is very poor in the absence of either of them.

The growth of coffee appears to be affected when exposed, and hence it requires a certain degree of shade for better performance. The quality of "Monsooned Malabar Coffee" is better if monsoon behaviour is good over Malabar region.

9.4.10 Tea

Two growth periods are seen in tea (mid-March to mid-June and mid-August to mid-November). Night temperature and relative humidity are the most important factors which determine the productivity of tea in addition to rainfall. In Annamalai, night temperature occasionally drops below 5°C and humidity is below 10% when the dry atmosphere is prevalent. This situation leads to tea production at lower levels.

In tea, there is a time lag between the physiological function of tea bush and the vegetative harvest. It is about five weeks in the pruning year and four weeks in other years. As such the weather conditions of the previous month is generally correlated to the current month's harvest.

Irrigation during cool dry season (mid-January to mid- March) will not produce the desired effect on production as temperature is the limiting factor during the period. Irrigation during the hot and dry period (March to September) that occur frequently in Nilgiris will be beneficial in crop production. The occurrence of heat wave during March 2004 adversely affected tea crop upto 50 percent over Himachal Pradesh.

9.4.11 Cocoa

High rainfall has a malevolent effect in places where the crop is grown under waterlogged conditions, while summer rainfall has a benevolent effect on the yield. On augmenting the dry season's rainfall by the use of irrigation, the results have not always been improved because the water lost from the leaves is so high with low humidity that the roots cannot match the loss. Crop yields are very low with heavy shade and increases with increasing light up to the 50% level. Cocoa yield is affected by the presence or absence of fertilizer if the level of light is above 50%. With added fertilizer, yields increase almost up to full light, whereas in the absence of fertilizer the yields fall off. The theory has therefore been advanced that the light regime for optimum yield of cocoa is a function of its mineral nutrition. The loss of apical dominance in cocoa is seen constantly at a temperature of above 32°C. The loss of apical dominance implies the development of side shoots of the plant. High maximum temperatures (> 36°C) during February and

March, 2004 resulted in low pod yield in cocoa over the central region of Kerala.

9.5 EFFECT OF WEATHER ON PLANTATION CROPS

The effect of weather elements on plantation crops is summarised in the form of positive and negative signs based on the above discussion and presented in Table 9.5. The influence of seasonal rainfall on plantation crops is the only variable discussed in literature, while there is scant mention of other variables.

Table 9.5 Effect of weather elements on plantation crops

Weather elements	*Coconut*	*Arecanut*	*Cocoa*	*Pepper*	*Cashew*	*Rubber*	*Cardamom*	*Coffee*	*Tea*
Rainfall (monsoon)	–	–	–	+	+	+	–	+	+
Rainfall (summer)	+	+	+	–	–	+	+	+/–	+
Maximum Temp. (low altitudes)	–/+		–						
Minimum Temp. (high altitudes)		–						–	–
RH (%)	It leads to insect pest and diseases. Hence, there is negative effect on all the crops. Tea requires high relative humidity for better yield.								
Solar radiation	It is not adequate during the monsoon period and hence it affects crop production. High rainfall with low solar radiation during the monsoon and low rainfall with high solar radiation during summer adversely affect most of the plantation crop production in the humid tropics.								
Wind speed	Moderate in Kerala except in the Palakkad Gap region. Mundakan *rabi* rice is affected in the above region. Beneficial to crops where winds are moderate.								
Evaporation	Evapotranspiration is restricted during the monsoon due to heavy rains. Atmosphere is relatively saturated under the above conditions. Therefore, uptake of nutrients may be restricted.								

– negative effect + beneficial effect or no adverse effect

In general, growth is restricted during rainy season under water-logged conditions, while crop growth is better during summer if irrigation is provided and other environmental factors are not limiting.

9.6 CLIMATIC NORMALS FOR CROP PRODUCTION

Crop distribution, production and productivity are influenced by air temperature, rainfall and the other meteorological variables viz. sunshine, solar radiation, humidity and wind. It is also understood that every crop has its own cardinal points in terms of climatic requirements. If the crops are selected for cultivation on the basis of the optimum climatic requirements, it is likely that crop production can be maximised under rainfed as well as irrigated conditions. For example, rice is grown in vast areas across the country in different agroclimatic zones except for a few pockets. However, it gives better yield only in some agroclimatic zones due to prevalence of optimum climatic conditions. The production potential could be explained through crop-wise agroclimatic zones, whether the crop is well-suited with minimum inputs for maximising yields (refer to Chapter 12 under agroclimatic classification) or not. The agroclimatic indices viz. the moisture availability index (MAI), the index of moisture adequacy (IMA) and the climate indices (both thermal and moisture regimes) could be derived on the basis of climatic normals. Based on the above indices, suitable crops have been suggested by the India Meteorological Department (IMD) in different agroclimatic zones. Many attempts were also made by the scientists from different parts of the country to explain the climatic boundaries for various crops based on agroclimatic indices. For example, the whole country has been divided into six fruit-growing regions from the climate point of view with information on frost occurrence (see Fig. 9.7). Such information is useful to minimise the ill-effects of frost on fruit crops and to avoid the zones where climatic risk is involved with respect to frost. Likewise, all climatic maps for different weather variables could be examined to find out the suitable areas for a given crop with respect to climate. Therefore, the climatic normal maps are one of the most useful tools in the selection of the right crops for maximising the yield. Similarly, the maps depicting the spatial distribution of time of flowering in case of fruit crops like mango, grape and cashew can be prepared on the basis of Hopkins' bioclimatic law (refer Chapter 3 under temperature), which will be handy for selection of a suitable variety and prophylactic crop protection measures so as to sustain crop production. In the case of cashew, suitable varieties have been suggested on the basis of climatic types. These were drawn on the basis of climatic normals by the National Research Centre for Cashew, Puttur (Karnataka), in collaboration with the State Agricultural Universities (SAUs).

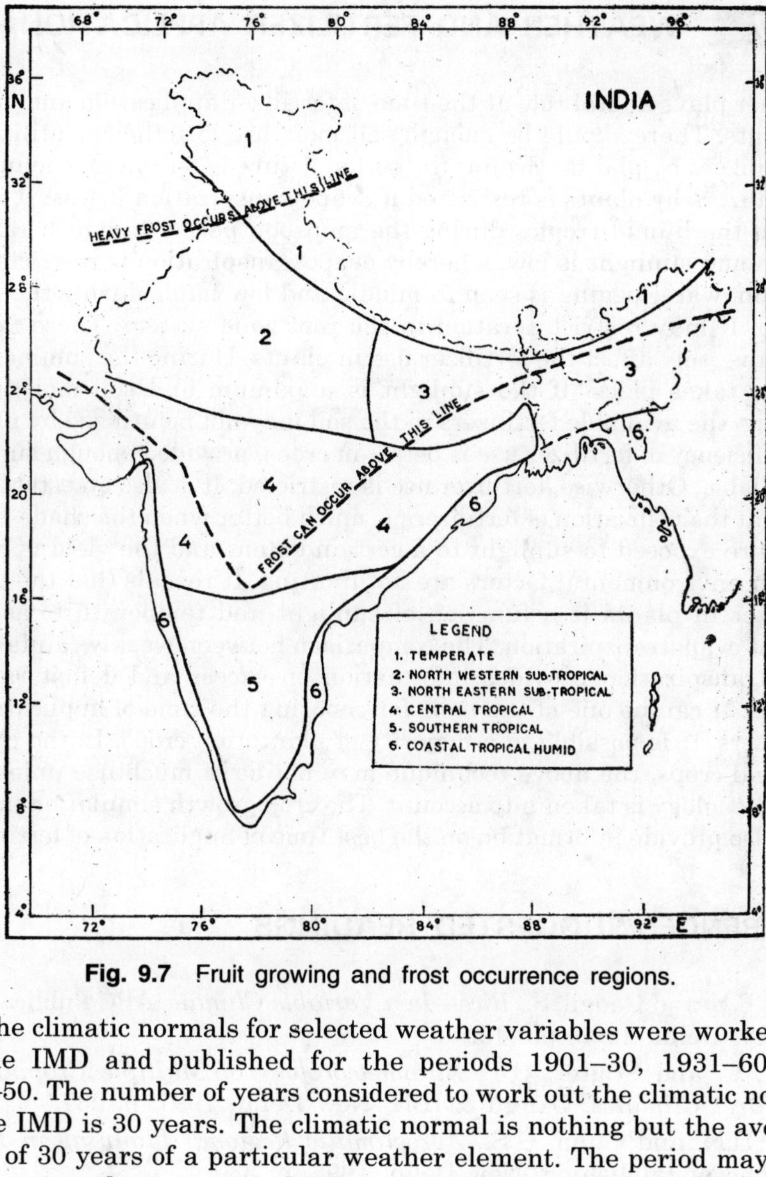

Fig. 9.7 Fruit growing and frost occurrence regions.

The climatic normals for selected weather variables were worked out by the IMD and published for the periods 1901–30, 1931–60 and 1901–50. The number of years considered to work out the climatic normal by the IMD is 30 years. The climatic normal is nothing but the average value of 30 years of a particular weather element. The period may be a week, a month or a year. In fact, the IMD prepared many climatic normal maps for different periods, depicting the spatial distribution of all major weather elements across the country such as rainfall, temperature, sunshine, evaporation, dewfall, and frost-prone areas and the like, in addition to the spatial distribution of different agroclimatic indices. They can be readily used and updated. Similar work can be taken up by the SAUs to meet the local needs of the States to draw detailed information on climate at the micro-level for each crop, and its tolerant weather limits, in terms of rainfall and temperature, so that crop yields can be maximized against climate risk.

9.7 WEATHER AND FERTILIZER APPLICATION

Weather plays a vital role at the time of fertilizer application and its use by plants. There should be enough soil moisture for efficient utilization of fertilizers by plants. Even if the soil moisture is optimum, the uptake of fertilizers by plants is restricted if evapotranspiration is less. It is the case in the humid tropics during the monsoon period in which rainfall is high and sunlight is low, whereby evapotranspiration is restricted. In addition, waterlogging is seen in middle and low lands during the rainy season. It may restrict aeration in the root zone system. These factors may adversely affect the fertilizer use in plants. During the summer, the reverse takes place. If the sunlight is maximum and soil moisture is limiting, the available fertilizers in the soil may not be utilized by plants. The efficiency of fertilizer use is better in crops, provided enough sunlight is available. Otherwise, fertilizer use is restricted. It is also an established fact that the utilization of fertilizers is much better when the shade-loving crops are exposed to sunlight to a certain extent and the yield is better if other environmental factors are not limiting. It reveals that the use of fertilizer in plants is a function of sunlight and temperature, as they restrict evapotranspiration. The comparison between weekly rainfall and evapotranspiration provides information on excess and deficit rainfall periods. It can be one of the tools for choosing the time of application of fertilizers. It is possible in a majority of plantation crops. In the case of seasonal crops, the above technique may not be of much use unless the crop phenology is taken into account. The crop growth simulation models may also provide information on the best time of application of fertilizers.

REFERENCES/SUGGESTED READINGS

Abrol, P.Y. and Gadgil, S., *Rice—In a Variable Climate*, APC Publications, New Delhi, 1999, p. 243.

Baldy, C. and Stigter, C.J., *Agrometeorology of Multiple Cropping in Warm Climates*, Oxford & IBH, New Delhi, 1997, p. 237.

Basu, D.N. and Guha, G.S., *Agroclimatic Regional Planning in India*, Concept Publishing, New Delhi, 1996, p. 330.

Chang, J.H., *Climate and Agriculture*, Aldine Publishing, Chicago, 1968, p. 304.

Kakde, J.R., *Agricultural Climatology*, Metropolitan Book Coompany, New Delhi, 1985, p. 387.

Lenka, D., *Climate, Weather and Crops in India*, Kalyani Publishers, Ludhiana, 1998, p. 490.

Mihara, Y. (Ed.), *Agricultural Meteorology of Japan*, University of Tokyo Press, Japan, 1974, p. 215.

Rao, B.V.R., *Operational Agricultural Meteorology*, Indian Society of Agronomy, Indian Agricultural Research Institute, New Delhi, 1988, p. 59.

Rao, G.S.L.H.V.P., *Weather and Rice: Sustaining Rice Productivity through Integrated Nutrient Management*, Kerala Agricultural University, Thrissur, 1993, pp. 36–46.

Rao, G.S.L.H.V.P., *Agrometeorology of Coconut: Six Decades of Coconut Research*, Eds. Aravindakshan, M., Nair, R.R. and Wahid, P.A., Kerala Agricultural University, Thrissur, India, 1988, pp. 81–93.

Rao, G.S.L.H.V.P., *Climate and Cashew*, Kerala Agricultural University, Thrissur, 2002, p. 100.

Rao, G.S.L.H.V.P. and Nair, R.R. (Eds.), *Agrometeorology of Plantation Crops*, Kerala Agricultural University, Thrissur, 1988, p. 187.

Seeman, J., Chirkov,Y.I., Lomas, J. and Primault, B., *Agrometeorology*, Springer-Verlag, Berlin, Heidelberg, 1979, p. 324.

Smith, L.P., *Methods in Agricultural Meteorology*, Elsevier Scientific Publishing, Amsterdam, 1975, p. 210.

Venkatraman, S. and Krishnan, A., *Crops and Weather*, Indian Council of Agricultural Research, New Delhi, 1992, p. 586.

Vitkevich, V.I., *Agricultural Meteorology*, Israel Programme for Scientific Translations, Jerusalem, 1963, p. 320.

Wilsie, P.C., *Crop Adaptation and Distribution*, Eurasia Publishing, New Delhi, 1974, p. 448.

WMO, *Agricultural Meteorology*, Technical Note No. 310, World Meteorological Organisation, Geneva, 1972, p. 357.

WMO, *Guidelines on Crop Weather Models*, Technical Note No. 50, World Meteorological Organisation, Geneva, 1983, p. 115.

QUESTIONS

1. Elucidate the following:
   (a) Effect of weather on plantation crops in the humid tropics.
   (b) Influence of altitude on distribution of crops.
   (c) Role of weather variables in environmental stress.
   (d) Climatic normals for crop production.
2. Write short notes on:
   (a) Weather and rice
   (b) Weather and coconut
   (c) Weather and coffee
   (d) Weather and cashew
   (e) Weather and fertilizer application
   (f) Climate and vegetation

3. Fill in the blanks:
   (a) ——— and ——— are the most important weather elements affecting crops.
   (b) Xerophytes are confined to ——— climates.
   (c) In black pepper, ——— is the main pollen vector.
   (d) Cashew is a photosensitive as well as——— crop.
   (e) Rice production is affected due to wind during ——— season.
   (f) ——— is the region where wind is a constraint to rice production during rabi.
   (g) Crop duration in rice is affected due to ———temperature.
   (h) ——— and ——— as weather elements are important in tea.
   (i) ———showers are important in case of cardamom.
   (j) ——— and ———showers are important for better coffee yield.
   (k) Cocoa is sensitive to ——— temperature, which results in apical dominance.

4. State true or false:
   (a) Most of the crops are included under mesophytes.
   (b) Biotic stress is the major component in environmental stress.
   (c) Atmospheric factors are categorized under abiotic stress.
   (d) Megatherms are favourable for tropical crops.
   (e) Microtherms are favourable for subtropical crops.
   (f) Cardamom is grown in the highlands of Kerala.
   (g) Solar radiation 45 days before flowering is important in case of rice.
   (h) Prolonged dryspell or drought during summer affects coconut yield during the same year.
   (i) Black pepper is grown in high ranges, where low temperature prevails throughout the year.
   (j) Rainfall during summer is beneficial to the quality of cashewnuts.

10 EVAPOTRANSPIRATION

In tropical countries like India, abundance or scarcity of moisture has a strong influence on plant growth, since temperature is not a limiting factor. Though rainfall is the main source for moisture supply to plants, plant growth does not depend on rainfall alone. But rainfall should balance the evapotranspiration of crops. Evapotranspiration is a combination of physical and biological processes through which water enters the atmosphere in the form of water vapour. The process by which water evaporates from soils and water surfaces is known as *evaporation*. The process by which water vapour escapes from living plants, principally the leaves, and enters the atmosphere, is known as *transpiration*. Evapotranspiration depends upon the availability of water, temperature and humidity of the air, wind movement, intensity and duration of sunlight, stage of development of the plant, type of foliage and nature of the leaves. Perennial crops like arecanut, cardamom, cocoa, coffee, coconut, rubber and tea require summer irrigation for obtaining better yield. It is also true in the case of biennials, annuals and seasonal crops that are grown during *rabi* and summer under moisture stress in rainfed conditions. Therefore, crop-wise evapotranspiration studies are useful tools in irrigation scheduling for effective water resources management. It is also important in nutritional management studies, since the nutritional uptake is the maximum when optimum soil moisture is available to the plant. It is one of the reasons why a particular fertilizer element responds positively when optimum soil moisture is available and negatively when soil moisture is not sufficient. Similarly, the uptake of nutrients when evapotranspiration is limited or below optimum due to near-zero saturation vapour pressure deficit in atmosphere, which is not uncommon during the rainy season extending up to three to four months in tropical monsoon climates, is yet to be understood fully. Heavy rain followed by prolonged dry spell is a common phenomenon in the tropical monsoon climates. In such an environment together with varied climates, crop- and site-specific evapotranspiration findings play a vital role for irrigation scheduling under scarce water resources management.

10.1 POTENTIAL EVAPOTRANSPIRATION

Potential evapotranspiration indicates the maximum limit for water that can be lost through evaporation and transpiration in the form of water vapour from a large vegetative or cropped surface if the soil moisture is not limiting. It depends upon a number of factors including temperature, incoming solar radiation, wind speed and saturation vapour pressure deficit. The concept of potential evapotranspiration (PET) was first introduced by Thornthwaite in 1943 and later by Penman in 1948. PET was defined by Thornthwaite as the evapotranspiration from a large vegetation-covered land surface with adequate soil moisture at all times. It is also considered the thermal efficiency (TE) for delineating climatic types based on the thermal regime. PET was defined by Penman as the evapotranspiration from the actively growing short green vegetation, completely shading the ground and never short of soil moisture availability. It is the maximum quantity of water capable of being lost as water vapour in a given climate by a continuous stretch of vegetation covering the whole ground when the soil is saturated (WMO, 1963). A number of empirical formulae are also available in literature for computing potential evapotranspiration, based on temperature, vapour pressure, aerodynamic and energy balance concepts.

10.1.1 Thornthwaite's Method

Thornthwaite, working in the humid eastern United States in 1948, assumed that an exponential relationship exists between mean monthly temperature and potential evapotranspiration. Thornthwaite proposed the following formula based on the above concept:

$$e = 1.6\left(\frac{10t}{I}\right)^{a}$$

where

e is the unadjusted PET (cm/month) month of 30 days each and 12 hours daytime

t is the mean monthly air temperature (in °C)

I is the annual or seasonal heat index, the summation of 12 values of monthly heat indices given by

$$i = (t/5)^{1.514}$$

a is an empirical exponent computed by using the equation

$$a = 0.000000675I^3 - 0.0000771I^2 + 0.01792I + 0.49239$$

The unadjusted values of e are corrected for actual daylight hours and days in a month.

10.1.2 Blaney–Criddle's Method

The relationship developed by Blaney–Criddle for the arid western United States in 1950 is as follows:

$$U = KF = \Sigma kf = \Sigma u = \frac{\Sigma ktp}{100}$$

where

U is the seasonal consumptive use of water by the crop for a given period (in inches)

K is the empirical seasonal consumptive use co-efficient for the growing season

F is the sum of the monthly consumptive use factors (f) for the growing season

u is the monthly consumptive use (in inches)

k is the empirical consumptive use crop co-efficient for the month, given by u/f

$$f = t \times \frac{p}{100}$$

t is the Mean monthly temperature, in °F

p is the monthly daylight hours expressed as a percentage of daylight hours of the year

10.1.3 Penman's Method

Penman (in England) in 1948 proposed an equation for evaporation from open water surface, based on combination of energy balance and sink strength, which is given below:

$$E_0 = \frac{\Delta Q_n + \gamma E_a}{\Delta + \gamma}$$

where

E_0 is the evaporation from open water surface (mm/day)

Δ is the slope of the saturation vapour pressure vs. temperature curve (de_a/dT) at the mean temperature T_a (in mm Hg/°C)

e_a is the saturation vapour pressure of the evaporating surface (e_s) in mm Hg at mean air temperature T_a (Here e_s is considered equal to e_a by assuming zero temperature gradient between surface(s) and air temperatures.)

T_a is the mean air temperature in K = 273 + °C

Q_n is the net radiation (in mm of water)

$$= Q_a(1 - r)\left(0.18 + \frac{0.55n}{N}\right) - \sigma T_a^{\,4}(0.55 - 0.092\, e_d)\left(0.10 + \frac{0.90n}{N}\right)$$

r is the reflection coefficient of evaporating surface, which is 0.06 for open water surface

Q_a is the Angot's value of mean monthly extra-terrestrial radiation in mm of water/day

n/N is the ratio between actual and possible hours of bright sunshine

σ is the Stefan–Boltzmann constant

e_d is the saturation vapour pressure of the atmosphere, in mm Hg, at dew point temperature = $\frac{RH_{\text{mean}}}{100}\, e_a$ in which RH is the mean relative humidity

γ is the psychrometric constant or ratio of specific heat of air to the latent heat of evaporation of water (0.49 for 0°C in mm Hg)

E_a is the aerodynamic component in which e_s is considered equal to $e_a = 0.35(e_a - e_d)(1 + 0.0098u_2)$

u_2 is the wind speed in miles per day at 2 metres height, equal to u_1(log 6.6/log h), in which u_1 is wind speed in miles per day at any other height h in feet.

10.1.4 Modified Penman's Method

Doorenbos and Pruitt (1975) proposed a modified Penman method, given below, for estimating fairly accurately the reference crop *ET*.

$$ET_c^* = W \cdot R_n + (1 - W)\, f(u) \cdot (e_a - e_d)$$

Radiation term + Aerodynamics term

where

ET_c^* is the reference crop evapotranspiration in mm per day (not adjusted)

e_a is the saturation vapour pressure in mb at mean air temperature in °C

e_d is the mean actual vapour pressure of the air in mb

$= e_a \times \frac{RH_{\text{mean}}}{100}$, in which, RH is relative humidity

$f(u)$ is the a wind-related function

$(1 - W)$ is the temperature- and elevation-related weighting factor for the effect of wind and humidity on ET_c

W is the temperature- and elevation-related weighting factor for the effect of radiation on ET_c, and

R_n is the net radiation (same as $Q_n = R_{ns} - R_{nl}$)

In which R_{ns} is the net incoming short-wave solar radiation = $R_A(1-a)\left(0.25+\frac{0.50n}{N}\right)$, in which R_A is same as Q_A or extra-terrestrial radiation expressed in equivalent evaporation in mm/day

n/N is the ratio between actual and possible hours of bright sunshine; and a is same as r or reflection coefficient; the value of which is taken as 0.25 for most of crops and which gives the conversion factor for R_A to R_{ns} for a given reflection of 25% and ratios of n/N, and

R_{nl} is the net long wave radiation = $f(t)\cdot f(e_d)\cdot f\left(\frac{n}{N}\right)$,

To determine ET_c, the unadjusted reference crop evapotranspiration, ET_c^* is adjusted for day and night time weather condition. This adjustment and the revised wind function term are the important variations made by Doorenbos and Pruitt to the original Penman method. The modified Penman relationship gives satisfactory values of crop ET.

10.1.5 Comparison of Potential Evapotranspiration

For a few stations in Rajasthan, the monthly potential evapotranspiration was computed by using the Thornthwaite, Penman, Blaney–Criddle and Christiansen methods. The Christiansen method overestimated the levels at all the locations throughout the year while the other three methods were relatively comparable. The Thornthwaite and Blaney–Criddle methods followed closely as the formulae were based on temperature. Though Penman's method is sound, it also gave higher values at Jaipur, Jodhpur and Ganganagar during the summer season due to high wind and solar energy (see Fig. 10.1).

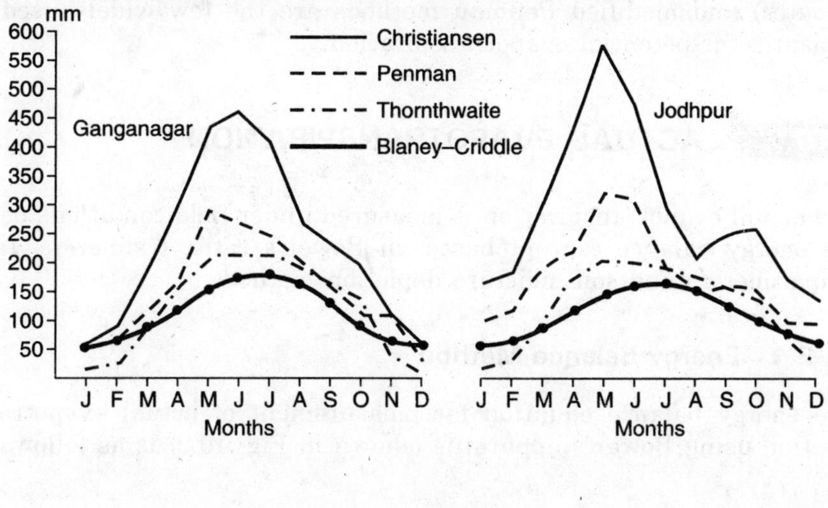

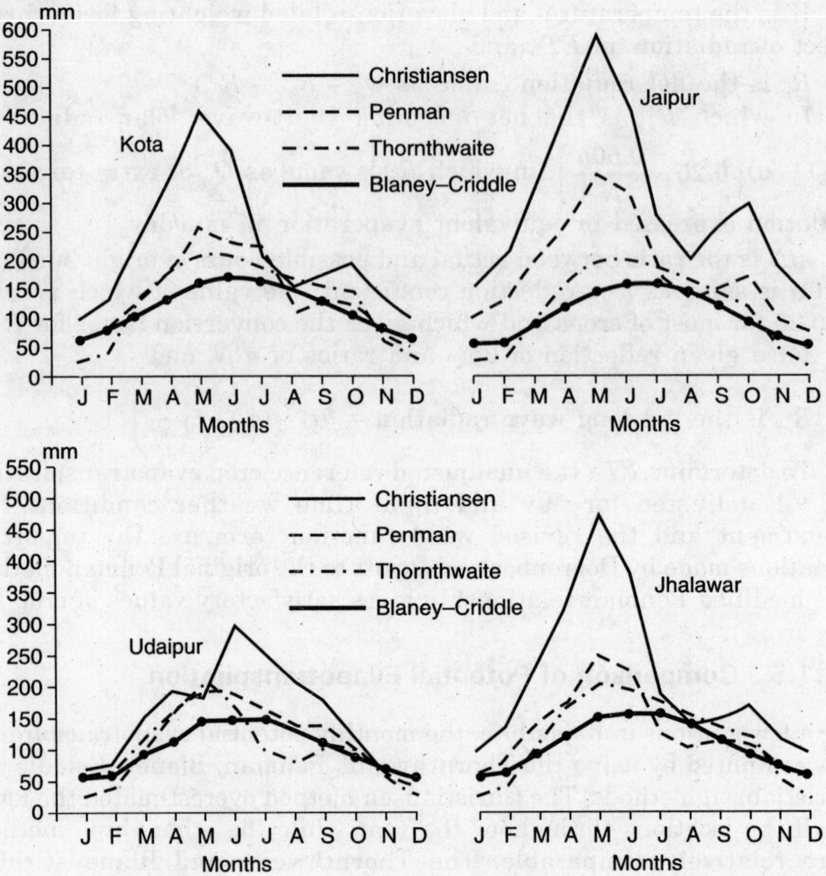

Fig. 10.1 Comparison of PET by different methods at selected stations in Rajasthan.

Similar results were noticed at many locations in India by several workers. The pan evaporation, Blaney–Criddle, Thornthwaite (coastal climates) and modified Penman methods are the few widely used as measures for potential evapotranspiration.

10.2 ACTUAL EVAPOTRANSPIRATION

The actual evapotranspiration is measured under field conditions using the energy balance concept based on Bowen's ratio, lysimeters, tank evaporimeters and soil moisture depletion methods.

10.2.1 Energy Balance Method

The energy balance equation for measurement of actual evapotranspiration using Bowen's apparatus (shown in Fig. 10.2) is as follows:

$$R_n = G + H + LE$$

where

R_n is net radiation
G is ground heat or soil heat flux
H is sensible heat
LE is latent heat of evapotranspiration

Fig. 10.2 Bowen's apparatus.

The above equation can be written as

$$R_n - G = H + LE$$

$$\frac{R_n - G}{LE} = \frac{H + LE}{LE}$$

$$\frac{R_n - G}{LE} = \frac{H}{LE} + 1$$

$$\frac{(R_n - G)}{LE} = \beta + 1$$

where

$$\beta \text{ (Bowen's Ratio)} = \frac{H}{LE},$$

where

$$LE = \frac{R_n - G}{1 + \beta}$$

$$\beta = \frac{H}{LE} = C_p \frac{\delta\varphi}{L\delta q}$$

where C_p is the specific heat of air at constant pressure, $\delta\varphi$ and δq are the vertical gradients of potential temperature and specific humidity, and LE is the latent heat of vaporisation.

The actual crop evapotranspiration (ET_c) or consumptive use can be computed by converting latent heat (in Wm^{-2}) to evapotranspiration using the following relationship:

$$LE = \rho_w L\, ET_c$$

where ρ_w is the density of water.

The energy balance components recorded for black pepper on a clear day (in January 1995) are depicted in Fig. 10.3. There is a steady increase

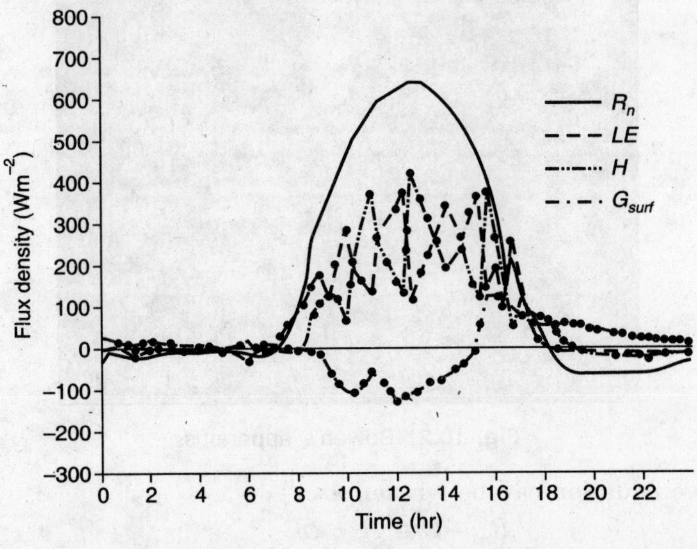

Fig. 10.3 Energy balance components in multiple cropping system.

in net radiation. It attains the maximum level (700 Wm^{-2}) at 1300 hrs, decreases rapidly in the afternoon and shows negative results during the night. LE was high (307 Wm^{-2}) while H was 75 Wm^{-2}, and the surface soil heat flux was –107 Wm^{-2} at 1000 hr, contributing 63%, 15% and 22 % of R_n respectively. This high soil heat flux density reduces the amount of energy available for partitioning into other fluxes, resulting in low rates of evapotranspiration during the morning hours when the vapour pressure deficit is less. After 1000 hr there was a decrease in *LE* up to 1100 hr, beyond which it increases further, reaching the maximum at around 1220 hrs. At this time, 60% of R_n was consumed by *LE*, 22% by *H* and 18% by *G*. In general, the partitioning towards *LE* is reduced when the soil moisture depletes to 10–12% by volume. *LE* reaches up to 80% if the soil moisture improves further from 10% to 18%. It reveals that most of the radiant energy available at the cropped surface is mostly used in evapotranspiration if enough soil moisture is available. If soil moisture depletes under rainfed conditions, the percentage of R_n increases towards sensible heat and decreases towards *LE*.

10.2.2 Lysimeters

Lysimeters are of two widely used types viz. weighing and volumetric. The India Meteorological Department has installed lysimeters at selected locations across the country on the basis of cropping systems to study evapotranspiration of different crops. In Kerala, lysimeters are installed at RARS Pattambi to study the actual evapotranspiration of paddy at different crop growth stages. A lysimeter in operation is shown in Fig. 10.4.

Fig. 10.4 Lysimeter.

Gravimetric or weighing lysimeter

An isolated block of cropped soil, representative of the field, is weighed first to determine the quantity of water added by precipitation or irrigation, and then after loss of water by ET is weighed again. The difference of weights in kilograms in two observations, multiplied by the conversion factor, gives ET in mm (see Table 10.1), that is,

ET (in mm) = [Difference in weights of two consecutive lysimeter observations (in kg × 0.6] + [Corresponding rainfall (in mm)]

Table 10.1 ET calculation using gravimetric lysimeter

Date	*Rainfall* (mm)	*Soil tank weight* (kg)	*Change in weight* (kg)	*ET* (mm/day)	*Remarks*
28-6-1999		1862.5			Non-rainy day
29-6-1999	0.0	1851.3	+11.2	11.2 × 0.6 + 0.0 = 6.72	
30-6-1999	5.2	1855.6	– 4.3	(–4.3 × 0.6) + 5.2 = 2.62	Rainy day
2-7-1999		1851.3			
3-7-1999	12.0	1865.3	–14.0	(–14 × 0.6) + 12 = 3.6	Rainy day

Volumetric lysimeter

Volumetric lysimeter is used to measure ET in terms of volume. It requires standing water (and hence is used for standing water crops) while gravimetric lysimeter does not require the same. The principle and function of volumetric lysimeter is not discussed here. On rain-free days, the water added to the reservoir drum (inflow) is measured with the help of a measuring jar. It is measured by volume in terms of cc. Then it is converted into mm by using the conversion factor 1 cc = 0.000694 mm. Thus, if the water added is 1000 cc, the loss of water due to ET would be 6.94 mm. The required equation is:

$$\text{ET} = [\text{Inflow} - \text{Outflow}] \times \text{Conversion factor} + \text{Rainfall}$$

Example:

Inflow = 3500 cc, Outflow = 2000 cc and Rainfall = 0.0 mm

Hence, ET = (3500 – 2000) × 0.000694 + 0.0 = 1.04 mm

10.3 REFERENCE EVAPOTRANSPIRATION

Reference evapotranspiration was coined by Doorenbos and Pruitt (1977). It is defined as the rate of evapotranspiration (ET_o) from an extensive surface of 8 to 15 cm tall green cover of uniform height, actively growing, completely shading the ground and not short of water. It can be estimated from the modified Blaney–Criddle method through a graphical technique as suggested by them. The modified formula is as follows:

$$ET_o = C[P(0.46\ T + 8)] \text{ (in mm/day)}$$

The CROPWAT software developed by FAO (Smith, 1992) can also be used to work out reference evapotranspiration. It is based on the modified Penman–Monteith formula. It was tested with the open pan evaporation method based on the data collected at Vellanikkara (10°31′N and 76°13′E) from 1984 to 2005. The ET_o underestimates the level from December to April and coincides in the remaining months when compared to that of open pan evaporation. This is due to abnormal seasonal fluctuations in saturation vapour pressure deficit and wind speed. The underestimate of reference evapotranspiration during winter can be attributed due to strong winds that blow through the Palakkad Gap in Kerala (see Fig.10.5). The wind speed is high across Thrissur and Palakkad districts from the middle of November to February. This may not be the case even at all the locations within Kerala. Hence, these studies are location-specific and need to be taken up at every location. The pan coefficients may also be developed for computing reference evapotranspiration based on pan evaporation, which will be easy from the users' point of view.

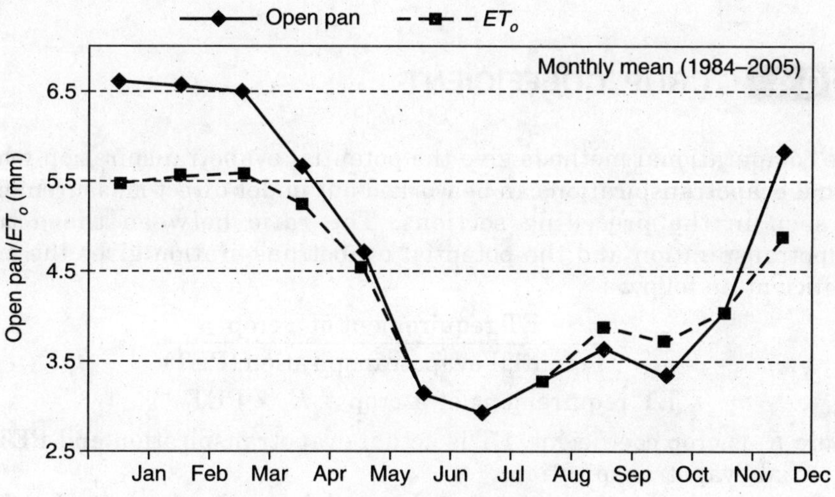

Fig. 10.5 Mean monthly reference evapotranspiration versus open pan evaporation (1984–2005) at Vellanikkara.

10.3.1 Pan Coefficient

Based on the pan coefficient, reference evapotranspiration can be worked out using pan evaporation. The pan coefficient works out as 0.83 for winter and 0.89 for summer under the climatic conditions that prevail over the Palakkad and Thrissur regions. The monthly pan coefficients vary between 0.8121 and 1.1054 (see Table 10.2).

The pan coefficient was worked out using the regression equation $Y = bx$, where Y is reference evapotranspiration, x is pan evaporation and b is a constant. The pan evaporation should be multiplied with the monthly pan coefficient for computing the monthly reference evapotranspiration, which can be used for scheduling irrigation in place of cumulative pan evaporation (IW/CPE). This helps in saving irrigation water, which is essential in areas where water is scarce.

Table 10.2 Monthly pan coefficients at Vellanikkara

Month	*Pan coefficient*	*Month*	*Pan coefficient*
January	0.8121	July	1.0173
February	0.8363	August	0.9987
March	0.8444	September	1.0809
April	0.8910	October	1.1054
May	0.9257	November	1.0100
June	1.0141	December	0.8302

10.4 CROP COEFFICIENT

The computational methods give the potential evapotranspiration while actual evapotranspiration can be worked out under direct measurements as seen in the preceding sections. The ratio between the actual evapotranspiration and the potential evapotranspiration gives the crop coefficient as follows:

$$K_c = \frac{\text{ET requirement of a crop}}{\text{Potential evapotranspiration (PET)}}$$

or

$$\text{ET requirement of a crop} = K_c \times \text{PET}$$

where K_c is crop coefficient, ET is actual evapotranspiration and PET is potential evapotranspiration.

The actual evapotranspiration obtained from a lysimeter under field conditions can be compared with the modified Blaney–Criddle formula suggested by Doorenbos and Pruitt (1975) for working out crop coefficients at various locations.

For many of the annual dryland crops, K_c is 0.3 to 0.6 from the time of sowing to its establishment. It increases to a maximum of 1.0 to 1.3 when the crop canopy covers almost the whole of the ground surface. When the crop attains physiological maturity, K_c declines to approximately 0.5. In the case of paddy, the seasonal crop coefficients vary between 1.0 and 2.0 depending upon the agroclimatic zone in which the crop is taken up. It also reveals that the crop coefficient varies from year to year during the same season *(kharif)* as weather variables vary. It is true in case of *rabi* also.

The monthly mean crop coefficients in case of young coconut palms is 0.82, varying between 0.6 in July and 0.96 in May (Rao, 1989). The actual evapotranspiration varied between 2.3 mm per day in June and 5.5 mm per day in May. The crop coefficient is 0.87–0.88 (in September and October), 0.78–0.85 (in November–February) and 0.87–0.96 (in March–May) and the least (0.60–0.68) from June to August.

10.5 ACTUAL AND POTENTIAL EVAPOTRANSPIRATION

Several workers attempted to compare the actual evapotranspiration (lysimetric data) with potential evapotranspiration (computational methods) for different crops across the country by using the concept of Doorenbos and Pruitt (1975). As an example, the studies taken up at Pattambi (10°48′N, 76°12′E, 25 m AMSL) for paddy (Thriveni) by Rao and Alexander (1988) are reported here. It reveals that the reference crop evapotranspiration (ET_o) computed by the Blaney–Criddle method was 24 % higher during the first crop (*kharif*) and 18% lower during the second crop (*rabi*). Radiation and Penman methods gave lower ET_o than the measured value by up to 18% (see Fig. 10.6). The reference evapotranspiration (ET_o) using pan evaporation was lower than the measured ET by 5–47%. Therefore, it is necessary to adjust the reference

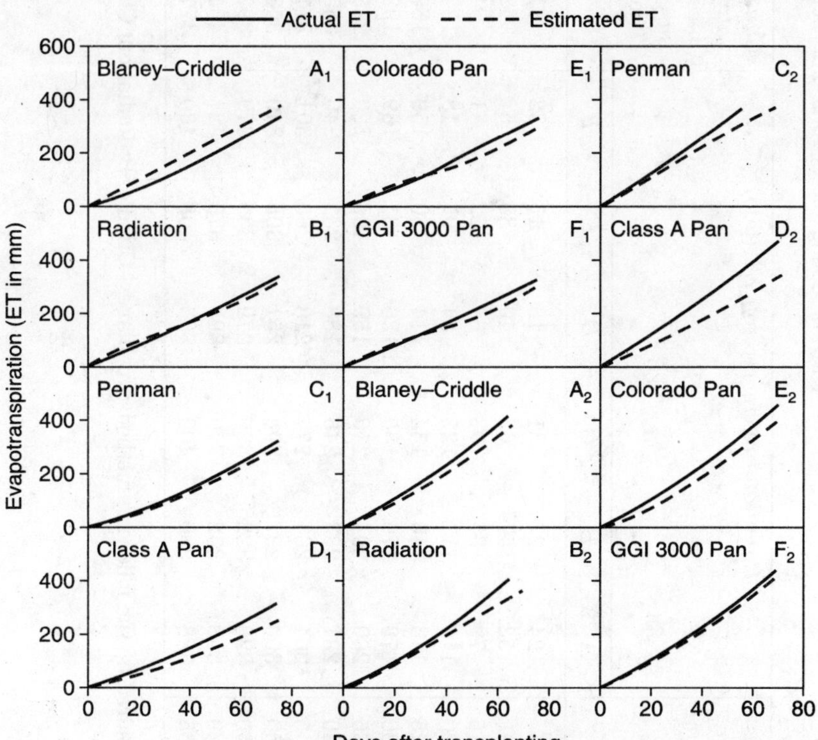

$A_1 - F_1$: ET during frist crop season; $A_2 - F_2$: ET during second crop season

Fig. 10.6 Actual and estimated evapotranspiration.

Table 10.3 Evapotranspiration and crop coefficients of the rice variety 'Thriveni' during the second crop season at RARS, Pattambi

Days after planting	*Cumulatie reference crop ET (mm)**						*Cumulative measured ET(mm)*	*Crop coefficients**					
	A	*B*	*C*	*D*	*E*	*F*		*A*	*B*	*C*	*D*	*E*	*F*
1–5	24	27	31	18	15	32	31	1.29	1.15	1.00	1.72	2.07	0.97
6–10	52	57	62	40	38	57	62	1.19	1.09	1.00	1.55	1.63	1.09
11–15	79	85	92	60	62	81	94	1.19	1.11	1.02	1.57	1.52	1.16
16–20	110	117	123	84	95	118	127	1.15	1.09	1.03	1.51	1.34	1.08
21–25	143	146	154	114	130	159	163	1.14	1.16	1.06	1.43	1.25	1.03
26–30	173	176	185	136	161	189	193	1.12	1.10	1.04	1.42	1.20	1.02
31–35	200	204	216	159	196	220	228	1.14	1.12	1.06	1.43	1.16	1.04
36–40	231	233	246	188	231	240	265	1.15	1.14	1.08	1.41	1.15	1.06
41–45	264	264	277	218	266	301	311	1.18	1.18	1.12	1.42	1.17	1.03
46–50	300	297	308	251	305	340	353	1.18	1.19	1.15	1.41	1.16	1.04
51–55	328	322	339	279	339	379	392	1.20	1.22	1.16	1.41	1.16	1.03
56–60	358	352	370	308	376	421	434	1.21	1.23	1.17	1.41	1.15	1.03
61–65	388	380	400	335	408	460	477	1.23	1.26	1.19	1.42	1.17	1.04

*Measured by the following methods: A: Blaney–Criddle; B: Radiation; C: Penman; D: Class A pan; E: Colorado pan; F: GGI 3000 pan.

crop ET_o with suitable crop coefficients for estimation of ET at Pattambi as illustrated in Table 10.3 (Rao and Alexander, 1998).

10.6 RAINFALL AND POTENTIAL EVAPOTRANSPIRATION

Based on rainfall and potential evapotranspiration, the water availability periods can be worked out. They are useful for selection of suitable crops. Cocheme and Franquine (1967) developed the above concept under the semi-arid and arid climates (such as in West African Countries). For the majority of the crops, the period during which rainfall exceeds the PET is best suited to be the maximum growth period. Of course, this is the period during which surplus rainfall leads to run-off, and in the absence of drainage facilities the crops are subject to waterlogging. Therefore, rainfall is detrimental to crops. The early vegetative phase including sowing period and the later phase, that is, the ripening period, can be adjusted when rainfall is less than PET (rainfall should be at least 50% of PET). The above conditions hold good in the arid and semi-arid belts of the country, where crops are grown only under low-rainfall regimes in rainfed conditions. However, the climatic conditions are totally different in the humid tropics, where the surplus water during the first crop season (*kharif*/monsoon) leads to waterlogging which is detrimental to the crops. The second and third crops (*rabi* and summer) suffer due to soil moisture stress, and crop failure is a common phenomenon during the above seasons if irrigation is not assured. The periods during which irrigation is required or the periods of surplus water could be determined on the basis of precipitation and potential evapotranspiration (see Fig. 10.7). Open pan evaporation is used in place of potential evapotranspiration at RARS, Pilicode (12°12′N and 75°30′E).

It reveals that the accumulation of moisture in soils starts with the onset of pre-monsoon showers or monsoon rains, and reaches its field capacity with the progress of rains by the first or second week of June depending on the onset of monsoon and types of rain. Torrential rains or downpour of rains are unfavourable for agriculture as most of the water produces run-off without penetrating into the soil and also leads to soil erosion. It is common in the humid tropics, where rainfall is very high during the monsoon. The above situation is noticed from June to August and sometimes extends to September. It leads to waterlogging, and hypodermal thickening at root tips is noticed due to lack of aeration. In the absence of rains or intermittent rains during the off-season, the accumulated soil moisture is used for plant growth during October and November depending on soil type. Thereafter, crops experience moderate to severe soil moisture stress in the humid tropics from December to May if premonsoon showers fail between March and May. It is difficult to raise crops during the above period without irrigation support. The choice of crop and successful crop harvest depend on the distribution of rainfall and its quantity under rainfed conditions. Frequent and abundant rainfall, particularly at harvest time, may definitely be undesirable and cause losses in yield. Similarly, excessive rains in the growing season retards development and favours the outbreak of various plant diseases. At each location, such rainfall and evaporation curves can be drawn graphically to find out the favourable and unfavourable periods for crops in relation to moisture availability.

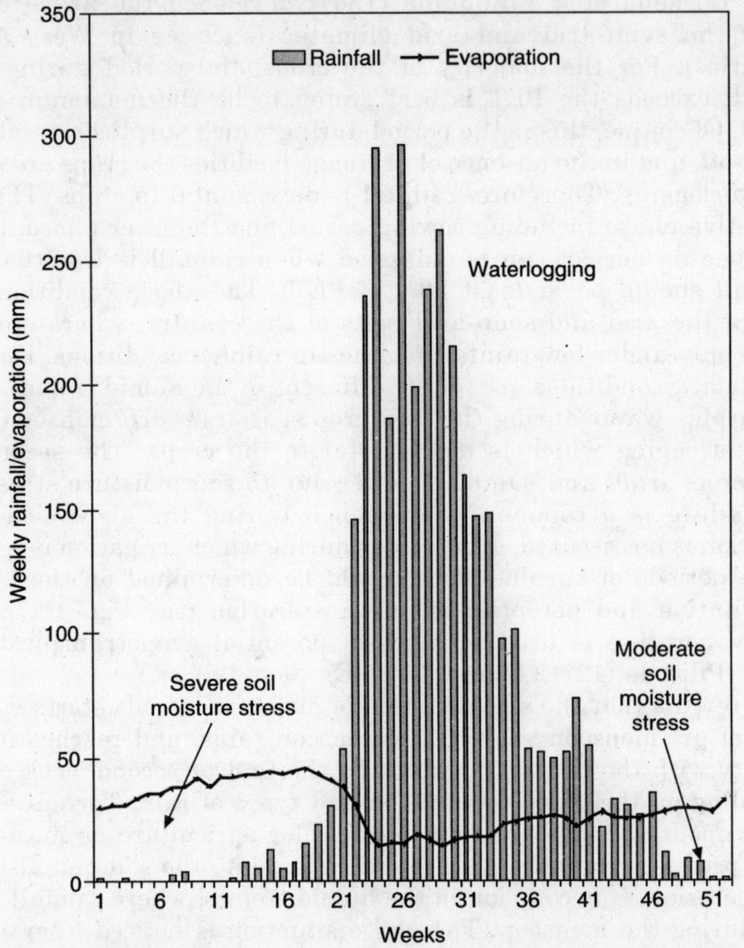

Fig. 10.7 Weekly rainfall versus pan evaporation at RARS, Pilicode (1983–1998).

10.7 THORNTHWAITE'S WATER BALANCE

Thornthwaite (1948) developed a book-keeping water balance procedure based on precipitation and potential evapotranspiration, which was later revised by Thornthwaite and Mather in 1955. The basic equation governing the water balance concept is

$$P = E + \Delta S + RO$$

where P is precipitation, E is evapotranspiration, ΔS is change in soil moisture and RO is run-off.

During the rainy season, when precipitation is more than potential evapotranspiration, the potential evapotranspiration is nothing but actual evapotranspiration. In the above situation, the excess rain reaches the soil and becomes soil storage, which is dependent on moisture holding capacity (or field capacity) of soils. If any excess precipitation over and above the field capacity of soils and potential evapotranspiration occurs, it manifests itself as surface run-off. In the absence of rain or in case of little rain during the off-season, the potential evapotranspiration is more than precipitation. In such situations, the soil moisture loss and precipitation, if any, is known as the *actual evapotranspiration*. During such occasions, soil moisture is drawn from the soil storage. The soil moisture loss is limited to the available soil moisture and its supply to plants, which is a complex matter. Hence, the actual evapotranspiration is defined as the precipitation together with soil moisture available to meet the water need or potential evapotranspiration if it is more than precipitation. The difference between potential evapotranspiration and actual evapotranspiration is known as the *water deficit*. In this water balance procedure, precipitation is treated as income and potential evapotranspiration as expenditure, while soil moisture stored in soils is a reserve from which moisture can be withdrawn (within certain limits) in times of deficient precipitation. The water surplus is defined when precipitation is more than the potential evapotranspiration and the field capacity of the soils. The excess water, after meeting the above two demands, becomes surface run-off (see Fig. 10.8). Using the water balance parameters, several indices such as index of moisture adequacy, humidity index, aridity index, moisture index, etc. could be worked out. The above climatic indices based on moisture regime are widely used for delineating climatic zones, defining climatic droughts as well as for explaining crop yields based on the index of moisture adequacy. Palmer (1965) developed a two-layer soil moisture budgeting procedure based on CAFEC (Climatologically Appropriate For Existing Conditions) precipitation and CAFEC evapotranspiration. Later, Baier and Robertson (1966) developed a six-layer soil moisture versatile budgeting. Saxton et al. (1974) developed a SPAW model for estimating soil moisture based on daily precipitation and open pan evaporation. On similar lines, Frere and Popov (1979) developed a cumulative water balance based on 10 days-period/weekly precipitation and potential evapotranspiration, taking into account the *crop coefficient factor*, which varies from crop to crop and within the crop from stage to stage. The water requirement satisfaction index (WRSI) is a good measure to explain yield variability. FAO developed a software for computing irrigation requirement based on WATBAL.

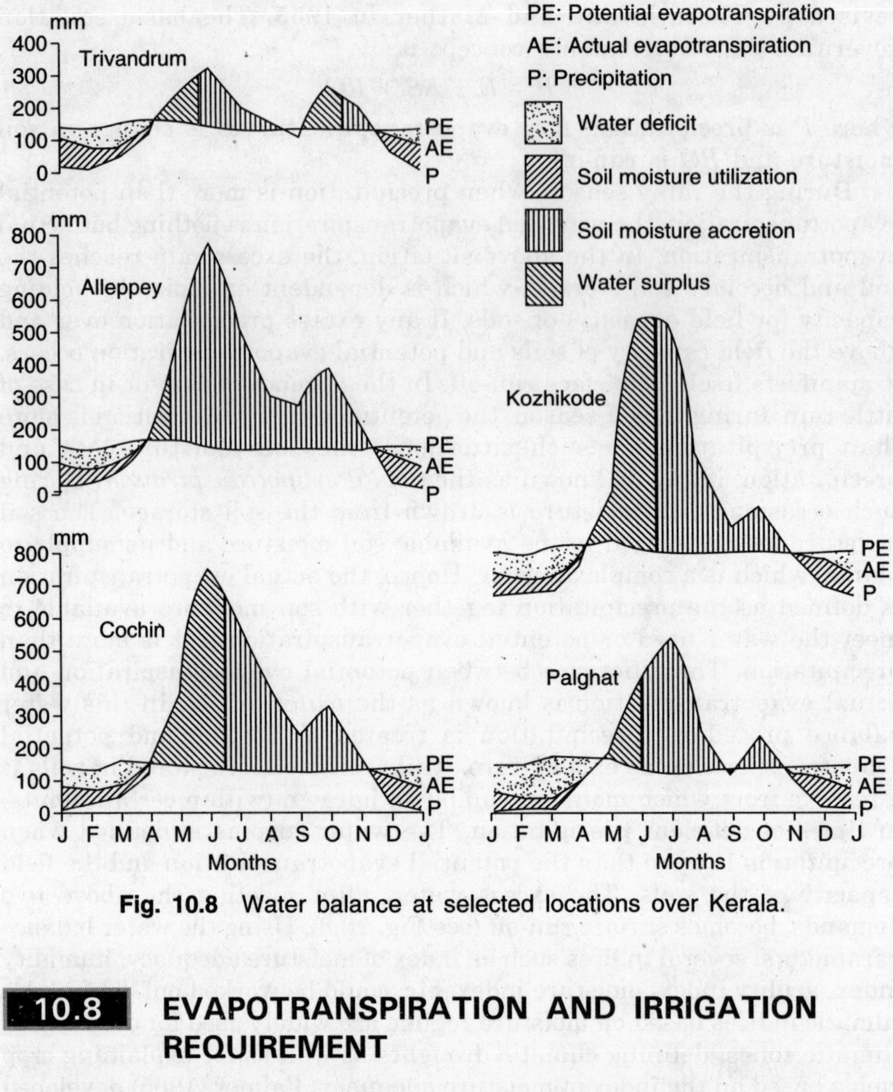

Fig. 10.8 Water balance at selected locations over Kerala.

10.8 EVAPOTRANSPIRATION AND IRRIGATION REQUIREMENT

Irrigation is practiced from December to May in the tropical humid climates in case of plantations, where water resources are very scarce. The reference evapotranspiration is the one universally accepted for computing water requirement of crops. Though it is universally accepted, pan evaporation is taken into account in irrigation scheduling due to its easy access. Moreover, computation of reference evapotranspiration need data on weather variables such as maximum and minimum temperatures, wind speed, sunshine and relative humidity, etc., which are recorded and maintained properly at only a few locations. Similar is the case in several universally accepted formulae for computing potential evapotranspiration. Several workers (Subramaniam and Rao, 1980; Subramaniam et al.,

1983; Rao et al., 1985; Rao and Vamadevan, 1986; Kumar et al., 1987; Rao and Rao, 1995; Jadhav et al., 1999 and Kingra et al., 2002) computed various empirical formulae and discussed merits and demerits in comparison with open pan evaporation. Though the lysimetric method is better for computing evapotranspiration when compared to empirical formulae under field conditions, it is very expensive and immobile in addition to being location- and crop-specific. Moreover, water savings may not be significant under field conditions between lysimetric and reference evapotranspiration, if irrigation schedule is based on them. Hence, pan coefficients can be developed as seen in Section 10.3.1 to work out reference evapotranspiration. From the crop reference evapotranspiration (ET_o), the actual evapotranspiration (ET) can be worked out on the basis of crop coefficient. From the actual evapotranspiration, irrigation water can be quantified. One such example is provided here for coconut. The irrigation requirement in case of coconut varies from 1106 litres (minimum) in December to 1488 litres in April, while in May it is 1391 litres per month (see Table 10.4).

Table 10.4 Irrigation requirement of coconut

Month	*ET_o* (mm)	*Evapotranspiration* (mm) ($ET_o \times 0.75$)	*Irrigation requirement* (l) ($\pi r^2 h$)
December	3.79	2.84	1,106.0
January	3.95	2.96	1,154.0
February	4.56	3.42	1,204.0
March	5.01	3.76	1,464.0
April	5.26	3.95	1,488.0
May	4.76	3.57	1,391.0
			Total = 7,807.0

ET_o: Reference evapotranspiration; *r*: Radius of coconut basin–2 metres; 0.75: Crop coefficient of coconut.

Irrigation requirement is low in May due to the influence of pre-monsoon showers, which makes atmosphere moist and thereby decreases the evapotranspiration rate. The irrigation quantity can be applied equally at weekly intervals. This rational climatic approach in irrigation scheduling is highly economic under scarce water resources mangement. Coconut growers can very well adopt the irrigation quantities during summer as recommended above, wherever water is available, which will improve coconut yield significantly, as the yield per palm is well below the normal under rainfed conditions due to prolonged dry spell from November to May.

REFERENCES/SUGGESTED READINGS

Baier, W. and Robertson, G.W., 'A new versatile soil moisture budget', *Can. J. Pl. Sci.*, **46**, 299–315, 1966.

Blaney, H.F. and Criddle, W.D., '*Determining water requirements in irrigated areas from climatological data*', Technical publications No. 96, Soil Conservation Service, United States Department of Agriculture, p. 48, 1950.

Christiansen, J.E., 'Pan Evaporation and Evapotranspiration from climatic data', *J. Irrig. and Drain*', **94**, 243, 1968.

Cocheme, J. and Franquin, P., *An agroclimatology survey of a semiarid area in Africa, South of Sahara*, Technical Bulletin No. 86, Food and Agricultural Organisation/World Meteorological Organisation, 1967.

Doorenboss, J. and Pruitt, W.O., *Crop water requirements*, Irrigation and Drainage Paper No. 24, Food and Agricultural Organisation, Rome, p. 179, 1975.

Frere, M. and Popov, G., *Agrometeorological Crop Monitoring and Forecasting*, Food and Agricultural Organisation, Rome, 1979, p. 64.

Hargreaves, G.H., *Water requirements manual for irrigated crops and rainfed agriculture*, Pub. No. 75-D-159, EMBRAPA and Utah State University, United States, 1971.

Jadhav, J.D., Mokashi, D.D., Shewale, M.R. and Patil, J.D., 'Relationship between PET and mesh covered pan evaporation at Solapur', *J. Agromet.*, **1(1)**, 93, 1999.

Kingra, P.K., Kaur, P. and Hundal, S.S., 'Estimation of PET by various methods and its relationship with mesh covered pan evaporation at Ludhiana', *J. Agromet.*, **4(2)**, 143–149, 2002.

Kumar, K.K., Kumar, K.R. and Rakhecha, P.R., 'Comparison of Penman and Thornthwaite methods of estimating potential evaportranspiration for Indian conditions', *Theor. Appl. Climatol.*, **38**, 140–146, 1987.

Michel, A.M., *Irrigation Theory and Practices*, Vikas Publishing, New Delhi, 1978, p. 801.

Nair, R.R., Pillai, G.R. and Rajagoplan, A., *Water management in coconut: Six decades of coconut research* (Eds. Aravindakshan, M., Nair, R.R. and Wahid, P.A.), Kerala Agricultural University, Thrissur, 1988, pp. 94–102.

Nelliat, E.V., and Padmaja, E.K., 'Irrigation requirement of coconut and response to levels of fertilizer under irrigated condition during the early bearing stage', *PLACROSYM–I* (Ed. Nelliat, E.V.), Proceedings of the First Annual Symposium on Plantation Crops, Indian Society of Plantation Crops, Kasaragod, pp. 196–199, 1978.

Palmer, W.C., *Meteorological drought*, Research Paper No. 45, United States Department of Commerce, Weather Bureau, 1965.

Penman, H.L., 'Natural evaporation from open water, bare soil and grass', Proceedings of the Royal Society of London, **193**, pp. 120–145, 1948.

Rao, A.S., 'Water requirements of young coconut palms in a humid tropical climate', *Irrig. Sci.*, **10**, 245–249, 1989.

Rao, A.S. and Alexander, D., 'Prediction of evapotranspiration and grain yield of rice (*Oryza sativa* L. cv. Thriveni) in a humid tropical climate', *Int. J. Biometeorol.*, **32**, 81–86, 1988.

Rao, A.S. and Vamadevan, V.K., 'Evaporation rates from different Pans under humid tropical conditions', *Mausam*, **37(3)**, 325–328, 1986.

Rao, A.S., Varadan, K.M. and Vamadevan, V.K., 'Evaporative index for scheduling irrigation in plantation crops', *Mausam*, **36(2)**, 205–208, 1985.

Rao, G.S.L.H.V.P., *Water management in coconut*, Proceedings of the National Seminar on water management for plantation crops, March 4, 1993, Ed. Satheesan, K.V., Centre for Water Resources Development and Management, Kozhikode, pp. 65–70, 1994.

Rao, K.B. and Rao, G.S.L.H.V.P., 'Vegetable cultivation in Rajasthan: A study on prediction of crop water requirements', *Ind. Geogr. J.* **70**(2), 104–106, 1995.

Saseendran, S.A. and Jayakumar, M., 'Effect of rainfall in coconut based land use of Kerala', *Agrometeorology of Plantation Crops*, Eds. Rao, G.S.L.H.V.P. and Nair, R.R., Kerala Agricultural University, Thrissur, pp. 36–46, 1988.

Saxton, K.E., Johnson, H.P. and Shaw, R.H., 'Modelling evapotranspiration and soil moisture', *Trans. Am. Soc. Agric. Engrs.* **17**(4), 673–677, 1974.

Smith, M., '*CROPWAT—A computer programme for irrigation planning and management*' Irrigation and Drainage Paper No. 46, Food and Agricultural Organisation, Rome, 1992.

Subrahmanyam, V.P., *Water Balance and Its Applications*, Andhra University Press, Waltair, p. 120, 1982.

Subramaniam, A.R. and Rao, A.S., 'Potential Evapotranspiration over Maharashtra from climatic data using different methods', *J. Instn. Engrs.*, **61**(1), 142–145, 1980.

Subramaniam, A.R., Rao, K.B. and Rao, G.S.L.H.V.P., 'Consumptive use over Rajasthan using different climatic methods', *Madras Agric. J.* **70**(12), 771–774, 1983.

Thornthwaite, C.W., 'Problems in the classification of climates', *Geogr. Rev.*, **33**(2), 233–255, 1943.

Thornthwaite, C.W., 'An approach towards rational classification of climates', *Geogr. Rev.*, **38**(1), 55–94, 1948.

Thorthwaite, C.W. and Mather, J.R., '*Instructions and Tables for computing potential evapotranspiration and water balance*', Water balance publications in climatology, New Jersey, **10**, 3, 1955.

WMO, *Guide to Agricultural Meteorological Practices*, World Meteorological Organization, Geneva, 1963, p. 134.

QUESTIONS

1. Define the following:
   (a) Potential evapotranspiration
   (b) Actual evapotranspiration
   (c) Reference crop evapotranspiration
2. Distinguish between:
   (a) Potential and actual evapotranspiration
   (b) Reference evapotranspiration and potential evapotranspiration
   (c) Thornthwaite and Blaney–Criddle methods
   (d) Penman and modified Penman methods
   (e) Pan and crop coefficients
3. Write short notes on:
   (a) Energy balance method
   (b) Lysimeters
   (c) Thornthwaite's water balance
   (d) Rainfall and potential evapotranspiration
   (e) Emperical PET formulae
4. Fill in the blanks:
   (a) ——— is considered a form of thermal efficiency (TE).
   (b) ——— is the year in which the concept of PET was first introduced by Thornthwaite.
   (c) Penman PET formula is based on —— and —— components.
   (d) ——— is the main difference between the Thornthwaite and Blaney–Criddle formulae.
   (e) ——— is the pan coefficient during summer over the Palakkad and Thrissur regions.
   (f) ——— regression equation is used to compute the pan coefficient.
   (g) ——— is the crop coefficient in case of young coconut palms.
   (h) ——— PET formula gives higher values.
   (i) ——— is the energy balance equation.
   (j) ——— is known as Bowen's ratio.
   (k) ——— and ——— types are seen in lysimeters.
   (l) ——— is the basic equation governing Thornthwaite's water balance.

5. Match the following:

(a) Temperature	Doorenbos and Pruitt
(b) Energy balance	Blaney–Criddle
(c) Crop coefficient	Thornthwaite
(d) Sink strength	Bowen Ratio
(e) Reference evapotranspiration	Penman

6. Elucidate the role of evapotranspiration in irrigation scheduling.

11 DROUGHT

Drought is a relative term used universally with reference to deficiency of rainfall when compared to normal rainfall of a given location. There is a shortcoming in understanding drought, as a period or periods of dryness is related to lack of rain and exists only when rainfall alone is less than normal. To overcome this, many attempts were made to define drought on the basis of either rainfall and temperature, or rainfall and water need, or consumptive use. The unique attempt in this direction was made by Thornthwaite (1948) who used the water balance concept for defining drought. He defined drought on the basis of the aridity index. Later on, the late Prof. Subrahmanyam, who was a student of Prof. Thornthwaite, widely used the concept of water balance in climatology. The delineation of climatological zones over India using the above concept was the first of its kind in India. His school of water at the Andhra University, Waltair widely used the above concept and the intensity of drought was worked out for the country as a whole and for different States separately. The Drought Research Unit, India Meteorological Department, Pune prepares aridity anomaly charts for studying drought intensity on meteorological subdivision basis. The Australian Bureau of Meteorology also defined drought on the basis of rainfall and minimum water need for a particular purpose. Van Bavel (1953) defined the intensity of drought on the basis of the number of times the soil reservoir is reduced to its lower limit. Several workers also adopted an ecological approach and defined drought as a state of deficient moisture conditions which produce a lasting adverse effect upon plants, particularly those of economic importance. In some other countries, it is thought that a month is dry when its mean rainfall (in mm) is less than twice its mean temperature (in degrees Celsius). This classification is widely used for defining natural vegetation and bioclimates across the world. Subrahmanyam et al. (1963) introduced the index of moisture adequacy for explaining crop distribution across India. Palmer (1965) described drought with an elaborate and comprehensive technique for computing the severity of drought on the basis of current and anticedent rainfall, evapotranspiration and soil moisture, by defining CAFEC (Climatically Appropriate For Existing Conditions) precipitation. Jagannathan (1968), George et al. (1973) and Rao (1979) studied the intensity and frequency of droughts using Palmer's technique. Sly (1970) derived the climatic index based on initial soil moisture at the beginning of the crop growing season, rainfall during the season, and irrigation requirement to meet the crop demand, and later simplified this as the ratio between rainfall during the crop season and the water requirement of a crop. All the above studies reveal that several attempts have been made to understand the drought and its effects on crops in and outside the country.

11.1 TYPES OF DROUGHT

Drought can be classified as meteorological, hydrological, agricultural and physiological, depending upon the deficiency of rainfall or soil moisture stress. However, physiological drought can also occur due to waterlogging during monsoon in the humid tropics, where rainfall is very high (see Fig. 11.1). In general, the occurrence of drought is noticed in major parts of India during *kharif* when monsoon is delayed or it fails, since crops are grown during the above period under rainfed conditions. The *kharif* crops raised during monsoon period may experience soil moisture stress, depending on the number of rainless days associated with the break of monsoon. The success of crops during the first crop season depends on monsoon behaviour. The India Meteorological Department defined drought on the basis of rainfall and the area affected

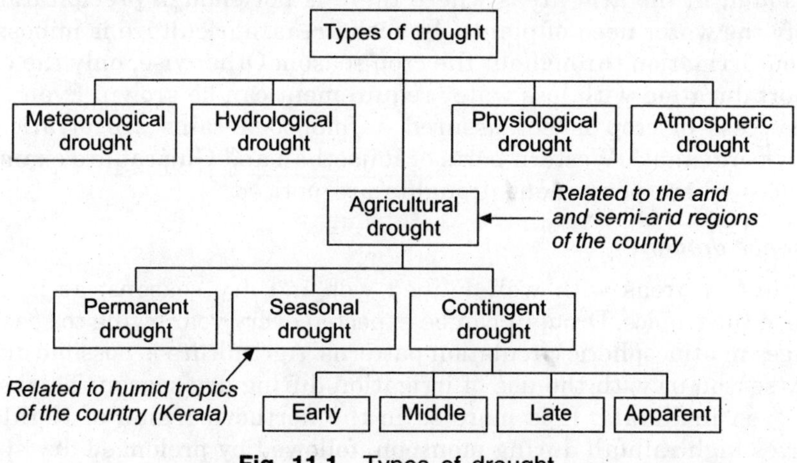

Fig. 11.1 Types of drought.

due to drought during the monsoon season. If the rainfall over the country is less by 10% of the long period average and 20% of the cropped area is affected due to drought, it is known as an *all-India drought*.

11.1.1 Meteorological Drought

A drought is said to be a meteorological drought if the rainfall deficiency is more than 25% of the normal rainfall over an area. It becomes *moderate drought* if the deficiency is between 26 and 50% of normal. If the deficiency of rainfall is more than 50%, it is known as *severe drought*.

11.1.2 Hydrological Drought

The meteorological drought, if prolonged, results in hydrological drought with marked depletion of surface water and consequent drying of reservoirs, lakes, streams and rivers, cessation of spring flows and fall in the groundwater table. Hydrological drought may be reflected in depleted snowmelt due to poor snowfall in an earlier season, and this may result in curtailment of power generation and thus affect industry as well as agriculture.

11.1.3 Agricultural Drought

It occurs when soil moisture and rainfall are inadequate during the growing season to support healthy crop growth in maturity, and causes extreme stress and wilting.

Agricultural drought constitutes a great hazard to agriculture in both the temperate and tropical regions of the world. There are three different types of agricultural droughts.

Permanent drought

It is found in the arid areas where there is not enough precipitation to satisfy the water need of plants. In such areas, agriculture is impossible without irrigation throughout the crop season. Otherwise, only the crops of short duration with less water requirement can be grown. Even then, the success of crop is not assured as monsoon rains are erratic and highly perdictable. Western parts of Rajasthan and Gujarat are examples of places where permanent droughts are noticed.

Seasonal drought

It occurs in areas with well-defined wet and dry seasons, as in most parts of the tropics. Drought can be expected every year owing to seasonal changes in atmospheric circulation patterns. Agriculture is possible during rainy season or with the use of irrigation during dry season. This is the case even in Kerala. It is more so in the northern region of Kerala. It receives high rainfall during monsoon, followed by prolonged dry spells.

Contingent drought

It is characteristic of the humid and sub-humid areas and occurs over a period of time when the rains fail. Contingent drought constitutes a serious hazard to agriculture because of its unpredictability (e.g. monsoon breaks). Depending upon the arrival of monsoon and its behaviour during the crop season, the contingent drought can further be divided into early-, mid- and late-season droughts. If the monsoon fails during the early, middle or late part of the crop season, the drought is known as early-, mid- or late-season drought respectively. The late-season drought is also known as *terminal drought*. All the droughts under this category are related to monsoon, unlike seasonal drought, which is defined on the basis of rainfall distribution during the off-season.

11.1.4 Apparent Drought

This is another type of drought related to agricultural drought. Normally in the arid and semi-arid climates, crops which require less water are chosen for cultivation. Under a given rainfall amount, they may come up well. In the case of crops with high water requirement, they fail to yield if chosen in place of crops with low water requirement. It means that the right kind of crop can survive apparent drought. The seasonal climate forecast may prove handy to farmers to avoid apparent droughts.

11.1.5 Physiological Drought

It occurs when plant functions fail physiologically due to waterlogging or water deficit. In fact, the definition of physiological drought should be based on both water deficit and excess water, unlike in other types of drought.

11.1.6 Atmospheric Drought

It can be defined on the basis of vapour pressure deficit over the crops or in the lower atmosphere, which may lead to no rainfall for prolonged periods. When the atmosphere is dry, and warm temperature prevails alongwith, dry wind, plants wilt even if irrigation is provided due to *advection effect*. Such a situation may lead to atmospheric drought.

11.2 ALL-INDIA DROUGHT AND FOODGRAINS PRODUCTION

Crops are selected across the country to fit with normal rainfall conditions. If there is any deviation in normal rainfall towards deficit, it will adversely affect the crop's growth, development and production under rainfed conditions. The delay or early break in monsoon (early season drought) is what adversely affects the sowing area of different crops across the country, by which the cropped area is reduced. Monsoon failure during July and August (mid-season drought) is highly detrimental to Indian foodgrains production, as seen during the monsoons of 1987 and 2002. The all-India droughts of 1987 and 2002 were due to monsoon failure, and the area affected due to drought was more than 20% across the country. This resulted in a sudden decline in agricultural production despite the adoption of improved technologies to mitigate the effect of drought. The effect of drought on the total foodgrains production of the country in 1987 was less by about 34 million tonnes (138.41 million tonnes in 1987–88 as against 172.18 million tonnes in 1988–89). Similar was the trend in 2002–03 also. Several workers have studied the incidence of drought over India by using different criteria. Most of them classified drought on the basis of deficiency of rainfall during the monsoon period. Mooley (1984) studied the origin, incidence and impact of droughts over India and identified 1877, 1899, 1918, 1972 and 1987 as the phenomenal drought years. Such drought years across the country were 26 in number as per the India Meteorological Department, and they are listed as follows: 1877, 1891, 1899, 1901, 1904, 1906, 1907, 1911, 1913, 1916,1918, 1920, 1926,1930, 1941, 1951, 1965, 1966, 1968, 1972, 1974, 1979, 1982, 1986, 1987 and 2002.

11.2.1 Relevance of All-India Drought Over Kerala

The *kharif* foodgrains production across the country is dependent on the monsoon rainfall and its distribution under rainfed conditions (see Fig. 1.4). In contrast, the northeast monsoon is important in case of Tamil Nadu and Jammu & Kashmir. In the tropical monsoon climates where monsoon is assured, it is the summer rainfall which is critical in the case of plantation crops as they are perennial in nature, and the availability of soil moisture round the year is very important for survival of the perennial crops.

In Kerala, the monsoon rainfall was less by 35% during monsoon 2002. It was a year with one of the lowest monsoon rainfall amounts. The water level in major hydel power generating reservoirs of Kerala was very low. The low hydel power generation cost the government heavily during the lull monsoon of 2002. The second and third crops were adversely affected due to no irrigation water, though the first crop paddy under wetland was not affected. However, reports indicated that 20% was the grain loss during *kharif* due to lull monsoon 2002. The effect of drought on plantation crops in Kerala was not felt during 1987 and 2002 due to break in monsoon during July. At the same time, the total plantation crop production was severely affected due to unprecedented drought that occurred in Kerala during summer 1983 and 2004. It reveals that the perennial crops like coconut depend upon summer showers during March, April and May for better yield. Thus, the term *drought* or *agricultural drought* with reference to coconut should be defined on the basis of rainfall received or available soil moisture during summer months viz. March, April, and May instead of total monsoon rainfall, even if it is less like in 1987 and 2002, which may not affect crops in the humid tropics due to soil moisture stress. In contrast, excess soil moisture may be a problem during monsoon.

11.3 WATER BALANCE AND ARIDITY

Thornthwaite defines drought as a condition in which the amount of water needed for transpiration and direct evaporation exceeds the amount available in the soil. The climatological drought can be better defined in terms of aridity index, which is the ratio of water or soil moisture storage deficiency (WD) to water need or potential evapotranspiration (PET) of a given location.

$$Ia = \frac{WD}{PET} \times 100 \qquad \text{where Ia is aridity index}$$

The index of aridity is an appropriate measure of dryness of a place. In order to understand whether a region is becoming drier, less dry or wet, one must know the trend of the annual aridity index. Yearly water balances must be computed using Thornthwaite and Mather's (1957) book-keeping water balance procedure, given by Subrahmanyam (1982), for computing the aridity index. The late Prof. V.P. Subrahmanyam did a pioneering work in this direction in the 1960s. He had done his D.Sc under Prof. Thornthwaite from USA, a reputed international climatologist. The whole idea of Thornthwaite's water balance concept was introduced in India by Prof. Subrahmanyam during his tenure at Andhra University, Waltair, Vishakhapatnam (Andhra Pradesh).

The annual aridity index (Ia in %) was worked out from 1871 to 2000 for the State of Kerala. It showed a marginal increasing trend (2.34%) in the last fifty years, though such a trend was not noticed for the study period as a whole. The trend towards aridity has increased annually as well as seasonally except in the monsoon towards the north of Kerala (see Fig. 11.2(a) and (b)). This was more so in case of summer (Rao, 1986).

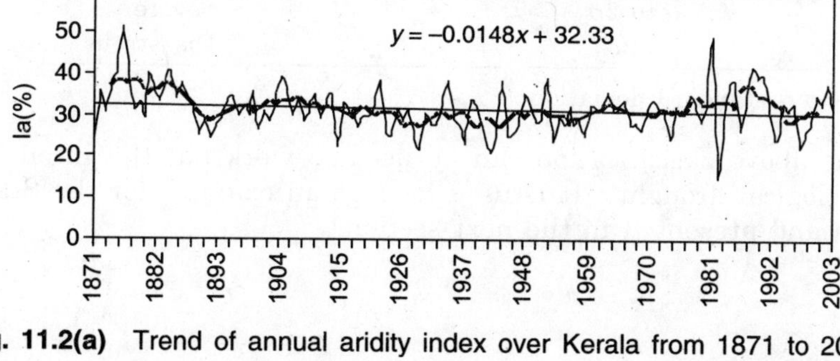

Fig. 11.2(a) Trend of annual aridity index over Kerala from 1871 to 2003.

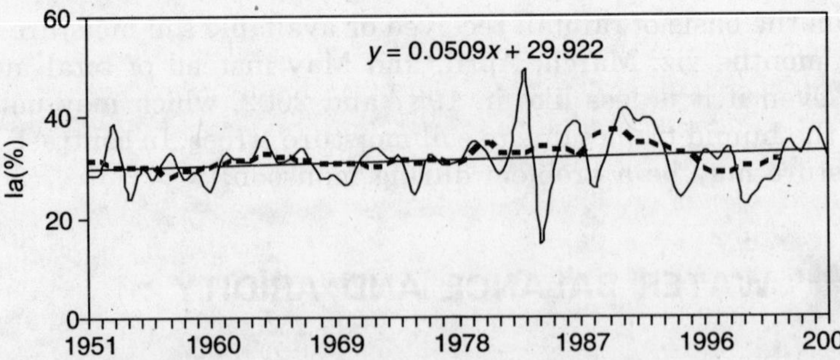

Fig. 11.2(b) Trend of annual aridity index over Kerala from 1951 to 2005.

Such analysis could be taken up at each location or for a region to understand the inter-annual variations and trends in dryness.

11.4 ARIDITY AND DROUGHT

Aridity is a measure of dryness or water deficiency. It is an index based on water deficiency over water need, while drought is a temporary phenomenon defined on the basis of rainfall deficiency. Aridity is a permanent character climatologically, where rainfall is very poor, while drought is only a temporary one and is noticed due to intermittent dry spells. Nevertheless, aridity index can be used to define intensity of drought.The occurrence and intensity of droughts can be worked out as per the procedure given by Subrahmanyam and Subramaniam (1964) and later modified by Subrahmanyam and Sastry (1971). It was classified, on the basis of the percentage departure of aridity index from the median, as Moderate, Large, Severe and Disastrous. The scheme is as shown in Table 11.1:

Table 11.1 Departure of aridity index from median

Departure of Ia from median	*Drought intensity*
$<\frac{1}{2}\sigma$	Moderate
$\frac{1}{2}\sigma$ to σ	Large
σ to 2σ	Severe
$< 2\sigma$	Disastrous

σ = standard deviation

The above classification can be used to work out the intensity of climatological droughts. It is used here as an example for the State of Kerala and presented in the next section.

11.4.1 Drought over Kerala

The State of Kerala experienced 70 drought years out of 135 (in the period 1871–2005), of which thirteen were moderate, twelve each were large and severe, and thirty-three were disastrous droughts. Though the decade 1881–90 experienced more droughts (nine), followed by 1971–80 (seven), the number of disastrous droughts was more (six) in 1981–90, indicating that the intensity of drought was high in 1981–90 (see Table 11.2). None of the years fell under any category of drought other than disastrous during the given decade. There was a gradual decline in number of drought years from 1901–10 (six) to 1951–60 (two) and increase from 1961–70 to 1991–2000. It varied between seven (1971–80) and five (1991–2000).

Table 11.2 Occurrence and intensity of droughts over Kerala from 1871 to 2005

	Intensity and occurrence of drought					
Decade	*Moderate*	*Large*	*Severe*	*Disastrous*	*Total*	*(%)*
1871–80	0	0	1	5	6	60
1881–90	0	2	2	5	9	90
1891–00	0	1	1	2	4	40
1901–10	2	0	2	2	6	60
1911–20	1	1	1	2	5	50
1921–30	2	1	0	1	4	40
1931–40	0	1	1	1	3	30
1941–50	1	0	0	2	3	30
1951–60	0	1	0	1	2	20
1961–70	3	1	1	1	6	60
1971–80	3	2	1	1	7	70
1981–90	0	0	0	6	6	60
1991–00	0	1	2	2	5	50
2001–05	1	1	0	2	4	40
Total	13	12	12	33	70	52
Drought intensity(%)	18.6	17.1	17.1	47.2	–	–

The percentage occurrence of disastrous droughts was more (47.2%) across the State, followed by moderate (18.6%), large and severe (17.1% each) droughts. It revealed that the intensity of drought was relatively high whenever droughts were noticed, as the majority of the years fell under the category of *disastrous*. Also, the intensity of drought was high in 1981–90, followed by 1871–80 and 1881–90. It appears that the intensity of drought was more in the recent decades, especially in 1981–90, as the aridity index was high. A similar trend was noticed between 1871 and 1890.

11.5 INDEX OF MOISTURE ADEQUACY AND AGRICULTURAL DROUGHTS

The aridity index is a measure of dryness, while the index of moisture adequacy is a measure of soil moisture availability. The former explains climatological droughts, while the latter explains agricultural droughts. The ratio of actual evapotranspiration (AET) to potential evapotranspiration (PET) is known as the *index of moisture adequacy* and is expressed in percentage (Subrahmanyam et al. 1963).

$$Ima = \frac{AE}{PE} \times 100$$

where *Ima* is the index of moisture adequacy. The same index was introduced as R.Index by Yao in 1969. The index can be used for determination of crop distribution. High values of the index at a given location indicate success of crops with high water requirement, while low values favour crops with low water requirement. Similarly, high values for a longer period indicate success of long-duration crop and vice versa. For millets, the index may vary from 40% to 60% and 80–100% in case of rice. Accordingly, crops can be chosen on the basis of spatial and temporal distribution of the index of moisture adequacy. The weekly indices can be worked out at each location on the basis of climatological data and crop choice can thus be made. The above index can also be used to define agricultural droughts with reference to crops. For the first time, Subramaniam and Sastry (1979) classified agricultural droughts on the basis of the index of moisture adequacy. It was modified by Rao (1979) for western Rajasthan.

11.5.1 Agricultural Droughts with Reference to Coconut

Most of the plantation crops suffer due to soil moisture stress during summer. In case of coconut, the nut yield is affected in the following year up to 30–50%, depending upon variety as well as crop management, due to severe soil moisture stress during summer. The following criteria were developed (Rao, 1991) for assessing agricultural droughts with reference to coconut based on the index of moisture adequacy (see Table 11.3).

Table 11.3 Mean index of moisture adequacy (Ima) and intensity of drought

Intensity of drought from December to April	*Mean Ima* (%)
No drought	< 30
Moderate	25–30
Large	20–25
Severe	15–20
Disastrous	< 15

Based on the above criteria, the intensity of agricultural droughts was worked out for the period from 1942 to 1997 by using the climatological data at RARS, Pilicode. Out of 41 agricultural drought years, 11 each fell under moderate and large, 12 severe and 7 under disastrous. It showed that 46.2% of the years fell under severe and disastrous droughts (see Fig. 11.3).

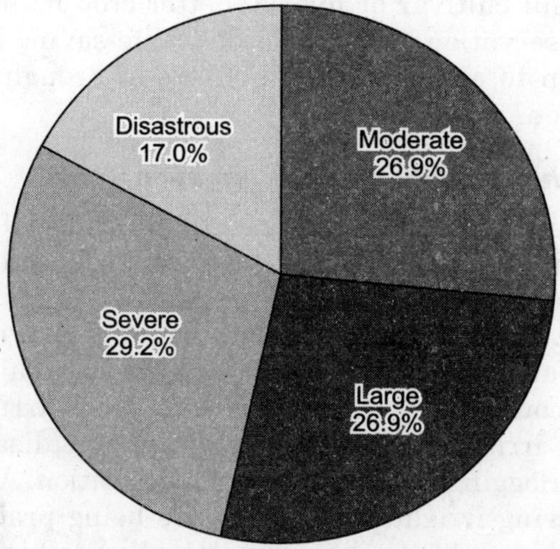

Fig. 11.3 Occurrence and intensity of agricultural droughts.

From the above, it is evident that the coconut crop, that is grown in the lowlands and midlands of Kasaragod district, experiences severe soil moisture stress almost in half of the years. It was probably one of the reasons why the fluctuations in coconut production were noticed very often. In the recent past, the occurrence of disastrous droughts in the summers of 1979, 1983, 1989, 1997 and 2004 adversely affected coconut production to a large extent.

11.5.2 Drought Characteristics in Coconut

Coconut palms show the following characteristics under severe soil moisture stress depending upon the duration and intensity of drought (Rao, 1994):

1. Withering and mortality in the case of young seedlings under poor management.
2. Drooping, wilting and drying of the lower whorl of leaves.
3. Breakage of leaves at the petiole or just above it.
4. Spindle leaf breaking, which leads to mortality in the case of senile palms under conditions of poor management.
5. Abortion of spadices, which starts from October or November.
6. Button shedding and immature nut fall.

7. Nut size decline
8. Decline in nut yield in the subsequent year by up to 50% depending upon type of crop management.

11.5.3 Can Drought Effect be Alleviated?

Choice of a coconut cultivar or hybrid, better crop management, *in situ* soil moisture conservation, and irrigation or life-saving irrigation during summer will help to alleviate the ill-effects of drought and to sustain coconut production.

Mitigating drought through summer irrigation

The nut yield in coconut can be increased significantly if coconut gardens are irrigated during summer. This is the reason why coconut productivity is higher in the neighbouring States of Kerala, viz. Tamil Nadu, Karnataka and Andhra Pradesh, where coconut is grown under irrigated conditions. In coarse textured soils having less available water, better yield response could be achieved through flood irrigation or basin irrigation. Flood irrigation, however, has its own disadvantages as it may lead to waterlogging, salinization and alkalization. As water is scarce, several water-saving irrigation methods are being practised in coconut plantations. Studies indicate that drip irrigation is highly effective and profitable in areas with limited water availability. If there is no water available for irrigation during summer months as is the case in many parts of the northern districts in Kerala, the question is: How to mitigate the effect of drought in coconut gardens under rainfed conditions? In this context, drought management under rainfed conditions plays an important role, and it can be considered a part of water management in water-scarce areas.

Mitigating drought through better management

Increase in nut yield can be achieved in conditions involving water scarcity through the adoption of a package of better management as detailed below:

Selection of planting material

Choose seedlings with good agronomic trails. Vigorous seedlings can withstand the effect of drought.

Depth of planting

Plant seedlings in pits of 1 metre depth. Deep planting has the following advantages: It produces more roots, and surface roots are prevented. Nutrient uptake is more. Whatever soil moisture is available in the root depth during summer, it can be effectively utilized by the roots since the subsurface evaporation is much low when compared to that of surface evaporation. Soil temperature at the root depth will be lower when compared to that of surface depths and mechanical damage or uprooting due to wind can be minimised.

Age and time of planting

Transplant 1-year-old seedlings immediately after the receipt of summer showers in April-May, or with the onset of monsoon. Seedling vigour can also be known if the age of the seedling is 9–12 months.

Shade

Young coconut seedlings require shade during summer months. Therefore, provide adequate shade during the first 2 years.

Fertilizer application

Apply fertilizers in two split doses under rainfed conditions (one-third in April-May immediately after premonsoon showers and two-third in September-October).

Interculture

In laterite and red sandy loam soils, give two ploughings or diggings in May-June and September-October and one raking in January. In areas where surface run-off is more, form mounds in September-October and level them in November-December.

Husk burial for moisture conservation

Burying fresh or dried coconut husk around the palms is recommended for moisture retention. Husk can be buried either in linear trenches taken 3 metres away from the trunk between rows of palms or in circular trenches taken around the palms at a distance of 2 metres from the trunk. The trenches may be half a metre long and equally deep. The husks are to be placed in 2 or 3 layers with their concave surface facing upwards and then covered with soil. The beneficial effect of husk burial is said to last for 5 to 7 years.

Green manure and cover crops

Grow green manure and cover crops to avoid surface run-off and to retain soil moisture.

Mulching

Mulch the basins in October-November with locally available tree leaves to control surface evaporation and to reduce soil temperature.

Leaf cutting

It is found that the cutting of lowermost leaves during summer may be beneficial to coconut. One leaf cut in alternative months, starting from January to May, may be practiced.

Disease management

In the northern districts of Kerala, where unimodal rainfall and prolonged drought situation are noticed, stem bleeding is a severe disease. Studies indicate that symptom expression is more in drought years. The affected portion may be removed by chiseling and the wounds dressed with hot coal tar. Application of calixin at the rate of 25 ml in 25 litre of water thrice in a year as soil drenching is found to control the disease. Application of neem cake at the rate of 5 kg/palm once in a year is also recommended to check the disease.

All the above management practices along with summer irrigation are recommended for increasing nut yield in coconut. Even without irrigation, the effect of drought could be mitigated with the above management practices.

11.6 PREDICTION OF DROUGHT

Is it possible to predict drought? The answer may be 'no', if it is related to a crop and is site-specific. The occurrence of drought is common in one or the other region across the country every year, and crops are affected due to soil moisture stress. The Drought Research Unit, India Meteorological Department, Pune prepares aridity anomaly charts to know the effect of drought on area spread under cultivation. Similarly, the National Remote Sensing Agency, Hyderabad monitors drought situation through the data obtained from remote sensing satellites. Based on the Normalized Vegetation Index, the temporal and spatial distribution of crop coverage can be obtained. The information so obtained is useful to know the past status, but is in no way useful in predicting drought. Of course, it is useful for mitigating the ill-effects of drought. Several research articles are also published on drought-prone areas and prediction of drought. These studies related to drought in India and abroad can not provide clues or methods of drought prediction and are therefore of no practical significance. The seasonal climate forecasts on rainfall, which are being issued by the developed countries, are also not dependable region-wise. However, efforts are being made by the international climatological community to develop precise seasonal climate forecasts. The long-range weather forecast issued by the IMD is found reasonably good except for a few years. Of course, it was a total failure during monsoon 2002. Even if it is correct, it gives information only on total rainfall of monsoon ahead. It fails to give information on time of occurrence of rainfall events. Also, information given on region-wise total rainfall is undependable. The short- and medium-range weather forecast of rainfall may be useful for assessing droughts, but the accuracy of weather forecast on specific locations is in doubt and the response time available to farmers is rather limited for combating drought. As drought prediction needs information crop- and site-specific information, the relevance of short- and medium-range weather forecast may be tested on experimental basis, wherever it is relevant. Considering all the aspects, probably proactive monsoon management is the right answer to mitigate the ill-effects of drought, which were discussed in length under *monsoon* in Chapter 8.

REFERENCES/SUGGESTED READINGS

ANGRAU, *Crop Strategies, Contingency Plans and Management*, Acharya N.G. Ranga Agricultural University, Rajendranagar, Hyderabad, 2002, p. 128.

CWRDM, *Water Atlas of Kerala*, Centre for Water Resources Development and Management, Kozhikode, 1995, p. 82.

George, C.J., Sastri, K.S.R. and Rentala, G.S., *Incidence of droughts*, Meteorological Monograph, Technical Bulletin No. 5, Division of Agricultural Meteorology, India Meteorological Department, Pune, 1973.

Jagannathan, P., '*Proceedings of symposium on cropping pattern in India*', A pilot study of drought for selected centres of India, Indian Council of Agricultural Research, New Delhi, 1968, pp. 239–250.

Mooley, D.A., 'Origin, Incidence and impact of droughts over India and remedial measures for their mitigation', *Sadhana*, **19**(4), 597–608, 1984.

Palmer, W.C., *Meteorological drought*, Research Paper No. 45, United States Department of Commerce, Weather Bureau, 1965.

Park, M., 'The effect of drought at Puttalam, Sri Lanka', *Trop. Agricst*, **78**, 11–17, 1932.

Rao, G.S.L.H.V.P., 'Some studies on agroclimatic aspects of Rajasthan', Ph.D thesis, Andhra University, Waltair, p. 245, 1979.

Rao, G.S.L.H.V.P., 'Effect of drought on coconut production', *Indian Cocon. J.*, **17**(8), 11–12, 1986.

Rao, G.S.L.H.V.P., 'Agricultural droughts with reference to coconut', *J. Plantation Crops*, **18**(supplement), 166–170, 1991.

Rao, G.S.L.H.V.P., 'Water management in coconut', *Proceedings of the national seminar on water management for plantation crops,* March 4, 1993, Ed. Satheesan, K.V., Centre for Water Resources Development and Management, Kozhikode, pp.65–70, 1993.

Rao, G.S.L.H.V.P., 'Drought management in coconut', *Indian Cocon. J.*, **24**(11), 6–8, 1994.

Rethinam, P., 'Management of drought situation in coconut plantations', *Indian Cocon. J.*, **18**(6), 3–5, 1987.

Singh, R.S. and Rao, B.V.R., *Agricultural Drought Management in India—Principles and Practices,* Central Research Institute for Dryland Agriculture, Hyderabad, 1988, p. 42.

Singh, V.P. and Singh, R.K. (Eds.), *Rainfed Rice: A Source Book of Best Practices and Strategies in eastern India,* International Rice Research Institute, Philippines, 2000, p. 293.

Sly, W.K., 'A climatic moisture index for land and soil classification in Canada', *Can. J. Soil Sci.*, **50**, 291–300, 1970.

Subrahmanyam, V.P., *Water Balance and its Applications*, The Andhra University Press, Andhra University, Waltair, 1982, p. 102.

Subrahmanyam, V.P., Rao, B.S. and Subramaniam, A.R., 'Moisture adequacy in relation to the distribution of a few crops in India', *Proceedings of IUGG,* California, pp. 462–467, 1963.

Subrahmanyam, V.P. and Subramaniam, A.R., 'Application of water balance concepts for a climatic study of droughts in South India', *Indian J. Met. & Geophys*, **15**, 394–402, 1964.

Subrahmanyam, V.P. and Sastry, C.V.S., 'New techniques in drought analysis', *Ann. Arid Zone,* **10**(2 and 3), 1971.

Subramaniam, A.R. and Sastry, A.S.R.A.S., 'A rational approach on agricultural droughts of Andhra Pradesh', *Proceedings of the International symposium on "hydrological aspects of droughts",* Indian Institute of Technology, New Delhi, 1979.

Thornthwaite, C.W., 'An approach towards a rational classification of climate', *Geogr. Rev.*, **38**, 55–94, 1948.

Thornthwaite, C.W. and Mather, J.R., '*Instructions and tables for computing potential evapotranspiration and the water balance*', Centreton, N.J. Laboratory of Climatology Publications, 1957, pp. 185–311.

Van Bavel, C.H.M., 'A drought criterion and its application in evaluating drought incidence and hazard', *Agron. J.*, **45**(4), 167–172, 1953.

Yao, A.Y.M., 'The R-index for plant water requirement', *Agrl. Met.*, **6**, 259–273, 1969.

QUESTIONS

1. Define the following:
   (a) All-India drought
   (b) Meteorological drought
   (c) Agricultural drought
   (d) Hydrological drought
2. Write short notes on:
   (a) Water balance and aridity
   (b) Aridity and drought
   (c) Drought characteristics in coconut
   (d) Types of drought
   (e) Relevance of all-India drought over Kerala
   (f) Climatological droughts
   (g) Agricultural droughts
3. State true or false:
   (a) The occurrence of drought is seen in India when the monsoon is delayed.

(b) Permanent drought is a feature of the arid areas.
(c) Meteorological drought if prolonged results in hydrological drought.
(d) Contingent drought is a feature of the semiarid areas.
(e) The effect of all-India drought during 1987 and 2002 was felt on the plantation crops of Kerala.
(f) Aridity index is the ratio between water surplus and water need.
(g) The IMD introduced the concept of water balance in India.
(h) The nut yield in coconut is not affected after the drought period is over.
(i) Plantation crop production in Kerala is influenced due to summer drought.
(j) Drooping of lower leaves during summer is an immediate symptom of drought in coconut.

4. Fill in the blanks:
(a) Meteorological drought if prolonged results in ———.
(b) Arid areas experience ——— drought.
(c) 2002 was ——— year.
(d) The *kharif* foodgrains production is dependent on ———.
(e) In Kerala, the rainfall was less by —— during monsoon 2002.
(f) In Kerala, ———was an unprecedented drought year with reference to plantation crops.
(g) In areas with limited water availability, ——— irrigation is highly effective and profitable.
(h) ——— is a severe disease in northern districts of Kerala.

5. Elucidate the following:
(a) Impact of all-India drought on Indian foodgrains production.
(b) Agricultural droughts with reference to crops.
(c) Drought characteristics in coconut and its management under rainfed conditions.

12 AGROCLIMATIC CLASSIFICATION

Climate varies from one location to another and from season to season. Although no two locations have identical climates, it is possible to demarcate and bring them into one zone, within which climate is broadly uniform. Such demarcation is usually known as *delineation of climatic zones*. Alternatively, locations can be categorised under different climatic types. That means if an area or region is categorised under a particular climatic type, it possesses uniform characteristics with respect to climate. The climate may be in relation to rainfall or temperature or both, depending upon the climate variables chosen for delineating climatic zones. Various regions have been identified in terms of climatic types to understand the different climatic patterns that exist across the world. The climatic regions in essence are the natural vegetation regions, which are defined climatologically. For example, the tropical and temperate climates support different types of natural vegetation, since there is a contrast in temperature and rainfall regimes. Identification of climates is one of the areas in the field of climatology and agricultural climatology. It helps in identification of *isoclimates* or *homoclimates*, which in turn help in the transfer of agrotechnology as well as suggesting new cropping systems. In this direction, the International Crops Research Institute for Semiarid Tropics (ICRISAT) is a classic example along with the Central Arid Zone Research Institute (CAZRI), where technologies are generated in tune with climatic types. The semi-arid and arid are the two climatic types classified by Thornthwaite in 1948, on the basis of the moisture regime, the details of which will be discussed later. Several climatologists, agrometeorologists and bio-climatologists have attempted to classify climates in relation to agriculture.

12.1 CLIMATIC CLASSIFICATION

Climatic classification is based on two approaches viz. *genetic approach* and *generic* or *empirical approach*. In the genetic approach, classification is based on climatic controls such as general circulation, net radiation and moisture fluxes. In the generic approach, classification is based on the observed climatic elements or their effects on the phenomena, usually vegetation or humankind. Because the controls of climate are far more difficult to measure than the climatic elements, all the climatic classification schemes have therefore adopted the empirical approach. Koppen (1900), De Martonne (1928), Emberger (1955) and Majundar (1964) demarcated climate types on the basis of empirical relationship between rainfall and temperature. Stenz (1946), Thornthwaite (1948), Troll (1965) and Hargreaves (1971) introduced the rainfall and potential evapotranspiration concept for delineating climates. However, Koppen and Thornthwaite classifications are universally accepted for delineating climatic zones. Troll and Hargreaves climate classifications are mostly used to define crop-growing seasons and stress periods respectively.

12.1.1 Koppen's Climatic Classification

The most widely used system of climatic classification in its various forms is that of Wladimir Koppen (1846–1943), a Russian-born meteorologist who devoted most of his life to a scientific career in Germany. He is known as the 'father of modern climatology'. He devised his first climatic classification (1900) on the basis of vegetation. The vegetation zones were climatologically defined and later (1923) revised with greater attention to temperature, rainfall and their seasonal characteristics. The Koppen climatic classification has five major climatic types recognised on the basis of temperature and rainfall. They are designated by capital letters as follows:

A: Tropical forest climates; hot seasons
B: Dry climates
C: Warm temperate rainy climates; mild winters
D: Cool forest climates; severe winters
E: Polar climates

A group of undifferentiated high-range climates is represented by the symbol H for the above five major climatic types. All the five major climates are further subdivided into 24 climatic types on the basis of seasonal temperature and rainfall characteristics.

Tropical forest climates: 3 types

Af: Tropical rainforest. Hot; rainy seasons
Aw: Tropical savanna. Hot; seasonally dry (usually winter)
Am: Tropical monsoon. Hot; seasonally excessive rainfall

Dry climates: 4 types

BSh: Tropical steppe. Semi-arid; hot
BSk: Mid-latitude steppe. Semi-arid; cool or cold
BWh: Tropical desert. Arid; hot
BWk: Mid-latitude desert. Arid; cool or cold

Warm temperate rainy climates: 7 types

Cfa: Humid subtropical mild winter. Moist seasons; long hot summer
Cfb: Marine. Mild winter; moist seasons, warm summer
Cfc: Marine. Mild winter; moist seasons, short cool summer
Csa: Interior mediterranean. Mild winter; dry hot summer
Csb: Central mediterranean. Mild winter; dry, short warm summer
Cwa: Subtropical monsoon. Mild, dry winter; hot summer
Cwb: Tropical upland. Mild, dry winter; short warm summer

Cool forest climates: 8 types

Dfa: Humid continental. Severe winter; moist seasons, long hot summer
Dfb: Humid continental. Severe winter; moist seasons, short warm summer
Dfc: Subarctic. Severe winter. Moist seasons, short cool summer
Dfd: Subarctic. Extremely cold winter. Moist seasons, short summer
Dwa: Humid continental. Severe dry winter; long hot summer
Dwb: Humid continental. Severe dry winter; warm summer
Dwc: Subarctic. Severe dry winter; short cool summer
Dwd: Subarctic. Extremely cold dry winter; short cool summer

Polar climates: 2 types

ET: Tundra; very short summer
EF: Perpetual ice and snow

The major categories (i.e. A, B, C, D, E) are based mainly on temperature criteria, and precipitation is also compared with potential evapotranspiration in A and B types. The major climatic types are defined as follows:

A. The coldest month has a mean temperature greater than 18°C, which is critical for the survival of certain tropical plants. The mean annual potential evapotranspiration is lesser than the mean annual precipitation.
B. Mean annual potential evapotranspiration is greater than the mean annual precipitation. There is no water surplus, hence no permanent streams originate here.
C. The coldest month has a mean temperature between –3°C and 18°C. Warmest month has a mean temperature greater than 10°C. The 10°C summer isotherm correlates with the poleward limit of tree growth, and the –3°C isotherm indicates the equatorward limit of permafrost.

D. The coldest month has a mean temperature below –3°C and the warmest month has a mean temperature greater than 10°C as in case of 'C'.

E. The warmest month has a mean temperature less than 10°C. The ET warmest month has a mean temperature between 0°C and 10°C. EF warmest month has a mean temperature less than 0°C.

The subdivisions of each major category are made with reference to the seasonal distribution of precipitation and additional temperature characteristics and detailed as follows:

Seasonal distribution of precipitation

f = No dry season, wet all the year (A, C, D)
m = Monsoonal, with short dry season and heavy rain during the rest of the year (A)
w = Summer rain (A, C, D)
S = Summer dry season (B)
W = Winter season (B)

Additional temperature characteristics

a = Hot summer, warmest month has mean temperature greater than 22°C
b = Warm summer, warmest month has mean temperature less than 22°C
c = Short cool summer, less than 4 months have mean temperature greater than 10°C
d = Very cold winter, coldest month has mean temperature less than –38°C

In the arid regions (BW and BS), the following subscripts are used:

h = hot, mean annual temperature greater than 18°C
k = cool, mean annual temperature less than 18°C

Rainfall-temperature relationship

Wladimir Koppen also suggested the following limits for the boundary between desert and steppe, and dry and tree climates for regions having maximum rainfall in summer, which is the case in Kerala too:

Boundary between dry and tree climates

$$r = 0.44\ t - 3$$

Boundary between steppe and desert

$$r = \frac{0.44t - 3}{2}$$

where t and r are mean annual temperature (in °F) and mean annual precipitation (in cm), respectively.

Koppen climatic classification lacks sub-humid category and his classification was criticised by Thornthwaite, who described it as unsystematic as it was based on a patchwork of irregular rules and definitions. However, Koppen classification explains natural vegetation fairly well, as his climatic regions are in essence the vegetation regions climatologically defined.

12.1.2 Thornthwaite's Climatic Classification

Thornthwaite (1948) proposed a rational classification of climates based on the concepts of potential evapotranspiration and water balance. It is found useful in diverse fields such as ecology, agriculture and water resources development. His classification is based on two main climatic indices, namely the moisture index and the thermal index (which is nothing but potential evapotranspiration). It was further improved by Thornthwaite and Mather in 1955. The four criteria chosen for climatic classification by them are as follows:

1. Moisture adequacy represented by values of the moisture index;
2. Thermal efficiency represented by the values of potential evapotranspiration;
3. Seasonal distribution of moisture adequacy; and
4. Summer concentration of thermal efficiency.

Moisture index and climatic types

The moisture index (Im) derived for delineating climatic zones is as follows:

$$Im = \frac{100\,S - 100\,D}{PE} = \frac{S - D}{PE} \times 100$$

where S is the annual water surplus, D is the annual water deficit and PE is the annual potential evapotranspiration. They can be computed by using the book-keeping water balance procedure (Thornthwaite and Mather, 1957). Using the moisture index (*Im*), different climatic types are derived, starting from Perhumid (A) to Arid (E). The same is given in Table 12.1.

Table 12.1 Moisture index (*Im*) versus climatic types

Symbol	*Climatic type*	*Moisture index (Im) in %*
A	Perhumid	100 and above
B_4	Humid	80 to 100
B_3	Humid	60 to 80
B_2	Humid	40 to 60
B_1	Humid	20 to 40
C_2	Moist sub-humid	0 to 20
C_1	Dry sub-humid	–33.3 to 0
D	Semi-arid	–66.7 to –33.3
E	Arid	–100 to –66.7

The seasonal moisture adequacy (see Table 12.2) is determined for the moist climates (A, B and C_2) based on the values of the aridity index (Ia), which is as follows:

$$Ia = \frac{D}{PE} \times 100$$

where D is water deficit and PE is potential evapotranspiration.

Table 12.2 Seasonal moist climates versus aridity index

Seasonal moist climates for A, B and C_2	*Aridity index*
r–little or no water deficit	0–10
s_1–moderate summer deficit	10–20
w_1–moderate winter deficit	10–20
s_2–large summer deficit	above 20
w_2–large winter deficit	above 20

Dry climate versus humidity index

The seasonal moisture adequacy for the dry climates is determined (see Table 12.3) based on the values of the humidity index (Ih) and is given as follows:

$$Ih = \frac{S}{PE} \times 100$$

where S is the water surplus and PE is potential evapotranspiration.

Table 12.3 Seasonal climates versus humidity index

Seasonal dry climates (C_1, D and E)	*Humidity index*
d–little or no water surplus	0–16.7
s–moderate winter surplus	16.7–33.3
w–moderate summer surplus	16.7–33.3
s_2–large winter surplus	above 33.3
w_2–large summer surplus	above 33.3

The values of water surplus and water deficit are obtained by using the water balance book-keeping procedure (Thornthwaite and Mather, 1957) based on precipitation and potential evapotranspiration. The potential evapotranspiration can be estimated on the basis of weather variables using empirical formulae. The moisture provinces are delimited on the basis of the computed values of the moisture index. The thermal provinces are similarly defined by using the annual potential evapotranspiration (PE), which is regarded as an index of thermal efficiency (TE).

Thermal efficiency and climatic types

On the basis of thermal efficiency (TE), the climatic types were derived by Thornthwaite (see Table 12.4).

Table 12.4 Thermal efficiency and its summer concentration versus climatic types

Climatic type	*Thermal efficiency* *PE* (cm)	*Type*	*Summer concentration*
A: Megathermal	114 and above	a′	Below 48.0
B_4: Mesothermal	99.7–114.0	b′	48.0–51.9
B_3: Mesothermal	85.5–99.7	b′	51.9–56.3
B_2: Mesothermal	71.2–85.5	b′	56.3–61.6
B_1: Mesothermal	57.0–71.2	b_1	61.6–68.0
C_2: Microthermal	42.7–57.0	c_2	68.0–76.3
C_1: Microthermal	28.5–42.7	c_1	76.3–88.0
D: Tundra	14.2–28.5	d′	Above 88.0
E: Frost	Below 14.2		

12.1.3 Troll's Climatic Classification

Troll's classification (1965) is based on length of moisture availability periods. Based on precipitation and potential evapotranspiration, the humid periods have been defined. A humid month is defined as one in which the mean annual precipitation exceeds mean potential evapotranspiration. This classification was found satisfactory to explain the vegetation zones of tropical Africa and South Africa, and it is as follows:

1. Tropical rainy climate with 9.5 to 12 humid months

2(a). Tropical summer–humid climates with 7 to 9.5 humid months

(b). Tropical winter–humid climates with 7 to 9.5 humid months

3. Wet dry tropical climates with 4.5 to 7 humid months

4(a). Tropical dry climates with 2 to 4.5 humid months (in summer)

(b). Tropical dry climates with 2 to 4.5 humid months (in winter)

5. Tropical semi-desert climates with less than 2 humid months

The above classification can be used for defining crop-growing seasons also.

12.1.4 Hargreaves' Classification

The climatic indices are derived on the basis of the Hargreaves (1971) method, which considers the Moisture Availability Index (MAI). It is nothing but a ratio of assured precipitation (precipitation value is expected with 75% probability) and potential evapotranspiration.

$$\text{MAI} = \frac{\text{Assured precipitation (P)}}{\text{Potential evapotranspiration (PET)}}$$

Based on the monthly MAI, the crop-growing periods were worked out and classified into four climatic types viz. very arid, arid, semi-arid and wet-arid, and their suitability to agriculture was given. Later, Hargreaves redefined the MAI versus moisture deficit (see Table 12.5).

Table 12.5 Moisture availability index (MAI) versus moisture deficit

MAI	*Category*
0.00–0.33	Very deficient
0.34–0.67	Moderately deficient
0.68–1.00	Deficient
1.00–1.33	Adequate moisture
> 1.34	Excessive moisture

The above classification is widely used by the India Meteorological Department, and crop-growing periods were delineated across the country on the basis of weekly MAI at different probability levels.

12.2 CLIMATIC ZONES OF INDIA

Thornthwaite's climatic classification is widely used in India by several workers for delineating climatic zones. The entire Western Ghats region and the north-eastern States fall under the humid climates while the northwest parts comprising Gujarat, Rajasthan, Haryana, Punjab, extreme north of Jammu and Kashmir, and some areas in Rayalaseema of Andhra Pradesh, fall under the arid climates. The central tract extending from north to south of the country falls under semi-arid tract (see Fig. 12.1).

12.2.1 Climatic Classification of Kerala

As per Koppen's classification, the entire State of Kerala falls under the tropical monsoon climates (Am). As per Thornthwaite's climatic classification, the State of Kerala falls under the perhumid (A) and humid (B_1–B_4) climatic types, except the southernmost pockets of the State and the eastern part of the Palakkad region, which falls under the moist sub-humid (C_2) climatic type. The State as a whole experiences megathermal climate (A′) based on thermal efficiency, which shows that the crop growth is not inhibited by temperature but is governed by rainfall alone. As per Troll's classification, the State of Kerala falls under the wet dry tropical climates with 4.5 to 7 humid months. The entire State of Kerala has been divided into a total of 20 agroclimatic zones with five climatic and seven soil types, as per Hargreaves' classification, and different cropping systems are suggested in each agroclimatic zone.

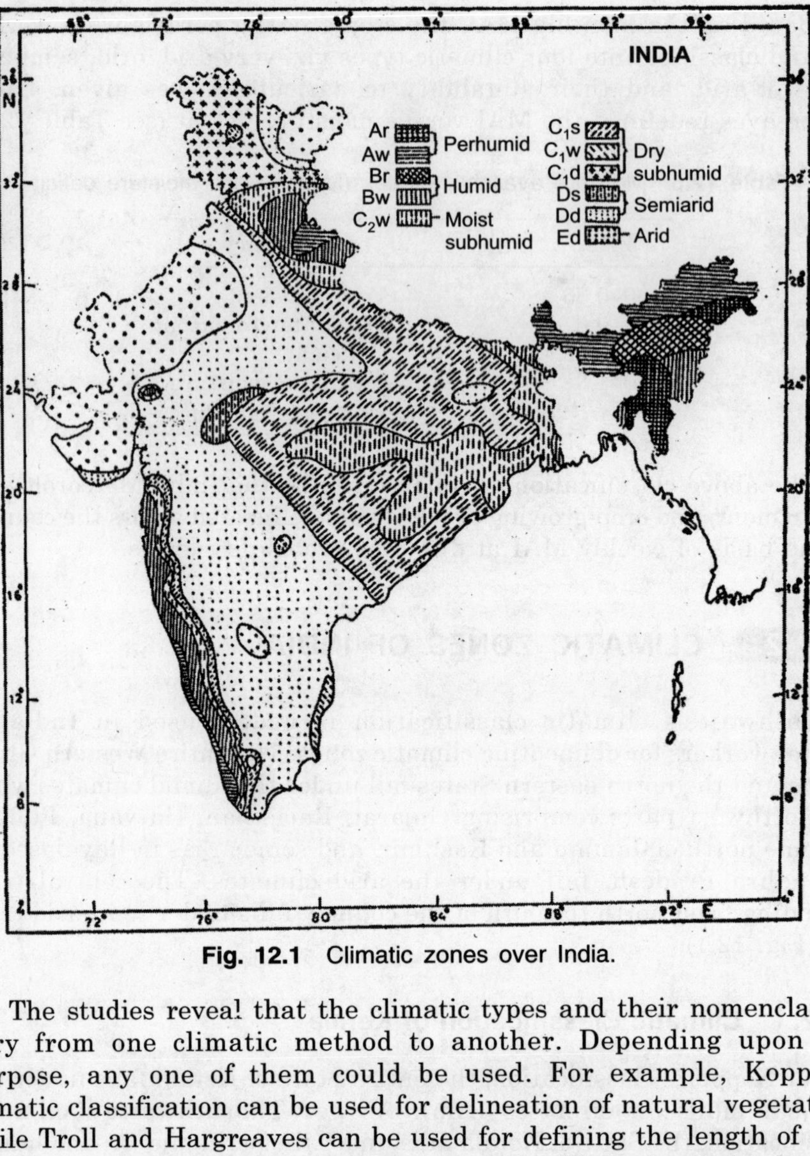

Fig. 12.1 Climatic zones over India.

The studies reveal that the climatic types and their nomenclature vary from one climatic method to another. Depending upon the purpose, any one of them could be used. For example, Koppen's climatic classification can be used for delineation of natural vegetation, while Troll and Hargreaves can be used for defining the length of crop growing seasons and the crop stress periods. Thornthwaite's climatic classification can be used for delineating climatic zones and studying climate variability based on moisture index. It has wider applications when compared to other methods.

12.3 SOIL CLIMATIC ZONES

The climate zones are delineated by using empirical relationships between either rainfall and temperature or rainfall and potential evapotranspiration. Though Thornthwaite used the water balance concept (wherein soil moisture is taken into account) for deriving different climates, they fail to explain different cropping systems as soil type and crops are not considered in his climatic scheme. Based on the above concept, Krishnan and Singh (1968) demarcated the soil climatic zones by superimposing the moisture index and mean temperature on the soil map of India, showing major soil types. The distribution of crops was explained in various climatic zones on the basis of soils and climate regimes, which include moisture and temperature indices. This scheme is definitely an improvement from the agricultural point of view, as it explains cropping systems based on climate and soil regimes.

12.4 AGROECOLOGICAL ZONES

Murthy and Pandey (1978) demarcated the whole country into eight agroecological regions on the basis of analysis of the individual attributes viz. physiographic, climatic, soil, crop and agricultural regions. Panabokke (1979) prepared a technical note for FAO on the "Agroecological zones of south and southeast Asia". This may be a useful tool for the development and transfer of appropriate agro-technologies for improving rainfed agriculture in different agroecological zones. He defined agroecological zones on the basis of rainfall regimes and soil grouping.

12.4.1 Definition of Agroecological Zone

It is defined as a major area of land that is broadly homogeneous in respect of its rainfall regime, and is made up of a grouping of soils that reflect broad similarities in respect of soil profile development.

12.4.2 Rainfall Regimes

These are annual sequential rainfall patterns that are broadly similar, both in respect of their seasonalities and amounts, and that could be grouped into agronomically significant categories.

12.4.3 Soil Grouping

A functional (non-taxonomic) grouping of soil units have been used in the FAO–UNESCO soil map of the world. Each grouping that consists of one or more such soil units represent, as far as possible, a similar degree of soil profile development.

Based on the above concept, six rainfall regimes and seven soil groups were derived. They were grouped into six zones (rainfall and soil regimes) for development and transfer of agrotechnology. A model of agroecological zones is as follows:

Group 1–Mainly rainfed or irrigated rice (R1S1, R2S1, and R3S1).

R1–Adequate or surplus rainfall throughout the year with no significant intervening dry period. The annual sequential rainfall usually follows a unimodal pattern.

S1–Fluvisols and Gleysols. They are soil units represents by the FAO-UNESCO classification. It is nothing but as per equivalent soil order in soil taxonomy.

Like that, all the agroecological zones were brought into six groups. A high degree of transferability is possible in the agroecological zone "R5 S4" while a moderate degree of transferability is possible within "R5 S3" and "R6 S2" according to Panabokke, wherever such zones are demarcated.

12.5 AGROECOLOGICAL ZONES OF KERALA

The National Commission on Agriculture or NCA (1973) delineated the whole country into different rainfall zones as A,B, C, D, and E, and tried to explain different cropping systems based on rainfall. The classification of agroecological zones based on monthly rainfall is as follows:

A: above 30 cm rainfall per month
B: 20–30 cm rainfall per month
C: 10–20 cm rainfall per month
D: 5–10 cm rainfall per month, and
E: below 5 cm rainfall per month.

Three rainfall seasons of four months each were broadly defined on the basis of rainfall distribution. They are as follows:

(a) Pre-monsoon season—February to May
(b) Monsoon season—June to September
(c) Post-monsoon season—October to January.

A subscript number with the designation letters would denote the number of months for which rainfall is that particular quantum. The four monsoon months (June–September) are put in parantheses. Nine rainfall zones, five cropping patterns and one livestock pattern had been identified for the State of Kerala. In Kerala, the following agroecological zone can be cited as examples:

E2C1B1(A3B1)B1C1E2

It denotes that during the pre-monsoon season, February and March record a rainfall amount of less than 5 cm, April has 10–20 cm rainfall and May has 20–30 cm rainfall. June, July and August receive more than 30 cm rainfall and September 20–30 cm rainfall during the monsoon season. October and November receive 20–30 and 10–20 cm rainfall respectively during the post-monsoon season while December and January less than 5 cm rainfall. The total of subscript number should be 12, indicating number of months in a year. Similarly cropping pattern was identified, taking into consideration the percentage area of the coverage of different crops. The National Commission on Agriculture concluded that Kerala is predominantly a Plantation State. An innovative alternative approach was suggested on the basis of altitude, soil, topography, rainfall and principal crops grown in the State of Kerala for delineating agroecological zones (Nair, 1974). Based on the above, Kerala has been divided into thirteen agroecological zones (see Table 12.6). The criteria chosen for identified parameters are as follows:

12.5.1 Altitude

Kerala is divided into two types as Type I and Type II:
Type I: 0–500 metres above the mean sea level; low altitude zone; hot humid tropics, spread over the entire State.
Type II: above 1000 metres from the mean sea level; high alitude zone; cool humid temperate climate, mainly in the high ranges of Idukki and Wayanad districts.

The region between 500 and 1000 metres above the mean sea level is left out as the area is covered by evergreen tropical forests and is not of direct significance to crop production, except that forest growth helps to regulate climate.

12.5.2 Topography

Model I: Extensive valleys with level but raised garden lands.
Model IIa: Valleys less extensive. Hills with moderate gradients and slopes having mild gradients.
Model IIb: Valleys less extensive. Hills with moderate gradients and with egg-shaped hump.
Model IIc: Valleys less extensive. Hills with table tops. Steep slopes.
Model III: Narrow valleys and hills with steep gradients. Steep slopes.

12.5.3 Soil Type

1. Alluvial soils (spread over river banks)
2. Sandy soils (coastal areas)
3. Sandy loam soils (coastal areas)
4. Laterite soil with well-defined B-horizon (natural midlands)
5. Laterite soil without B-horizon (natural highlands)
6. Red soil (southernmost Kerala)
7. Black soil (Chittor taluk of Palakkad district)
8. Peat (kari) soil (Kuttanad)
9. Acid-saline soil (Pokkali and Kaipad areas)

Table 12.6 Agroecological zones of Kerala

Zone	*Name*	*Area (in km)*	*Proportion (in per cent)*	*Altitude*	*Rainfall*	*Soil*	*Topography*	*Principal Crops*
I	Onattukara	519	1.6	Low	Pattern I	Sandy loam	Model I	Rice, Coconut, Tapioca, Arecanut
II	Coastal sandy	1564	4.8	Low	Pattern I	Sandy loam	Model I	Rice, Coconut
III	Southern Midland	3224	10.0	Low	Pattern I	Laterite without B-Horizon	Model III	Coconut, Rice, Tapioca, Arecanut
IV	Central Midland	2666	8.2	Low	Pattern I&II	Laterite	Model IIa	Rice, Coconut, Tapioca, Arecanut, Banana
V	Northern Midland	3765	11.6	Low	Pattern II	Laterite	Model IIb	Coconut, Rice, Arecanut, Cashewnut
VI	Malappuram	4254	13.1	Low	Pattern II	Laterite	Model IIc	Rice, Arecanut, Coconut, Cashew
VII	Malayoram	8861	27.4	Low	Pattern I	Laterite without B-horizon	Model III	Rubber, Coconut, Pepper, Rice
VIII	Palakkad	1280	3.9	Low	Pattern II	Red loam	Model II	Rice, Cotton, Groundnut, Millets
IX	Red loam	317	1.0	Low	Pattern I	Red loam	Model III	Coconut,Tapioca, Rice
X	Chittoor Block	508	1.6	Low	Pattern II	Black soil	Model IIa	Rice, Sugarcane, Cotton, Groundnut, Millets
XI	Kuttanad	284	0.9	Low	Pattern I	Peat (Kari)	Model I	Rice, Coconut
XII	River Bank Alluvium	–	–	Low	Pattern I	Alluvium	Model I	Rice, Coconut, Sugarcane, Arecanut
XIII	Highranges	5140	15.9	High	Pattern I&II	Red loam	Model III	Tea, Coffee, Cardamom, Rice
	Total	32283	100.0					

12.5.4 Rainfall

Pattern I: Both the southwest and northeast monsoons are active and moderately distributed. Southwest monsoon with June maximum rainfall.

Pattern II: Only southwest monsoon is active with July maximum rainfall. Northeast monsoon is relatively weak.

The above rainfall pattern is seen from south and north of Thrissur at about 11°N latitude. Pattern I is seen in the south while Pattern II is seen in the north of Kerala. It is an extensive scheme and perhaps the best for delineating agroecological zones, as all the variables that are relevant are taken into account (see Fig. 12.2). The principal crops that

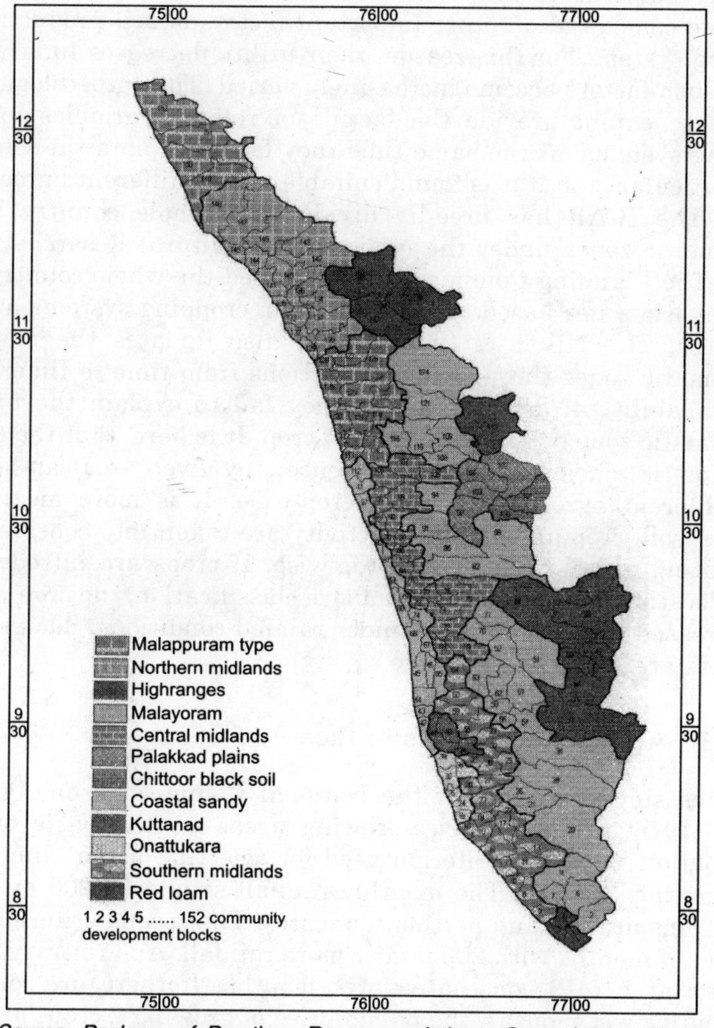

Source: Package of Practices Recommendations: Crops (12th edition), edited by A.I. Jose et al.

Fig. 12.2 Agroecological zones of Kerala.

are grown in each agroecological zone and other related details are also given in Table 12.6. The above classification can be further improved by superimposing moisture and thermal regimes in place of rainfall. On similar lines, the whole country has been divided into different agroecological zones based on the concept of Panabokke (Subramaniam, 1983).

12.6 AGROCLIMATIC CLASSIFICATION

The climatic classification based on standard climate indices such as average temperature and rainfall or monthly water balance may not take into consideration other important factors of soil productivity and cropping system. For this reason, their utility decreases in large areas, where such factors become increasingly varied. The agroecological zones to a large extent provide the factfile of related variables to explain cropping systems; at the same time they fail to explain the yield levels of a particular crop if it is found suitable in two different agroecological zones. The ICAR has broadly divided the whole country into 127 agroclimatic zones under the National Agricultural Resarch Project in 1980s. The Planning Commission has divided the whole country into 15 agroclimatic zones based on rainfall, soils, cropping systems and water resources. The NBSS & LUB (ICAR) also updates the mapping of agroclimatic zones through mid-corrections from time to time based on the availability of data. However, they fail to explain the productive agroclimatic zone or zones for a given crop. It is here, that the crop-wise agroclimatic zones will be handy. Hence, any given area can be defined into different agroclimatic zones crop-wise. It is more meaningful if climate, soil, crop and crop productivity are taken into consideration in delineating agroclimatic zones crop-wise. If crops are introduced in a given location on the basis of the above classification, the crop yields can be increased to a large extent under rainfed conditions. Such studies or concepts are scant in literature.

12.6.1 Agroclimatic Zones of Rice

On lines similar to that of the National Commission on Agriculture, Huke (1976) delineated rice-growing areas on the basis of rainfall distribution. The first criterion used for selecting major climatic zones was monthly rainfall. The monthly rainfall should be 200 mm to meet evapotranspiration and percolation losses. The second criterion was the number of months with 200 mm or more rainfall. An arbitrary boundary was set at 5 to 9 consecutive wet months. If there are less than 5 consecutive wet months, the possibilities of growing 2 crops are limited. If there are more than 9 consecutive wet months, the South Asian farmer is more likely to grow 2 crops of puddle rice. Based on the above concept, he grouped the rice-growing areas of India into three climatic regions. Based on the wet months criteria, four rainfall zones have been delineated. There are 4 subdivisions within the second zone and 2 in the third zone. Altogether, 10 zones have been defined. The fourth zone, comprising areas with less than 2 consecutive wet months, is not suitable for any type of agriculture unless supplementary irrigation is provided. Biswas (1982) and Sarkar and Biswas (1988) improved Hargreaves' climate classification to identify rice-growing areas. The whole country has been divided into 3 agroclimatic zones with respect to rice. The probability levels at which rice can be harvested successfully all the years have also been explained.

12.6.2 Agroclimatic Zones of Small Cardamom

The whole cardamom tract across the Western Ghats has been delineated into three agroclimatic zones based on climate indices viz. length of crop-growing season, available soil moisture (Ima), annual temperature range, soil type and productivity of cardamom. The production potential of cardamom over the Zone I is relatively better (>200 kgha^{-1}) when compared to that of Zones II and III across the Western Ghats, where the length of the crop-growing season is more than 300 days with annual Ima of more than 90%. The annual temperature range is relatively less (14.1°C) across Zone I. The production potential of small cardamom is low (100–150 kgha^{-1}) over Zone III (Karnataka) where the length of crop growing season was less than 250 days with annual Ima of 70–75%. The annual temperature range is also high (19.0°C), which may be detrimental in the case of cardamom (see Fig. 12.3).

Zone II falls under intermediary category (150–200 kgha^{-1}), where the length of the crop-growing season is more than 250 days with annual Ima varying between 80 and 85%. It is evident that Zones II and III can be made better in terms of production of small cardamom by improving *in situ* soil moisture conservation practices as it experiences unimodal rainfall. It is understood that Tamil Nadu (Thandikudi) and southern parts of Kerala (Pampadumpara) are relatively better suited for high cardamom yields as weather risk is less, whereas weather risk is more over the cardamom tract of Karnataka under the rainfed conditions.

12.6.3 Agroclimatic Zones of Cashew

A unique attempt has been made to demarcate "Agroclimatic zones for cashew" across the West and East Coasts of India on the basis of climate requirements, nutritional status of soils and cashew productivity (Rao, 2002). Cashew productivity can be improved towards the north across both the west and east coasts, while it may decrease towards inland plateau of peninsular India.

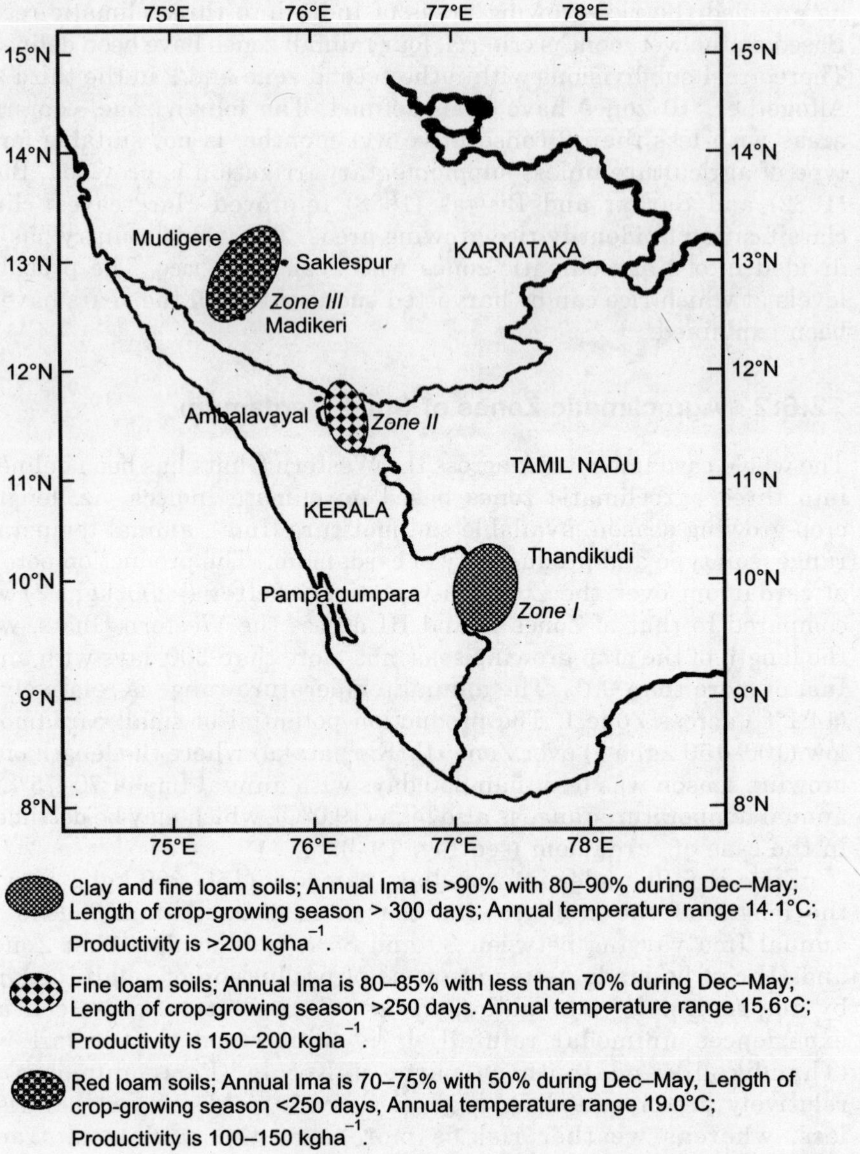

Fig. 12.3 Agroclimatic zones of small cardamom.

It is also seen that the extreme south of peninsular India across the west and east coasts may not be suitable for promoting cashew cultivation. At the same time, cashew cultivation could be promoted mostly by starting from the northern districts of Kerala to the Konkan region of Maharashtra along the West Coast. Across the East Coast, northern Tamil Nadu, Andhra Pradesh and Orissa are most favourable for cashew cultivation (see Fig. 12.4). West Bengal, northeastern states and the inland plateau of south peninsular India may not be conducive for obtaining better cashew yields. At Madakkathara in Kerala along the West Coast, high wind speed and maximum surface air temperature beyond 35°Celsius during the reproductive phase (February–April) may be detrimental for obtaining optimal yield and nut of good size and high quality in cashew. However, it needs further examination. The agroclimatological zones for cashew can be tested and improved further with more data gathered on weather, soil and genotype in detail, so as to suggest productive cashew areas across the country. It can be one of the tools for the planners while horizontal and vertical expansion of cashew programmes are taken up to sustain cashewnut production across the country.

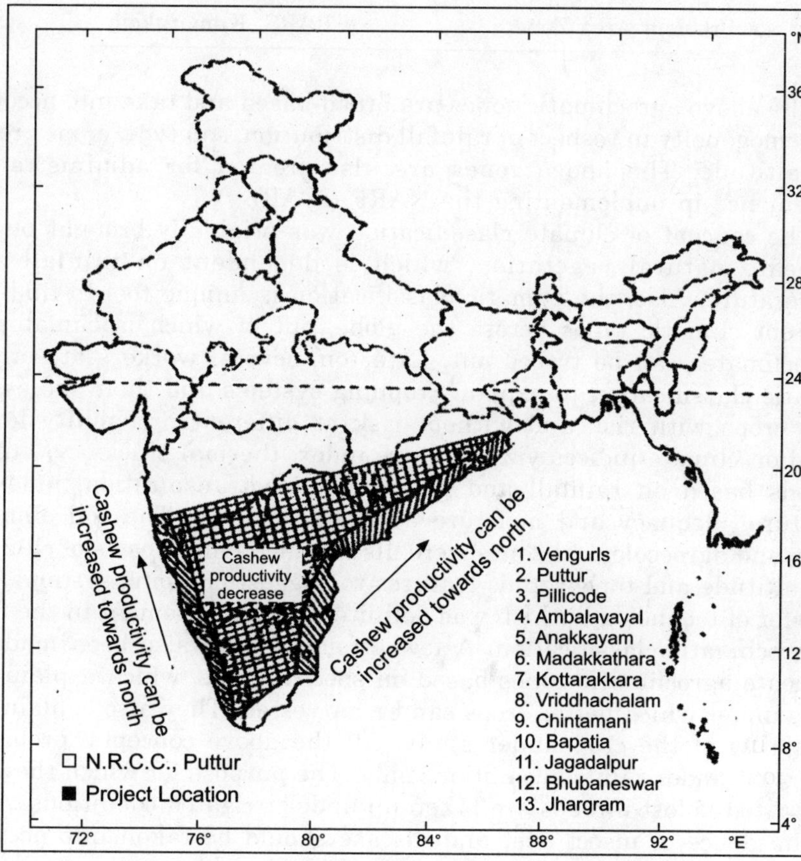

Fig. 12.4 Agroclimatological zones for cashew.

12.7 AGROCLIMATIC ZONES OF KERALA

Kerala has been divided into five agroclimatic zones under NARP (ICAR). The Northern Zone, Southern Zone, Central Zone, High Range Zone and

Problem Area Zone are the five agroclimatic zones spread across the State (see Table 12.7).

Table 12.7 Agroclimatic zones of Kerala (NARP, ICAR)

Name of the zone	*Headquarters*
Northern Zone	RARS, Pilicode
Central Zone	RARS, Pattambi
Southern Zone	Vellayani
High Range Zone	RARS, Ambalavayal
Problem Area Zone	RARS, Kumarakom

The above agroclimatic zones are broad-based and take into account the homogeneity in respect of rainfall distribution, soil type, crops grown and altitude. The above zones are also created for administrative convenience in implementing the NARP (ICAR).

The concept of climate classification was originally brought out to delineate natural vegetation, which is dependent on rainfall and temperature. Also, the climatic classification is unique tool to find out different climatic types across the globe, out of which isoclimates or homoclimates can be traced out. Later on, several workers attempted climatic classification to explain cropping systems and their success to grow crops with risk and without risk at different probability levels based on climate indices viz. moisture index, thermal efficiency, humid periods based on rainfall and potential evapotranspiration, index of moisture adequacy and moisture availability index. The soil climatic zones and agroecological zones were also defined on the basis of climate, soil, altitude and topographic features to suggest cropping systems and transfer of technology, which was definitely an improvement in the field of agroclimatic classification. A few attempts have also been made to delineate agroclimatic zones based on specific crops, which explain the areas under which better crops can be harvested. They also explain the suitability of the crop under study. All the above concepts, probably, hold good under rainfed conditions only. The purpose for which they are delineated is lost if crops are taken up under irrigated conditions. Also, the incidence of insect pest and disease should be taken into account while delineating agroclimatic zones crop-wise. However, studies on climatic or agroclimatic classification have not been seen in India for the last two decades. Hence, it is felt that agroclimatic classification crop-wise should be initiated by agrometeorologists with multi-disciplinary approach, as it has immense use in the field of agriculture to sustain production under rainfed conditions.

REFERENCES/SUGGESTED READINGS

Basu, D.N. and Guha, G.S., *Agroclimatic Regional Planning in India*, Concept Publishing, New Delhi, 1996, p. 330.

Biswas, B.C., 'Agroclimatic classification on the basis of moisture availability index and its application to dry farming tract of Gujarat', *Mausam*, **33**, 465–476, 1982.

Critchfield, H.J., *General Climatology*, Prentice-Hall of India, New Delhi, 1994, p. 453.

De Martonne, 'Une nouvella function climatologique l' indiced' aridite la', *Meteorologie*, pp. 449–458, 1928.

Emberger, L., 'Una classification biogeographique des climates', Recucil des travaux Fac. Sc. D l' univ. de Mon Pellier, Fasc. 7, 1955.

Hargreaves, G.H., *Water requirements manual for irrigated crops and rainfed agriculture*, Pub. No. 75-D-159, EMBRAPA and Utah State University, 1971.

Huke, R., 'Geography and climate of rice', *Climate and Rice*, International Rice Research Institute, Manila, pp. 31–50, 1976.

ICRISAT, *Climatic Classification: A Consultant Meeting*, International Crops Research Institute for Semi-Arid Tropics, Hyderabad, 1980, p. 153.

*Koppen, W., '*Versucheeiner klassification des klimate*', *Geogr. Zeitschs, Band* **6**, 593–611, 1900.

Koppen, W., *Die klimate der Erde* – Berlin, walter de Gruyter, 1923.

Krishnan, A. and Singh, M., 'Soil climatic zones in relation to cropping patterns', *Proceedings of symposium on cropping patterns in India*, Indian Council of Agricultural Research, New Delhi, pp. 172–185, 1968.

Majundar, N.C., 'A new climatic index and its application to sub-classification of arid zones', *Proceedings of symposium on problems of Indian arid zones,* Jodhpur, pp. 626–638, 1964.

Murthy, R.S. and Pandey, S. 'Delineation of agroecological regions of India', *11th Congress of the International Society of Soil Science,* Edmonton, Canada, 1978.

Nair, K.N.S., *Agroclimatic Zones and Cropping Patterns of Kerala*, State Planning Board, Govt. of Kerala, 1974.

NCA, *Rainfall and Cropping Patterns of Kerala*, National Commission on Agriculture, Govt. of India, New Delhi, 1973.

Panabokke, C.R., *Agroecological Zones of South and Southeast Asia*, FAO regional office for Asia and the Far East, Bangkok, 1979.

Rao, G.S.L.H.V.P., 'Agroclimatic zones', *Climate and Cashew*, Kerala Agricultural University, Thrissur, 2002, pp. 89–94.

Sarkar, R.P. and Biswas, B.C., 'New approach to agroclimatic classification to find out crop potential', *Mausam*, **39**, 343–358, 1988.

Singh, R.P., Singh, A. and Ramakrishna, Y.S., 'Cropping patterns for drylands of India—An agroclimatic approach', *Ann. Arid Zone*, **13**(2), 145–164, 1974.

Stenz, E., *The climate of Afghanisthan, its aridity, dryness and divisions*, Quart. Bull.Polish. Inst. Arts and Science, America, 1946.

Subramaniam, A.R., 'Agroecological zones of India', *Bioclim. Ser. B.* 32, *Arch. Met. Geophys.*, 329–333, 1983.

Thornthwaite, C.W., 'An approach toward a rational classification of climate', *Geographical Rev.*, **38**(1), 55–94, 1948.

Thornthwaite, C.W. and Mather, J.R., *The Water Balance*, Centreton, N.J. Laboratory of Climatology, New Jersey, 1955.

Thornthwaite, C.W. and Mather, J.R., *Instructions and tables for computing potential evapotranspiration and the water balance*, Centreton, N.J. Laboratory of Climatology Publications, **3**, 185–311, 1957.

Troll, C., 'Seasonal climates of the earth', *World Maps of Climatology*, Springer-Verlag, Berlin, p. 28, 1965.

QUESTIONS

1. Define the following:
   (a) Climatic zone
   (b) Agroecological zone
2. Distinguish between:
   (a) Genetic and generic approaches
   (b) Agroclimatic and agroecological zones
   (c) Troll's and Hargreaves' climatic classification
   (d) Climatic and agroclimatic classification
   (e) Thornthwaite's and Koppen's climatic classification
   (f) Koppen's and Troll's climatic classification
3. Write short notes on:
   (a) Climatic classification
   (b) Moisture availability index
   (c) Importance of agroecological zones
   (d) Agroclimatic zones of cashew
   (e) Moisture index and index of moisture adequacy
   (f) Climatic classification of Kerala
   (g) Agroecological zones of Kerala
   (h) Agroclimatic zones of rice

4. Fill in the blanks:
   (a) ——— is known as the father of modern climates.
   (b) ——— introduced the water balance concept.
   (c) ——— defined moisture availability periods based on P and PE.
   (d) ——— used assured rainfall for MAI.
   (e) The rainfall zone 'A' is defined by the NCA as that with rainfall amount of ——— per month.
   (f) Kerala falls under ——— climate as per Koppen's classification.
   (g) Central India mostly falls under ——— climate as per Thornthwaite's classification.
   (h) Southern parts of Kerala fall under ——— pattern of rainfall.
   (i) A high degree of transferability is possible in ——— agroecological zone as per Panabokke.
   (j) ——— suggested the innovative approach for delineating agroecological zones of Kerala.
   (k) Huke suggested ——— crops of puddled rice to the South Asian farmer.
   (l) Cashew productivity can be increased towards ——— across the west and east coasts of India.
5. Elucidate the following:
   (a) Koppen's climatic classification
   (b) Thornthwaite's climatic classification

13 WEATHER FORECASTING

The need for weather forecasting and its relevance in the field of agriculture are traditionally known to Indian farmers since time immemorial, as only on-time rains help in successful cultivation of crops. Despite effective weather services made available through different means, farmers do practice agricultural operations based on local myths and beliefs that are in vogue in different parts of the country. These are known as "weather- and agriculture-related idioms". Even today, farmers do pray to rain gods and goddesses in different forms, depending upon religion, for benevolent effects of weather so as to get good harvest. Recently also, during the all-India drought year 2002, when monsoon failed across the country, prayers were arranged by different religious groups for rains. It is high time that the weather scientists go through local beliefs that are in force in different parts of the country as well among different communities, and probe whether there is any scientific basis behind such local beliefs for agricultural operations, in addition to the regular weather forecasts that are being issued. Since its inception in 1875, the India Meteorological Department, which is the only agency in the field of weather forecasting, has strived to serve the agricultural community as a whole through different types of weather forecasting. In addition to the above, weather forecasting services to various purposes and organisations are issued by the IMD.

India has been divided into 36 meteorological subdivisions based on homogeneity in rainfall distribution. The aim of the department is to forecast daily weather meteorological subdivision-wise and communicate the same in time to the users, depending upon the type of forecasting. It has six Regional Meteorological Centres to serve the regional weather forecast needs. Most of the State Capitals in India also have the State Meteorological Centres to serve the weather forecast needs of the States, and these Centres issue Farmers' Weather Bulletins. The bulletins contain a forecast on weather elements valid for 36 hours, and an outlook for the subsequent two days. The weather forecasting services to various fields is depicted in Fig.13.1. The daily weather forecast is prepared for the entire country by the weather forecasting division, IMD Pune, and communicated to Mausam Bhavan, New Delhi for dissemination across the country (see Fig. 13.2). There is a vast improvement in the precision level of short-range weather forecasting with that of advent of supercomputers along with that of radar and satellite information.

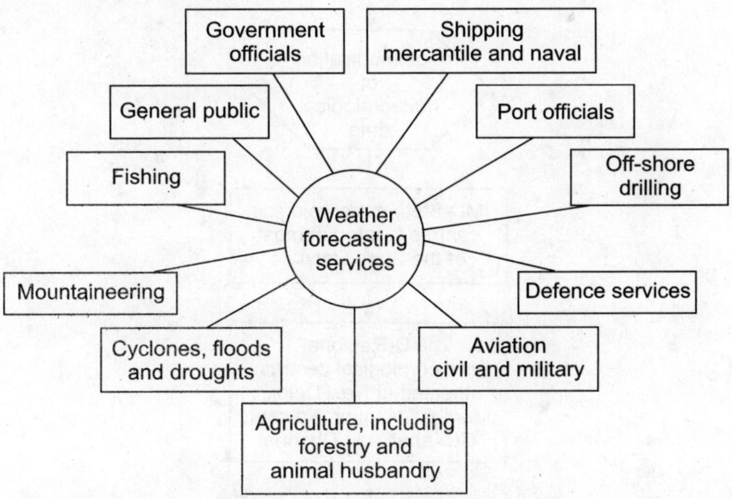

Fig. 13.1 Weather forecasting services.

13.1 WEATHER FORECASTING AND AGRICULTURE

The conditions of development, growth and yield of different crops depend on a chain of factors which involve seed, soil health, cultural and agronomic practices, nutrient management, agrotechniques, weather and farmer. Among different factors that influence crop production, weather plays a decisive role, as aberrations in it alone explain up to 50% variation in crop production. It is understood that reliable weather forecast can minimise crop losses to a considerable extent. It is also well-understood that Indian foodgrains production is expected to be as per the estimates of the Government of India, provided monsoon behaviour is normal. The loss in Indian foodgrains production is estimated to be around 20% due to bad monsoon 2002, characterised by failure in rainfall in July across the country. The troughs and peaks in foodgrains production continue to exist from year to year (refer to Fig. 1.4) despite technological improvement in agricultural inputs such as irrigation facilities, high-yielding crop varieties, adoption of plant protection measures and nutrient management of soils. The British saying "Indian agriculture is a gamble of the monsoons" still holds good inspite of technical advancement in the field of Indian foodgrains production, as only good monsoon yields good crop and vice versa. All these indicate that a sound mechanism has to

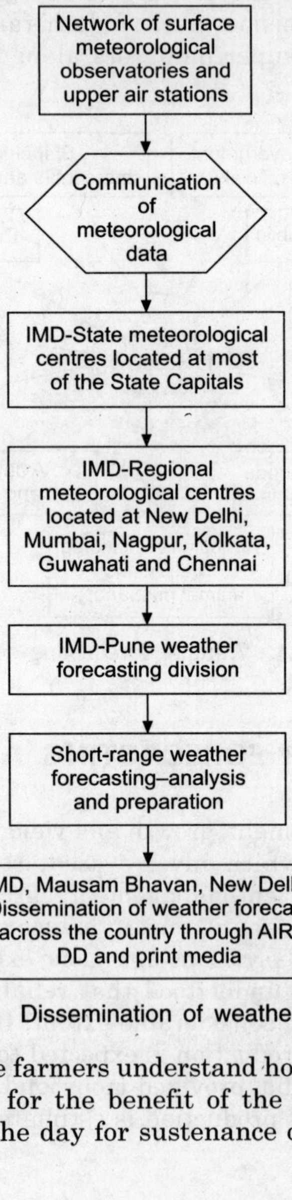

Fig. 13.2 Dissemination of weather forecast.

be developed to make the farmers understand how the services of weather agencies could be used for the benefit of the farmers. Weather-tuned farming is the need of the day for sustenance of agriculture production in the country. It is possible to sustain effectively if information on weather forecasting is made reliable and disseminated appropriately. Hence it is realised that better understanding of the weather and its variability on the overall effect of the different processes contributing to agricultural production is necessary to adopt appropriate cropping systems as well as soil and water management practices. It is in this context that the importance of weather forecasting is felt for the benefit of the farming community.

The objective of providing weather forecast is to disseminate information to the farmers on real-time basis, so that they can utilise the beneficial aspects of weather information to increase crop production, or to minimise the damage which may be caused directly or indirectly by unfavourable weather. Some of the crop pest and disease incidence also depends on weather. The crop output could be increased and recurring losses minimised if reliable forecast on incidence of pests and diseases is given in a timely manner on the basis of weather variables. Hence, weather forecast plays a key role in sustaining agricultural production and would help in accomplishing the task of increasing foodgrains production as planned by the Government of India. This could be possible if information is provided to the farmers, in advance, on the type of weather situations likely to be encountered and methods to be adopted for efficient management of inputs. Farmers need intensive training on various aspects of climate and weather which influence crop growth and its development. The institutions at national and state levels should take initiative in training the farming community on "Weather-tuned farming". One of the present effective tools in this direction is *weekly agromet advisory service*, the details of which are dealt with in the section entitled *Weather service to farmers*.

The economic importance and benefits of weather forecasts are given in Tables 13.1 and 13.2.

Table 13.1 Economic importance of weather forecasts in the protection of animals against hazards (WMO, 1987)

Hazard	*Rating of Importance*					
	Little	*Some*	*Much*	*Very much*	*No. of replies*	*Total score*
Cold stress /Wind chill	51	39	30	21	141	303
Heat stress	44	49	14	23	130	276
Water storage	62	32	40	15	149	306
Storms (wind, rain, snow)	55	34	29	19	137	302
Diseases	60	54	17	11	142	263

Table 13.2 Qualitative economic benefit of weather forecasts (WMO, 1987)

Rating	*Much (3)*	*Some (2)*	*Little (1)*	*None (0)*	*Total score*
Farmers	45.9	43.5	8.2	2.4	233
Market gardeners	45.1	35.5	19.4	–	226
Stud farmer	18.2	68.2	9.1	4.5	200
Commercial and local authority*	37.8	38.8	14.3	9.1	191

*Average of 10 sectors

13.2 TYPES OF WEATHER FORECASTING

The types of weather forecasting generated by the institutes in India, its validity and the present forecast skill are schematically presented in Fig. 13.3. All the four types of weather forecasting are one way or other useful from the agricultural point of view. However, the reliability of weather forecasting decreases with increase in forecast time and decrease in area, as the forecast models are not location-specific and the forecast is poor if it is for three days and beyond.

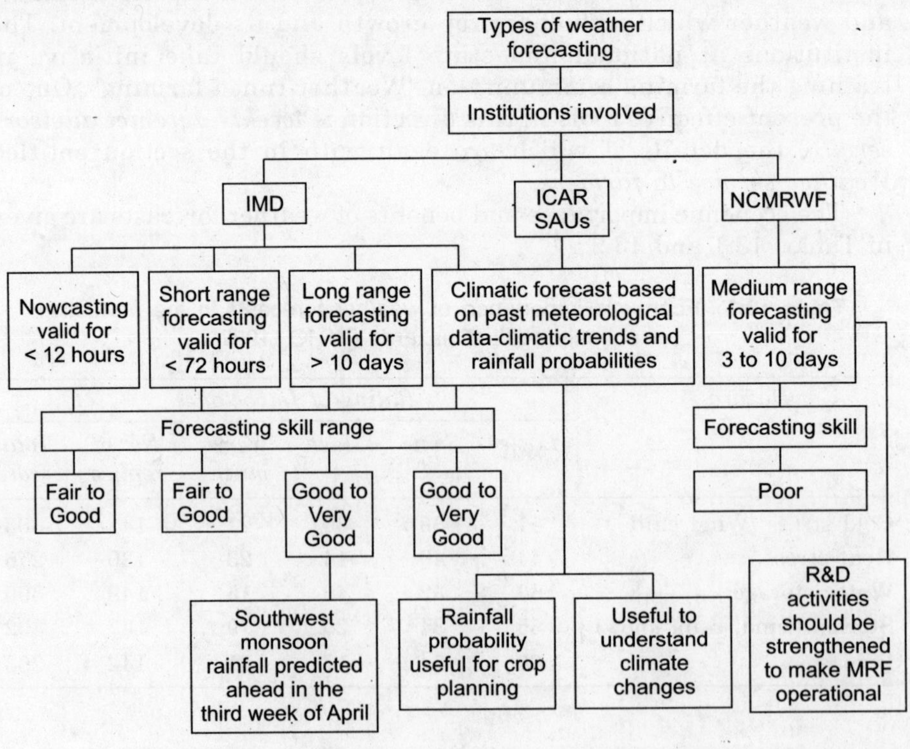

Fig. 13.3 Types of weather forecasting, its validity, skill and utility.

13.2.1 Nowcasting

This type of weather forecasting is issued for less than 12 hours. It is closely related to local weather phenomena like thunderstorms, duststorms, cyclones, and cold and heat waves. Nowcasting is also useful for outdoor functions.

13.2.2 Short Range Forecasting (SRF)

It comprises a forecast and warning of weather elements hazardous to agriculture which is valid for 36 hours, and an outlook for the subsequent 2 days. The short-range forecast includes cloud spread, rainfall distribution, heavy rainfall warnings, maximum and minimum temperatures, heat and cold waves, low-pressure areas, cyclone warning, hail or thunderstorm, duststorm, snow, frost and likelihood of maximum wind speed. The short-range forecast is issued twice a day on the basis of synoptic conditions. Though the short-range forecast is useful in weather-based agricultural operations, the reaction time to the farmers is too short for preventive measures against adverse weather. Thus, forecasting in the medium and long ranges is important.

13.2.3 Medium Range Forecasting (MRF)

It is a forecast and warning of weather elements hazardous to agriculture, which is valid for 3 to 10 days. This forecast includes cloud amount, rainfall, maximum and minimum temperatures, average wind speed, wind direction and weekly cumulative rainfall. The medium-range weather forecast is a challenging one to weather scientists, as it involves enormous numerical computations requiring expertise in weather science. The National Centre for Medium Range Weather Forecasting (NCMRWF) was established in New Delhi by the Government of India during 1988, to develop only medium-range weather forecasting for 3 to 10 days ahead and disseminate the same in time. This is a part of the programme for developing agrometeorological advisory services across the country based on medium-range weather forecasting.

13.2.4 Long Range Forecasting (LRF)

It is a forecast for more than 10 days, or a month or a season. The India Meteorological Department revived issuing the long-range weather forecast in 1988 on total monsoon rainfall of the country by the end of May. These forecasts can be used for predicting likely trends in foodgrains production of India before the beginning of the kharif season, as foodgrains production depends mostly on the distribution and amount of monsoon rain across the country. The cultivable cropped area depends on monsoon rain and its distribution. These forecasts can hold the foodgrain prices in check through buffer stock operations. In good monsoon years when prices fall, the government may step in and buy, and in bad years when prices tend to rise, it may unload a part of what it had purchased in the past. This will help in price stability, less financial loss, better distribution of the available foodgrains, and its import and export policies. With this in view, the crop weather watch group (CWWG) in the Agriculture Ministry, Government of India, closely follows the forecast and its impact on foodgrains production during the *kharif* season. The all-India drought during 2002 was managed well by the Government of India with an effective planning system of distributing foodgrains across the drought-affected States.

13.2.5 Climate/Agroclimate Forecasting

It requires past meteorological data for a good number of years (say 30–50 years). The trends in rainfall, its variability and the probability of distribution of rainfall over a season can be determined weekly by using the historical data of rainfall for a given location. This information is useful to crop planners and farmers, as crop growth periods can be adjusted under rainfed conditions depending upon rainfall probabilities. The climate trends will also help in understanding the impact of climate variability on agricultural production over a period of time. The periods of water surplus and deficit could be worked out for a particular location on the basis of past meteorological data (rainfall and evaporation) through the water balance procedure. This information can be used for irrigation scheduling during dry spells and for providing drainage in excess-rainfall or water-surplus periods. Analogous techniques can be tested for daily rainfall forecasting during the monsoon in the humid tropics location-wise, wherever long series of daily rainfall data are available and monsoon is stable, like in Kerala. This technique can be used by the end of July to predict rainfall during August and September. Global warming is another area of interest, which can be studied on the basis of Long Period Averages. Reports indicate that the year 1998 was the weather-related disaster year all over the world. It was attributed to global warming. The crop-weather relationships could also be well understood based on past biological and meteorological data, which in turn will help in mitigating the ill-effects of weather aberrations through better agronomic practices. In this direction, research programmes should be strengthened in the field of agricultural meteorology at the ICAR and SAU levels, with technical personnel having relevant subject background and commitment.

13.3 TOOLS OF WEATHER FORECASTING

Weather forecasting requires several basic inputs. They are: (i) area coverage under the network of meteorological stations, (ii) collection of meteorological data on cloud cover, sunshine, radiation, rainfall, temperature, wind and pressure, and (iii) upper air data on wind, humidity and temperature at various heights. These data, collected simultaneously across the country, are coded and plotted on surface and upper air charts. These charts are known as *synoptic charts*. The synoptic charts together with the satellite pictures on cloud spread are the basic materials needed to understand the initial state of the atmosphere. At the same time, the weather forecaster also needs information on the state and stage of crops to give weather forewarnings, for which the crop weather calendars become handy. In this context, students must have some knowledge of ground-based space tools, radar and satellites used for measuring upper air data. Also, one should know how the crop weather calendars come handy in operational weather forecasting with reference to crops.

13.3.1 Pilot Balloons

A small balloon is inflated with a gas (hydrogen or helium) lighter than air and released for measurement of wind direction and speed at different heights in the atmosphere by using a theodolite. The theodolite is an optical telescope which measures horizontal and vertical angles of the ascending balloons at known intervals of time. The path of the balloon is determined from the data observed through graphical techniques. The main difficulty with an optical theodolite is encountered when the sky is obscured, because the balloon is soon lost and can be no longer traced. This difficulty could be overcome by using radio-theodolites where radio signals are used to track the balloon. How the track of balloon is carried away by winds during monsoon is depicted in Fig. 13.4.

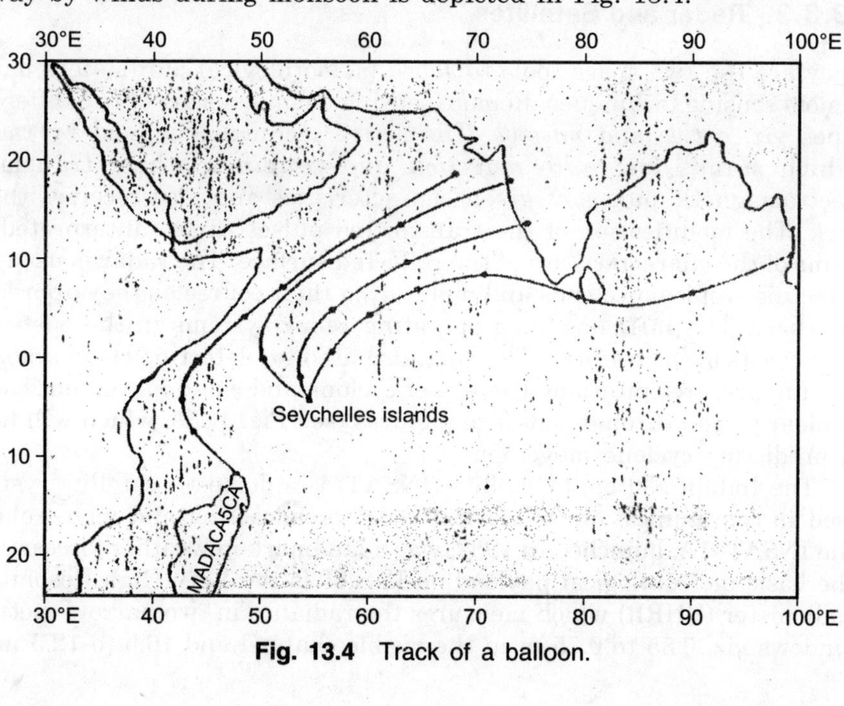

Fig. 13.4 Track of a balloon.

To measure temperature, humidity and pressure along winds at different heights in the atmosphere, meteorographs were attached in the past to the ascending balloons. The meteorograph is a combination of barograph, thermograph and hygrograph. The success of observations taken by using the meteorograph depends on its coming back to earth intact and reaching the sender. Once the balloon bursts due to its own volume expansion at greater heights, the meteorograph falls in unknown locations, which may not be accessible to the sender at all. Many times they do not reach the sender in time, and hence the utility of data recorded by meteorographs in weather forecasting is very much limited. To overcome this difficulty, radiosonde was introduced in place of meteorographs.

13.3.2 Radiosonde

The radiosonde is a device with sensitive sensors which transmit observations by radio signals to a ground receiver as the balloon ascends. It is carried aloft by means of a free balloon. The instrument measures temperature, humidity and pressure at various levels of the atmosphere through which it passes. The meteorological instruments contained in the radiosonde are designed to register temperature ranging from about +40 to –90°C, relative humidity from 15 to 100% and barometric pressure from 1040 to 40 hPa or less. The advantage of the radiosonde is that the data are immediately available and useful in weather forecasting.

13.3.3 Radar and Satellites

They are the two space tools to sense meteorological parameters using remote sensing techniques. Remote sensing techniques are of two different types viz. *active* and *passive*. The active technique is used by radar (which stands for *radio detection and ranging*), which transmits electromagnetic pulses of given characteristics and also receives them back. The modification of the transmitted pulses can be interpreted in terms of the characteristics of the reflected surface. The radar is used for detecting tropical cyclones and monitoring their course as they approach the coast. The IMD has been operating radar systems in the east and west coasts of the country. They are also used for detecting cloud amount and intense precipitation. The eye of cyclone and cloud movement could be clearly seen through satellite pictures (see Fig. 13.5), which will help in predicting cyclone movement.

The Indian National Satellite (INSAT) is a domestic satellite system used in communication, television, radio broadcasting and meteorology. The INSAT–1A, launched in 1982, was a landmark in weather forecasting. The basic meteorological payload on INSAT is the Very High Resolution Radiometer (VHRR) which measures the radiance in two narrow spectral windows viz. 0.55 to 0.75 m in the visible channel and 10.5 to 12.5 m in the thermal infrared band. The former is a measure of the solar radiance reflected from the earth's surface on cloud tops, that is, the brightness of these surfaces, whereas the thermal infrared measurement gives an indication of the temperature of the earth's surface and cloud tops, assuming them to be acting as blackbodies. Hence, the INSAT series give information on cloud spread, amount and information on the temperature of the earth's surface and cloud tops. This information is used in short-range and medium-range weather forecasting. INSAT-3A was launched on 10th April, 2003 for meteorology and telecommunication purposes.

Fig. 13.5 Eye of the cyclone.

13.3.4 Synoptic Chart

An enormous volume of meteorological data is being collected from all over the world, continuously round the clock, through various telecommunication channels. To assess, assimilate and analyse the vast data, they have to be suitably presented. For this purpose, the observations are plotted on maps in standard weather codes. These maps are called *synoptic maps* or *charts*. The surface and upper air charts are the two types of synoptic charts currently in use.

Synoptic charts display the weather conditions at a specified time over a large geographical area. The surface synoptic charts plotted for different synoptic hours (00, 03, 06, 09, 12, 15, 18 and 21 UTC) depict the distribution of pressure, temperature, dew point, clouds, winds, and present and past weather. In place of GMT, Universal Time Co-ordinate (UTC) is used. The upper air charts are also prepared at the standard pressure levels of the atmosphere (at different heights) of the atmosphere, wherein the pressure, wind and temperature are plotted. The surface charts together with the upper air charts provide a composite three-dimensional weather picture pertaining to a given time.

The surface synoptic charts are the most used charts. It contains the maximum number of observations with the largest number of parameters plotted, and often forms the base on which the pressure level charts are built up. The pattern of pressure distribution is brought out by drawing isobars, troughs, ridges, lows, highs, depressions, cyclones, cols, fronts and discontinuities. These systems are clearly marked and labelled by using appropriate symbols and colours.

The meteorological observations are plotted according to standard station models prescribed by the World Meteorological Organisation (WMO). The following are some of the codes used for plotting meteorological parameters in synoptic charts:

TTT	:	Dry bulb temperature in °C;
TdTdTd	:	Dew point temperature in °C (plotted in red colour);
TxTxTx	:	Maximum temperature;
TnTnTn	:	Minimum temperature;
ww	:	Present weather;
VV	:	Visibility;
PPP	:	Barometric pressure reduced to sea level;
ddd ff	:	Wind direction and speed;
C_L	:	Types of low clouds;
C_M	:	Types of medium clouds;
C_H	:	Types of high clouds (plotted in red colour);
N	:	Total amount of cloud;
h	:	Height of the base of low cloud;
Nh	:	Amount of cloud reported under h;
RRR	:	Amount of rainfall in mm (plotted in green colour).

There are small differences in the plotting model (see Fig. 13.6) depending upon the type of station viz. hilly, inland, coastal and ship stations.

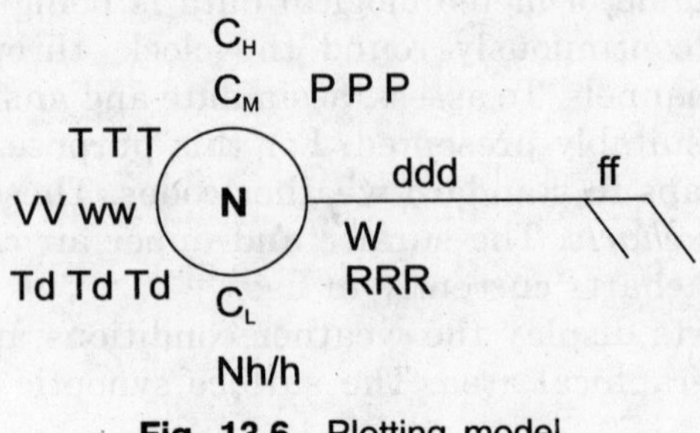

Fig. 13.6 Plotting model.

13.3.5 Crop Weather Calendar

In order to provide the farmer with an efficient weather service, it is essential that the weather forecaster should be familiar with the crops that are grown in a particular agroclimatic zone. The type of forewarning to be given, depending upon the state and stages of the crop, should also be known. In case of farmers, they should become familiar with weather bulletins and learn how to interpret them. To meet the above requirement, detailed information collected from the agricultural departments has been condensed by the IMD and presented in a pictorial form known as *crop weather calendar*.

The crop weather calendar consists of three parts viz. crop husbandry, climate requirements and weather warnings. The important crop phases like sowing, germination, transplantation (in the case of rice), tillering, elongation, flowering, grain formation and harvest are indicated under crop husbandry against the standard agrometeorological weeks in the lower part of the calendar (see Fig. 13.7). The normal monthly rainfall and number of rainy days are depicted in the middle of the calendar with weather requirements, and the uppermost portion of the calendar indicates the nature of the weather warnings to be issued in different crop periods. These crop weather calendars help the weather forecaster to see at a glance the type of weather warning that are to be issued, for a particular district in a given weather situation during a particular phase of the crop. They are of equal interest to the farmers for better crop management. However, they fail to produce the desired results, since they are not used as handy tools in practice and became obsolete with fast-changing crop varieties. Also, the crop weather calendars prepared district-wise may not be of much use at the micro-level. Hence, the crop weather calendars are to be prepared at the village panchayat level by the agrometeorological field units (AMFUs), established in different agroclimatic zones, through research and development activities taken up for major crops and varieties. They help in improving the quality of agroadvisory services based on medium range weather forecasting.

13.3.6 Crop Weather Diagram

It gives season-wise information on crop husbandry (tillage to harvest), actual weather and normal weather month- or week-wise, and information on pest and disease incidence. It may help in understanding favourable weather conditions that are responsible for better crop yields and vice versa, if they are prepared continuously for a number of years. Using the crop weather diagrams, attempts can be made for obtaining better crop yields through agronomic manipulations, and one can also predict crop yields qualitatively on the basis of weather conditions. It is also possible to select a suitable variety for a given location based on crop weather diagrams.

The main difference between crop weather diagram and crop weather calendar is that the former indicates the current state of information on

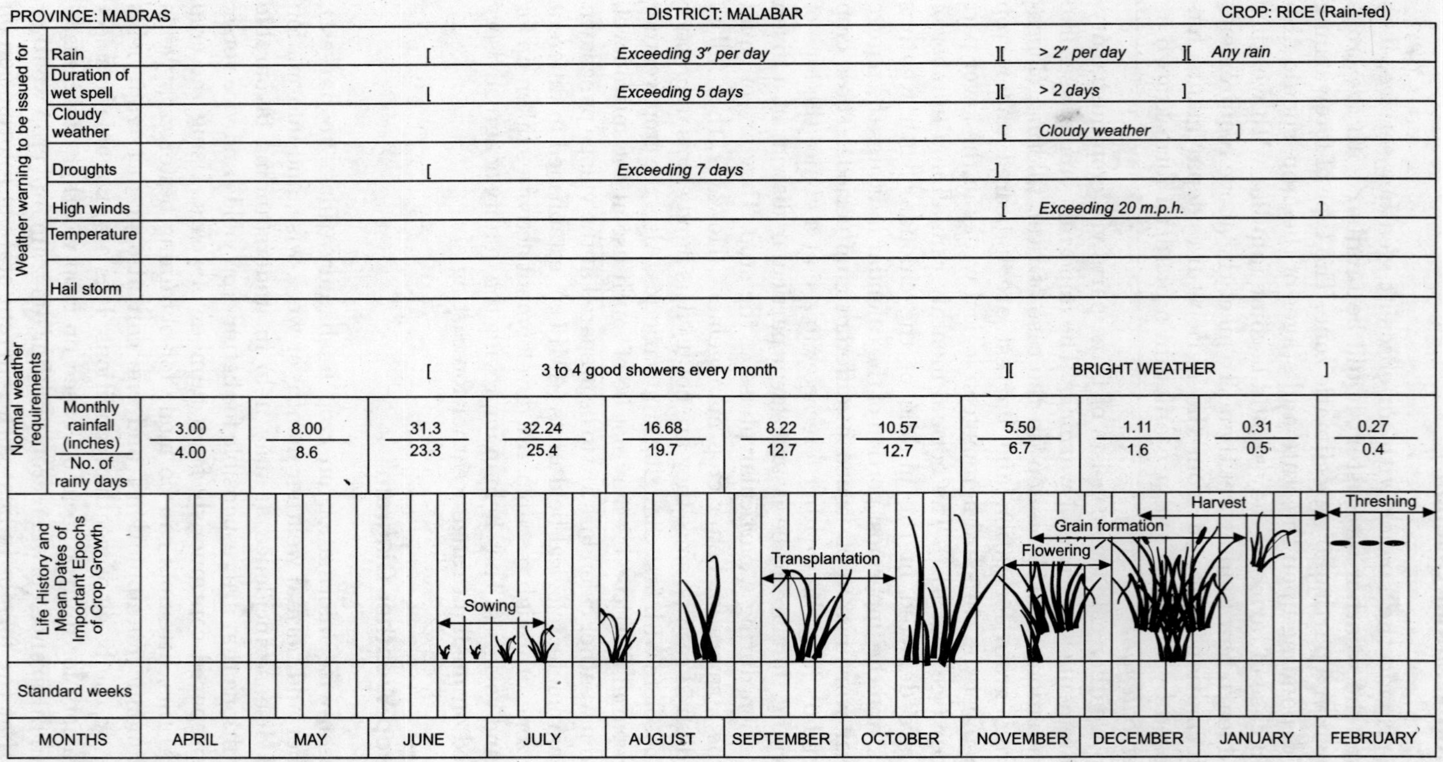

Source: Division of Agricultural Meteorology, IMD, Pune

Fig. 13.7 Crop weather calendar of paddy.

crops and weather, while the latter indicates the mean state of information on crops and weather—including forewarnings and the periods during which they are to be issued. The crop weather diagram can be a tool to assess the crop condition and its yield in relation to weather, while the crop weather calendar is a tool for a weather forecaster for providing efficient weather service. Of course, the mean state of crop weather diagrams should be a basis for improving the quality of crop weather calendars. Both are handy and useful in agroadvisory, which helps to improve crop yields through better agronomic practices. As an example, the crop weather diagram of cashew grown in the west coast is depicted in Fig. 13.8. It can be used in understanding the reproductive phase of different varieties and their variations from one season to another, if prepared every year. The qualitative assessment also can be made on crop yield, depending upon changes that take place in weather.

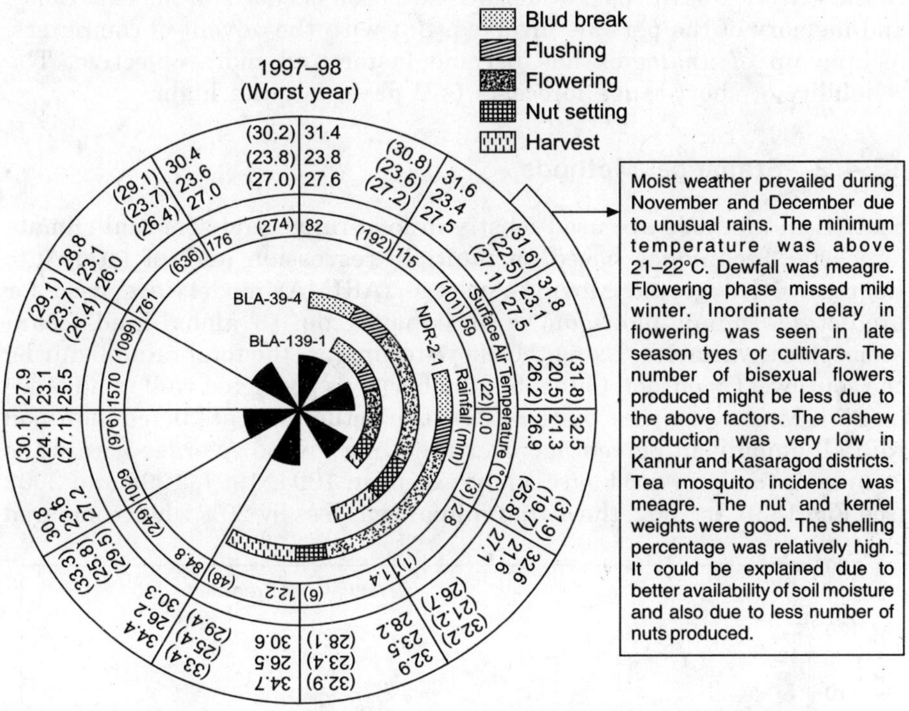

Fig. 13.8 Crop weather diagram of cashew at RARS, Pilicode.

13.4 METHODS OF WEATHER FORECASTING

The methods in use for weather forecasting can be divided into three broad categories viz. conventional or synoptic, statistical and numerical weather prediction.

13.4.1 Synoptic Methods

The state of atmosphere over an area is known at an instant through a set of meteorological variables viz. rainfall, maximum and minimum temperatures, and wind and pressure systems measured simultaneously at various locations. These observations are called *synoptic observations*. Using the observations recorded simultaneously, surface and upper air charts are prepared. These give the present state of atmosphere. The inferences on expected movement of weather systems are drawn by using the previous and present charts. In addition to the synoptic charts, satellite pictures also supply considerable information on cloud spread to the forecaster in issuing short-range forecast. Short-range forecasts are issued by using the past analogous situations on the presumption that a current situation will evolve on the lines of the past analogous one. Often, selection of the past analogous situation is based on the experience and memory of the persons involved. But with the advent of computers, picking up of analogies has become faster and more objective. The reliability of short-range forecasts (< 2 days) is quite high.

13.4.2 Statistical Methods

Statistical methods are used mostly in long-range and seasonal climate forecasts. Techniques based on multiple regression and, of late, Auto Regressive Integrated Moving Average (ARIMA) models are used for predicting Indian monsoon rainfall based on 16 global-land-ocean-atmospheric variables. Using the above technique, the total rainfall during the southwest monsoon (June–September) was predicted well in advance on 25th May every year by the IMD. The actual and predicted monsoon rainfall amount in percentage over the Long Period Average across the country is fairly in good agreement except in 1994, 1997, 1999 and 2002 (see Fig. 13.9). In 2002, the actual rainfall was less by 20% when compared

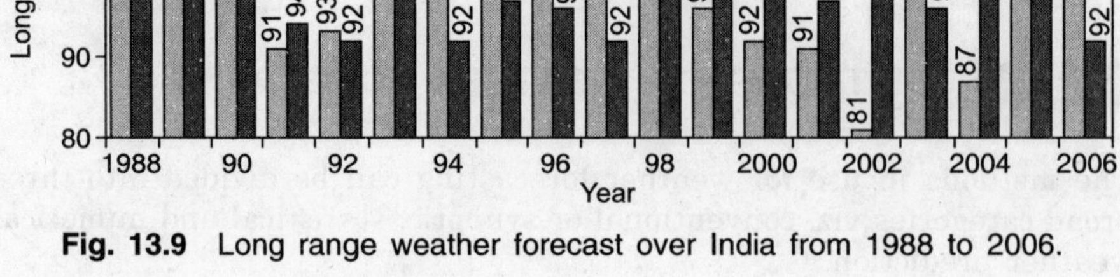

Fig. 13.9 Long range weather forecast over India from 1988 to 2006.

to the predicted rainfall. The long-range weather forecast was a failure for the first time in 2002, for which the causes were analysed and the model was improved by using 8 to 10 variables. The first forecast arrives in the 3rd week of April and the second in July.

The Long Period Average (LPA) rainfall for the period 1871–1984 during the southwest monsoon of the country is 853 mm, which is almost 80% of the annual rainfall. However, the ARIMA models fail on meteorological sub-division basis, though they give reasonably good results for the country as a whole. Therefore, a study was undertaken by Rao et al. (1994), to delineate the ambit of validity of myopic rainfall forecast for the country as a whole in agriculture planning and management over different meteorological subdivisions. It showed that the normal forecast is valid for the entire country except for western Rajastan, Saurashtra and Kutch regions, as the forecast reliability was less than 70%. The long-range weather forecast for the State of Kerala from 1988 to 2002 (see Fig. 13.10) showed that it needs improvement before the model is put in operation, as the model fails on regional scales. A new 8- or 10-parameter Power Regression Model is deviced in place of the 16-parameter model from 2003 onwards for estimating monsoon rainfall at the national level. It helps the IMD to advance the long range forecast by 40 days and generate a quantitative forcast update by mid-July, which will be useful for contingency planning.

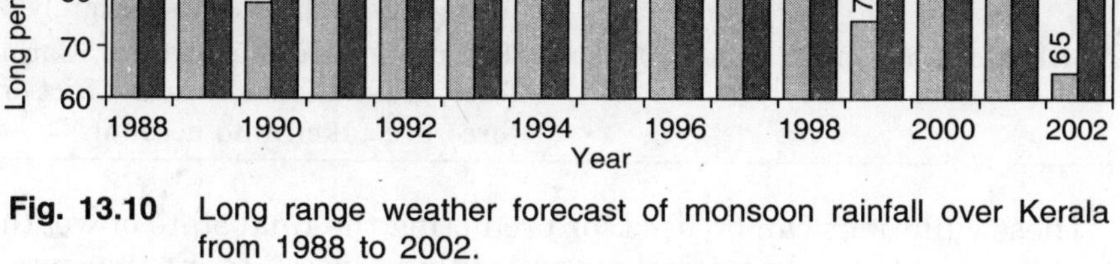

Fig. 13.10 Long range weather forecast of monsoon rainfall over Kerala from 1988 to 2002.

13.4.3 Numerical Weather Prediction (NWP)

The motion of atmosphere is primarily governed by distribution of radiation, which is uneven across land and ocean surfaces. Later, air motions are driven by several other forces viz. Coriolis force (due to earth's rotation), pressure gradient force (due to difference in pressure at two points), gravitational force (earth pulls air always downwards), and centrifugal and frictional forces which are acting very close to the surface of the earth. All the above forces can be expressed in terms of differential equations based on certain laws of physics. An integrated form of these equations is known as an *atmospheric model* which explains the evolution of atmosphere from a given state to some future state (Goswami, 1996 and 1997). The relevant laws of physics can be expressed in terms of seven equations, which govern the behaviour of seven atmospheric variables (see Table 13.3).

Table 13.3 Important atmospheric variables and equations

Variables	*Equations*
1. Pressure (P)	Gas law or equation of state. Relates temperature, pressure and density. $P = rRT$
2. Temperature (T)	First law of thermodynamics. Relates temperature changes to external heating or cooling, eddy diffusion and changes in pressure, external heating or cooling due to radiative balance, and latent heat released during cloud formation.
3. Density (r)	Continuity equation for air; expresses conservation of mass of air.
4. Water vapour (q)	Continuity equation for water vapour; expresses rate of change of water vapour in terms of source (evaporation), sink (precipitation) and diffusion due to turbulent eddies.
5. West to East component of wind (u)	Newton's second law Force = Mass × Acceleration
6. South to North component of wind (v)	Applied to West-East, South-North and vertical direction separately.
7. Vertical component of wind (w)	The pressure gradient force, Coriolis force, gravitational force and frictional force are taken into account.

These equations can be used for predicting the final state of weather through a purely mathematical process, taking into account the current state of weather in a fairly large number of synoptic stations of a given region. This is known as *numerical weather prediction*. The first attempt in this direction was made by L.F. Richardson, a British scientist in 1922. At present, this task is undertaken by the NCMRWF, Government of India, New Delhi by using supercomputers (CRAY–XMP), as the NWP requires enormous computations like 1.3 million points for the seven variables, and the forecast is to be produced well in advance. The NWP technique is used in medium-range weather forecasting. The numerical weather prediction has a practical limit up to 3 to 5 days in the tropical areas and less than 10 days in the middle latitudes. The NWP techniques are also applied to short-range forecast of meteorological parameters at the NWP unit of IMD, Pune. A CYPER–2000 supercomputer aids the computations. The dynamic models are also put in use for seasonal climate forecasts in various countries.

13.5 WEATHER SERVICE TO FARMERS

A reliable system of short-range and medium-range weather forecasts is needed for effective farm management practices such as cultivation of land, preparation of seed bed, planting, choice of crops in case of prolonged monsoon breaks, fertilizer management, harvesting of crops, post-harvest storage and transportation to markets. The inflicted wind damage is also not uncommon over the humid tropics (see Fig. 13.11). Through wind breaks and shelter belts, the damage can be minimised if the wind prone areas are demarcated.

Fig. 13.11 Inflicted wind damage during monsoon on 14-6-1996 at RARS, Pilicode.

The India Meteorological Department started issuing Farmers' Weather Bulletins (FWBs) on a regular basis in 1945, and later it was entrusted to Regional Meteorological Centres (New Delhi, Guwahati, Calcutta, Mumbai, Nagpur and Chennai) which contain a forecast of expected weather for the next 36 hours and an outlook for another 2 days. In due course, meteorological centres at State headquarters started issuing FWBs. All India Radio used to broadcast FWBs twice in a day in regional languages to reach the most remote corners. However, they are found to fall short of the requirements of farmers in terms of duration of forecast validity, capability for small areas and likely impact on agronomic operations. The crop weather calendars prepared for main crops lost their significance with the advent of fast-changing varieties. As they contain limited information, there was a need to frame them in relation to crops specific area-wise. Thus came the concept of Agrometeorological Advisory Services (AAS) which is in operation since 1977. However, the utility of the same is very much limited. Realising the defects and to make the AAS more meaningful, the National Centre for Medium Range Weather Forecasting (NCMRWF) took up the task in 1988 on the basis of medium-range weather forecasting.

13.5.1 Agromet Advisory Services (AAS)

The agroadvisory service based on medium-range weather forecasting (3 to 10 days) has been made operational by the National Centre for Medium Range Weather Forecasting (NCMRWF). The ICAR and State Agricultural Universities co-operate in this task. The NCMRWF has proposed to establish 127 agrometeorological field units (AMFUs) in all the agroclimatic zones which cover the entire country. 107 agroclimatic zones have been covered across the country till 2007. The agroclimatic zones were delineated by the ICAR under the National Agricultural Research Project (NARP). The AMFUs prepare AAS bulletins. These contain three parts. The first part of the AAS consists of weather events that occurred during the past week and the weather forecast for three days ahead. These forecasts are given to acquaint the farmer with the expected weather, and focus on cloud amount, rainfall, average wind speed, wind direction, maximum and minimum temperatures and weekly cumulative rainfall. The second part contains the factual information on varieties of crops, their state and stage, ongoing agricultural operations and incidence of insect pest and diseases of crops. The final and third part of the AAS bulletin provides information on agroadvisory. The bulletins are prepared by the Agrometeorological Advisory Board, consisting of scientists belonging to various disciplines and Agricultural Officers of the locality. Such AAS bulletins are being prepared and issued in English as well in local languages. The AMFUs functioning at several agroclimatic zones have made the weekly or biweekly AAS operational effectively. The NCMRWF provides information on MRF every Tuesday to all the established AMFUs and twice in a week (Tuesday and Friday) to only selected stations (AMFUs).

The IMD has also established Agromet Advisory Service Units (AASUs) at the State Meteorological Centres (SMC). The AASUs issue agromet advisories to the States on Monday and Thursday, based on short-range weather forecasting. These describe the weather over the State during the preceding week, the rainfall forecast valid for the next 48 hours and an outlook for 2 subsequent days. Information on crops is obtained from the State Agriculture Department. The schematic presentation of agroadvisory based on weather forecast is given in Fig. 13.12.

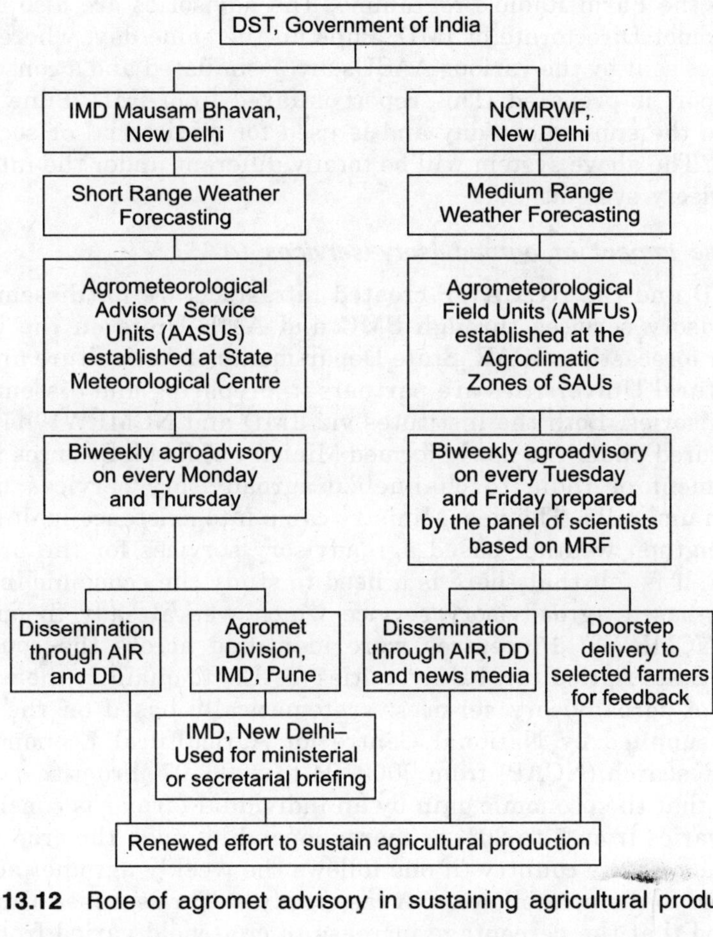

Fig. 13.12 Role of agromet advisory in sustaining agricultural production.

The above system is brought under the newly created Ministry of Earth Sciences (MoES) in which integrated agroadvisory services (IAAS) will be made operational with effect from the June 2008. It will be a five-tier system, taking into account the district as a unit for dissemination of agroadvisory service.

Dissemination of Agromet Advisory Services

The agroadvisory bulletins prepared weekly/biweekly by the AMFUs could be communicated on the same day or the next day to the selected farmers through a special messenger. The AIR, DD and the other communication facilities can be effectively used to disseminate the AAS. It is being done in several parts of the country. It is hoped that by the end of this decade, all the 127 agroclimatic zones will be covered under the scheme.

The agromet advisories prepared by the AASUs of the State Meteorological Centres are sent to the AIR stations for broadcasting through the Farm Radio Programme. The advisories are also faxed to the Agromet Directorate of IMD, Pune on the same day, where all the advisories sent by the various AASUs are assimilated and a consolidated brief report is prepared. This report is faxed from IMD, Pune to New Delhi on the subsequent day and is used for ministerial or secretarial briefing. The above system will be totally different under the integrated agroadvisory system.

Economic impact of agroadvisory services (AAS)

The IMD and the NCMRWF created infrastructure for disseminating agroadvisory services through SMC and AMF Units on the basis of weather forecasting. ICAR, State Departments of Agriculture and State Agricultural Universities are partners in preparing and disseminating agroadvisories. Both the institutes viz. IMD and NCMRWF have been restructured under the newly-formed Ministry of Earth Sciences (MoES), Government of India to channelize agroadvisory services under a common umbrella. The new Ministry came into existence in July 2006. To strengthen weather-based agroadvisory services for the benefit of farmers, it is felt that there is a need to study the economic impact of weather-based agroadvisory service which was already in operation under NCMRWF. 15 centres were identified across the country in various State Agricultural Universities (SAUs) to quantify the economic impact of agroadvisory services, systematically based on the master format supplied by National Centre for Agricultural Economics and Policy Research (NCAP) from 2003–04 to 2006–07. From the study, it is clear that the economic gain by an individual farmer is considerable and it varies from 5 to 20% or more, depending upon the crop and the season across the country (if one follows the weekly agromet advisory) compared to the non-followers. A case study at Thrissur (see Table 13.4) indicated that the percentage increase in crop yield varied from 6.4 to 19.0, depending upon crop and season, and 7.1 to 36.0 in terms of net return over non-AAS farmers. No doubt, the weather-based agroadvisory service has an overall beneficial effect, and farmers gain knowledge in addition to monetary benefit. At the same time, the benefits accrued against weather extremes through various farm packages, except in one or two cases like drought or strong winds, are not clearly brought out. Therefore, there is a need to bring out clearly the benevolent or malevolent effects of agroadvisory, in terms of occurrence of weather events and mitigation against adverse weather, rather than simple agronomic advisory.

Table 13.4 Impact of weekly AAS on crop yields from 2003–04 to 2005–06 at Thrissur, Kerala

Season	*Crop*	*Yield* (t/ha)		*Increase in yield over non-AAS farmers (in %)*	*Increase in net return over non-AAS farmers (in %)*
		AAS	*Non-AAS*		
Rabi 2003–04	Paddy	3.7	3.0	19.0	30.0
	Banana	31.8	27.9	12.2	11.5
	Coconut (nuts/ha)	13460	11025	18.1	13.2
Kharif 2004	Paddy	2.8	2.6	7.1	31.6
	Banana	22.8	20.5	10.1	7.9
Rabi 2004–05	Paddy	3.0	2.7	7.5	34.1
	Banana	31.3	29.3	6.4	11.9
Kharif 2005	Paddy	2.7	2.5	6.5	38.5
	Banana	25.3	22.4	11.2	7.1
Rabi 2005–06	Paddy	3.3	2.9	13.6	36.0
	Banana	27.8	24.5	11.8	12.3

Role of ICT in AAS dissemination

Dissemination of agromet advisory in time to the beneficiaries is very important. At present, under the “Experimental Agromet Advisory Services” project funded by MoES, the bulletins are handed over directly to the selected farmers. It is the only effective mechanism for getting feedback, so as to improve the quality of AAS. In addition, agro-advisory bulletins (AAB) have also been disseminated through print media (dailies), All India Radio and television programmes (local or regional cable TV, Doordarshan, etc.). Agencies like the IMD, NCMRWF, ICAR institutes and SAUs striving hard to disseminate bulletins by launching weather portals in order to reach the farmer community on near-real time basis. It has drawn the attention of a large segment of the user community. The web portals launched by various organizations across the country are shown in Table 13.5.

Table 13.5 Agroadvisory on various websites across the country

S. No.	*Website*	*Organization*
1.	www.ncmrwf.gov.in	NCMRWF, MoES, Noida
2.	www.indiaweatherwatch.org	Collaborative portal of DST, IMD, NCMRWF and IITM
3.	www.imdagrimet.org	IMD, Pune
4.	www.cropweatheroutlook.org	CRIDA, Hyderabad
5.	www.kau.edu	Kerala Agricultural University, Thrissur
6.	Department websites	Departments of Agricultural Meteorology in SAUs

In addition to the above, AMFU and District-level Crop Weather Watch Groups may attempt to link with Kissan Call Centres, District IT Cells, Virtual Universities, KVKs, NGOs and also Village Knowledge Centres (VKCs) and Village Resource Centres (VRCs), for effective online information. In this direction, the ICT tools will play a major role in the years to come. Extreme events like cyclones, floods, and heat and cold waves can be intimated through mobile Short Message Service (SMS) by connecting Agro-Meteorological Field Units with mobile network providers like BSNL, MTNL and private operators. With all the ICT technologies, if online messages are disseminated, the crop losses amounting to millions and billions of dollars can be avoided or minimised. It will promote the agricultural sector in a big way for the benefit of the farmers. It can be an essential in-built component of AAS in an agrarian country like India for sustenance of agricultural production.

The agroadvisory bulletins are prepared at the AMFUs by the panel of scientists on the basis of the "package of practices recommendations" developed by the SAUs for the respective zones. In the absence of a separate package based on R&D in operational agricultural meteorology, these are quite useful for agromet advisory survice. The essence of an agroadvisory is to create awareness on weather forecasting among the farmers and advise them on day-to-day farm operations for sustenance of agricultural production. This is possible if the interaction between weather variables on one hand and crop growth and its development on the other is well understood, for which crop growth simulation models have to be revalidated or developed for major crops. On similar lines, an interdisciplinary team should work to develop models on pest and disease dynamics. This needs strengthening of research and development activities of the AMFUs in every agroclimatic zone. In addition to that, the IMD in collaboration with the ICAR and SAUs should try to reorient the various research programmes going on under AICRP on agrometeorology in tune with AAS. Simultaneously, the skills of medium-range weather forecast must be improved further by the NCMRWF; as of now the forecast skill is rated in the range of *poor* to *fair*. The utility of agroadvisory based on medium-range weather forecast in the field of agriculture will be fruitful only if the forecast skill is improved for at least 3–5 days ahead, along with improvement in quality of AAS.

The crop weather watch groups (CWWGs) should be established at the district level, on similar lines to what we have today at the Centre, to integrate the various activities related to weather service to farmers at the micro-level. All the above related activities of weather forecasting should be brought under one organisation with commitment to improving "weather service to farmers", so as to minimise crop losses and increase farm output.

13.6 CHALLENGES IN WEATHER FORECASTING

Many times the weather forecast given by the weathermen has a question mark on its validity, as it is neither location-specific nor area-specific, and the present forecast skill is rated as *poor* for medium-range weather forecasting. Supercomputers came handy in solving equations much easier and faster. This will help to know the future state of the weather ahead, based on the initial state at the earliest. Then, why is the forecast skill poor for MRF? The answer is well documented in several articles (Das, 1981; Datta, 1989; Goswami, 1996 and 1997; Kulshrestha, 1995; and Raman, 1990). The reliability of weather forecasting is a complex issue, which depends on various components (see Fig. 13.13).

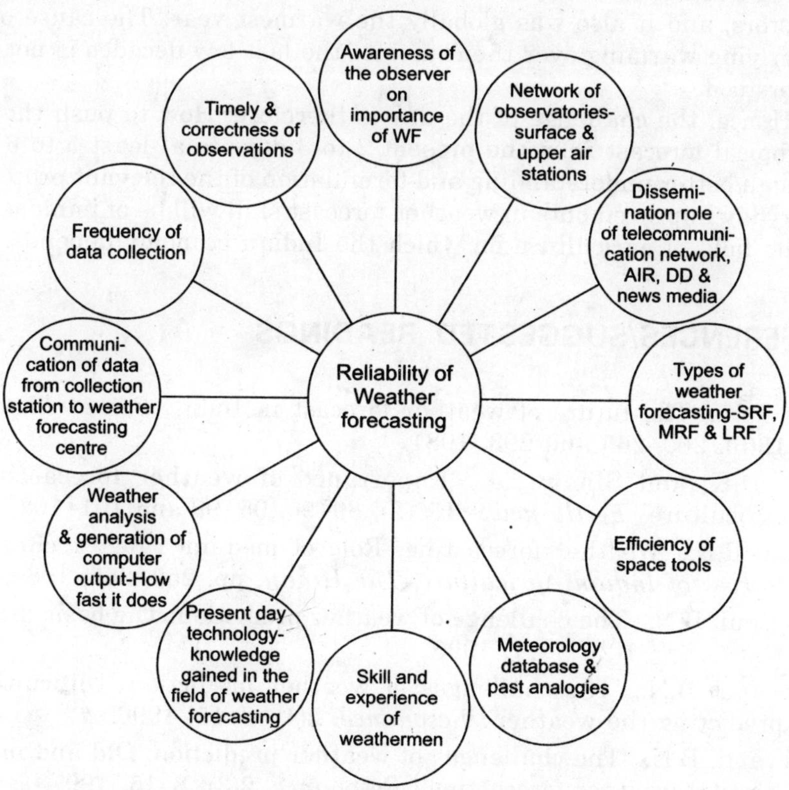

Fig. 13.13 Role of various components in weather forecasting.

Goswami (1996 and 1997) stated that a major effort is required to specify the initial state of weather, for which the observational system including instrumental errors and large data-void regions is to be taken into account. In addition, rainfall is the variable which dictates the atmospheric circulation in a tropical country like India. The knowledge gained at present is inadequate on how rains form and interact with atmospheric circulation in the tropics. It is amusing to see rainfall in a particularly small area not covering a few hundred square metres even during heavy monsoon in the humid tropics. How does this happen? How can this type of local circulation be related to global atmospheric circulation? It needs thorough understanding to develop a perfect atmospheric model for weather forecasting, which is very complex.

El Nino refers to an abnormal increase of sea surface temperature over the central and eastern parts of the equitorial Pacific Ocean of the coast of Peru. The role of the same in long-range weather forecasting is another area of research, as it has a negative correlation with the Indian summer monsoon, though not always true as that felt in 1997 and 2002. However, reports indicate that the heavy northeast monsoon rainfall across Tamil Nadu was attributed to the El Nino factor. The El Nino during 1997 was one of the factors which led to many parts of the world getting warmer in 1997 and 1998. 1998 was the year of weather-related disasters, and it also was globally the warmest year. The cause of the underlying warming over the course of the last few decades is not fully understood.

Hence, the challenge of the day is therefore: How to push the skill of tropical forecast from the present 2 to 3 days to at least 5 to 6 days through better understanding and formulation of the relevant processes? Everyday improvements in weather forecast skill will be of immense use in the field of agriculture on which the Indian economy depends.

REFERENCES/SUGGESTED READINGS

Das, P.K., 'The future of weather forecast in India', *Science Reporter*, **18**(6), 260–265 and 293, 1981.

Das, D.K. and Sindhu, J., 'Importanee of weather forecasting in agriculture', *Fertil. news*, **46**(12), 89–90, 95–99 and 101–102, 2001.

Datta, R.K., 'Weather forecasting: Role of medium range techniques', *Survey of Indian Agriculture, The Hindu*, pp. 252–255, 1989.

Goswami, B.N., 'The challenge of weather prediction: The basic driving', *Resonance*, **1**(2), 8–17, 1996.

Goswami, B.N., 'The challenges of weather prediction: Difficulties in predicting the weather', *Resonance*, **2**(1), 8–15, 1996.

Goswami, B.N., 'The challenges of weather prediction: Old and modern ways of weather forecasting', *Resonance*, **2**(3), 8–15, 1997.

IMD, *Agricultural Meteorology Division* (1932–57), Silver Jubilee of the Agricultural Meteorology Division, India Meteorological Department, Yeravada Prison Press, Pune, 1957, p. 42.

Kulshrestha, S.M., 'The monsoon game', *Science Reporter*, **32** (8), 9–14, 1995.

Pant, G.B., and Rupakumar, K., *Agrometeorological Data Monitoring and Management*, Indian Institute of Tropical Meteorology, Pune, 1995.

Parthasarathy, B., Munnot, A.A. and Kothawala, D.R., 'Regression model for estimation of Indian foodgrain production from summer monsoon rainfall', *Agric. and Forest Met.*, **42**, 167–182, 1988.

Raman, C.R.V., 'A new approach to rainfall climatology for agriculture planning', *Science Today*, July, 13–21, 1975.

Raman, C.R.V., 'Weather: Crucial role of forecasts', *Survey of Indian Agriculture, The Hindu*, pp. 207–211, 1990.

Rao, B.V.R., Katyal, J.C., Singh, J.B., Rao, A.V.M.S., Victor, U.S., Srivastava, N.N. and Vijaykumar, P., *Applicability of long range forecast of Southwest monsoon rainfall in different parts of India with special reference to Andhra Pradesh*, Central Research Institute for Dryland Agriculture, Hyderabad, p. 20, 1994.

WMO, *Economic benefits from agromet services*, Commission on Agrometeorology (CAgm.), Report No. 26, World Meteorological Organisation, Geneva, 1987.

QUESTIONS

1. What is the aim of weather forecasting in agriculture?
2. What is meant by Numerical Weather Prediction?
3. Distinguish between:
   (a) Nowcasting and SRF
   (b) AASUs and AMFUs
   (c) SRF and MRF
   (d) Crop weather calendar and crop weather diagram
   (e) MRF and LRF
   (f) Weather forecast and climate forecast
4. Expand the following:
   (a) ARIMA (b) CWWG
   (c) AASU (d) AMFU
   (e) NWP (f) SRF
   (g) MRF (h) LRF
   (i) UTC (j) FWB
   (k) NCMRWF (l) Radar
   (m) INSAT (n) SMC
   (o) IMD

5. Write short notes on:
   (a) Weather forecasting services
   (b) LRF and Indian foodgrains production
   (c) Types of weather forecasting
   (d) Methods of weather forecasting
   (e) Agrometeorological Advisory Services
   (f) Climate/agroclimate forecasting
   (g) Crop weather calendar
   (h) Synoptic chart and plotting model
6. Fill in the blanks:
   (a) Upper atmospheric wind is measured by ________.
   (b) ________ instrument is used to measure humidity in the upper atmosphere.
   (c) Farmers' Weather Bulletins are being issued since the year ________.
   (d) NCMRWF was established in the year ________.
   (e) Synoptic charts provide ______ state of atmosphere.
   (f) Td Td Td stands for______.
   (g) RRR stands for ________.
   (h) N stands for ________.
   (i) The sixteen-parameter model has been replaced by the IMD with the ______ parameter model.
   (j) Weekly agromet advisory service is provided by ______ units.
   (k) ________ come handy in solving equations much easier and faster.
   (l) NCMRWF provides MRF ________ a week to selected AMFUs.
   (m) Information on ________ is provided in the lower part of the crop weather calendar.
   (n) India has been divided into ______ meteorological subdivisions.
7. Match the following:

(a) SRF	CWWG
(b) MRF	AMFU
(c) LRF	AASU
(d) AAS	NCMRWF
(e) SMC	IMD

8. Elucidate the following:
   (a) Role of weather forecasting in agriculture
   (b) Weather service to farmers
   (c) Numerical Weather Prediction and challenges in weather forecasting

14 CROP SIMULATION MODELLING

The crop development process from sowing or transplanting to harvest responds to environmental fluctuations that are predominantly weather-induced. The important weather variables which drive crop growth and development are rainfall, solar radiation, vapour pressure deficit, and maximum and minimum temperatures. It is understood that the effect of the driving forces (weather elements) on crop development finally decides the total dry matter production under homogeneous environment. To quantify the crop-weather relationship, the necessary tools should be obtained through modelling. The modelling process may involve statistical, descriptive and explanatory methods. The former is known as *empirical statistical modelling* and the latter two as *dynamic simulation modelling*. The potential crop yields under field conditions in different agroclimatic zones could be analysed on the basis of simulation modelling, which in turn will help to identify the crop production constraints. In addition, simulation modelling can be a substitute to multilocational field trials for introducing variety in different agroclimatic zones, thus saving time and money. Hence simulation modelling is felt necessary for maximising agricultural production through better crop management practices. The tools required for application of the simulation models are (i) the required software for the crop chosen and (ii) a PC/AT with knowledge in computers and crops. The models can be revalidated under a given climate situation with a multi-disciplinary group comprising a crop modeller, a crop scientist with physiology background and an agrometeorologist. The only basic requirement to run the simulation models is a reliable database on weather and crop data, which includes data on soils, crop management practices and crop genetic coefficients. ICRISAT, Hyderabad was the first institute in India to introduce the crop simulation models. It started working in the 1970s, followed by IARI (ICAR), New Delhi.

14.1 EMPIRICAL STATISTICAL MODELS

The impact of weather elements on crop growth and yield is mostly understood in India through different statistical techniques. Based on the long series of data available on crop yields and climatological data, regression equations can also be developed for estimating crop yields. The widely used methods are correlation and regression techniques for studying crop-weather relationships and weather-based yield prediction. A simple regression equation of the following form helps to predict the dependent variable (crop yield) based on the independent variable (rainfall):

$$Y = a + bX$$

where Y is the dependent variable, X is the independent variable, and a and b are the regression coefficients. A relationship between water deficit and latex yield in rubber is worked out using the above and shown in Fig. 14.1. It reveals that the rubber yield is less if the water deficit is high.

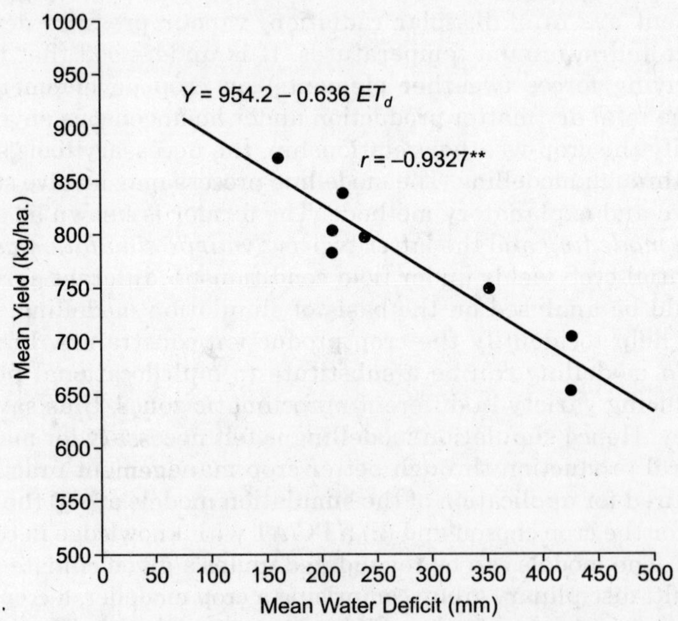

Fig. 14.1 Water deficit and rubber yield.

If two or more independent variables are involved, the multiple regression equation used is

$$Y = a + \sum_{i=1}^{m} b_i X_i$$

where Y represents the dependent variable, X_i represents several independent variables, and b_i represent regression coefficients.

Using the monthly rainfall figures from September to December in the multiple regression equation, the annual cashew production of the State of Kerala was estimated (see Fig. 14.2). The estimated and actual annual cashew production of the State was in good agreement, especially from 1986–87 onwards. The same was the case in cashew productivity. However, this model cannot be applied outside the State of Kerala.

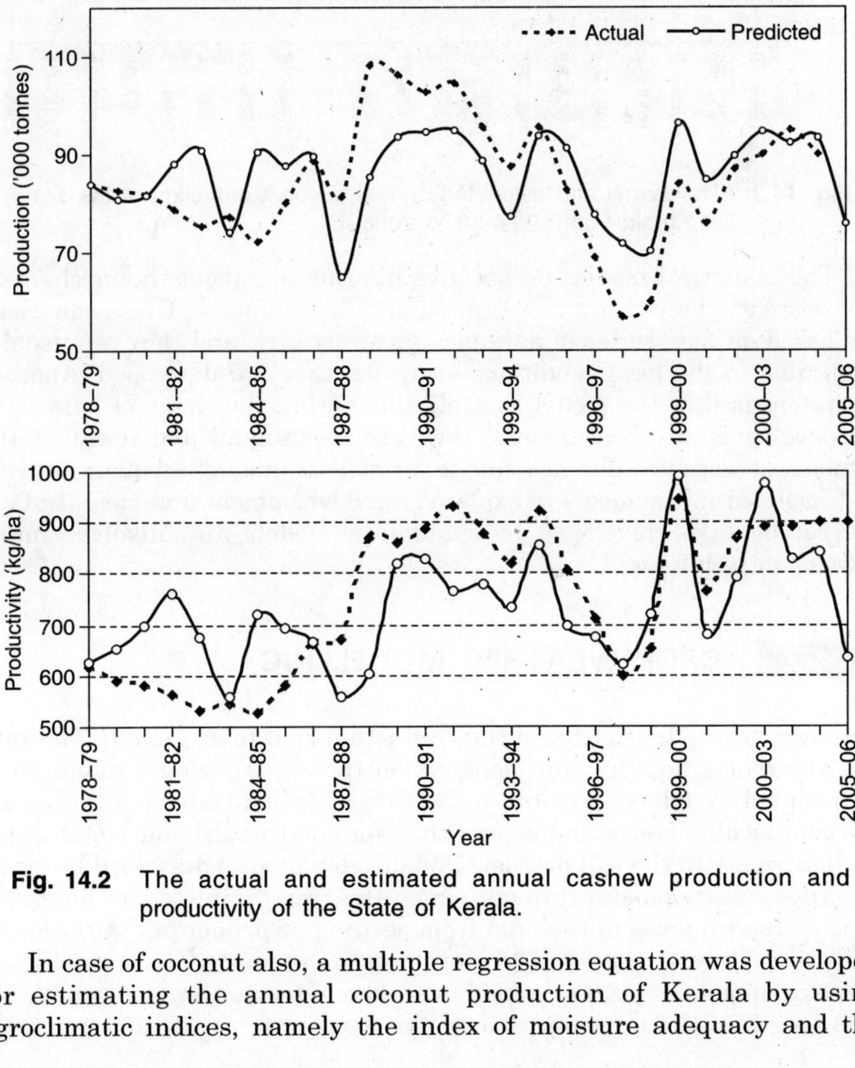

Fig. 14.2 The actual and estimated annual cashew production and productivity of the State of Kerala.

In case of coconut also, a multiple regression equation was developed for estimating the annual coconut production of Kerala by using agroclimatic indices, namely the index of moisture adequacy and the humidity index for 42 months before the harvest takes place. The same equation was put in operation till 2006–07 for estimating the annual coconut production of the State (see Fig. 14.3). It gave fairly good results.

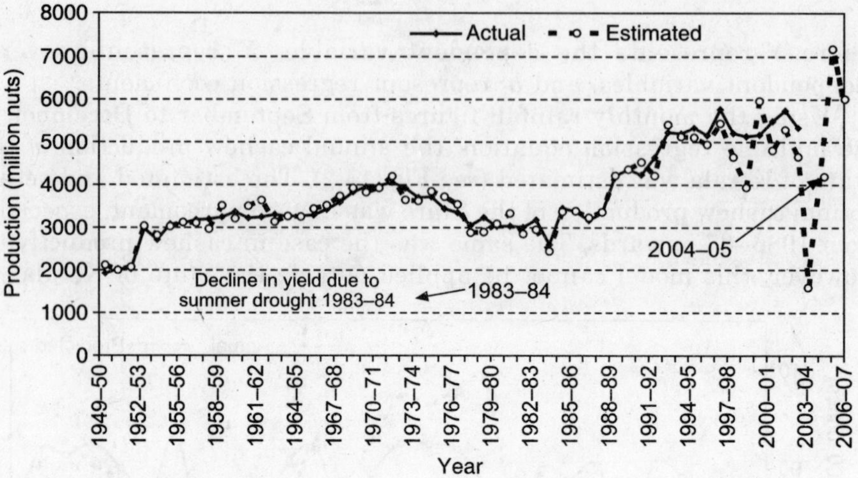

Fig. 14.3 The actual and estimated annual coconut prduction of the State of Kerala from 1949–50 to 2006–07.

The statistical models do not give information about the mechanism and process involved in plant growth and development. These equations are based on long series of data at a given locality, and they are usually applicable to the local conditions on which they are developed. Another limitation is that the model is applicable within the range of data used for developing it. Nevertheless, they are meaningful and useful if the response of crops to different weather variables is understood quantitatively and included in the model to explain crop development and yield. In fact, several modules developed in simulation models also involve simple regression techniques.

14.2 CROP WEATHER MODELLING

The increase in dry matter of the crop is referred to as *growth*. The rate of growth of a healthy crop depends on the rate at which radiation is intercepted by foliage, and/or on the rate at which water and nutrients are captured by root systems, and therefore on the distribution of water and nutrients in the soil profile. Crop development is described in terms of various phenophases through which the crop completes its life-cycle, that is, the progress of the crop from seeding or primordial initiation to maturity. Finally, the yield of crop stand is expressed as a product of three components, viz. the period over which dry matter is accumulated (the length of the growing period), the mean rate at which dry matter is accumulated, and the fraction of dry matter treated as yield when the crop is harvested.

It is understood that crop growth, development and yield depend upon Daily Thermal Time (DTT), the length of the day and the amount of solar radiation Photosynthetically Active Radiation (PAR) received by the crop.

$$\text{DTT} = \frac{\text{Max. daily temp.} + \text{Min. daily temp.}}{2} - \text{Base temp.}$$

where DTT is the daily thermal time accumulation.

The time needed for the crop to reach a development stage depends upon temperature measured above a base value (DTT), and for photoperiodically sensitive phases, such as flowering, the day length above a fixed base. In the absence of stress, the harvest index does not vary much from year to year for a specified cultivar or variety. Therefore, crop weather modelling is based on the principles that govern the development of a crop and its growing period based on temperature and/or day length. They are used to quantify the rate of crop growth in terms of radiation interception, water use and nutrient supply, which moderate the harvest index when the crops experience stress.

The basic information required to be generated for crop weather modelling includes (Miller et al. 1992):

(a) Crop phenology in relation to the temperature and day length
(b) Water use by the crop during different phenophases of crop growth
(c) The relationship between radiation interception, crop water use and total dry matter production
(d) Partitioning of dry matter into various plant components as influenced by water and nutrient availability, and
(e) The effect of weather parameters on biotic interference to crop growth.

14.3 DYNAMIC SIMULATION MODELS

A dynamic simulation model is a mathematical equation or set of equations which represent the behaviour of a system. These models simulate the day-to-day assimilation of photosynthetic material, based primarily on the exchange of energy and mass among the various growth processes taking place in a plant. Such models are very complex and require the knowledge of both biological and meteorological environments. They are useful for studying the physiology of crop growth and development. Unlike the empirical statistical model, a simulation model demands a large quantum of information with respect to interaction of the soil-plant-atmosphere continuum and predicts numerous parameters of crop production. The crop simulation model is not the total alternative to field experimentation, but it reduces the need for cumbersome field experiments considerably. Once the crop simulation model is validated or standardized for a particular crop or variety under a given environment, a lot of information on crop growth and productivity, as influenced by weather parameters, fertilizers, irrigation and soil parameters, can be generated within hours. The simulation models can generate information on different crop management and cultural practices viz. age of seedlings, optimum plant population, spacing, time of fertilizer application and its dosage, number of irrigations required during the crop season and yield expected in a given weather situation. A trial on time of planting of a variety or selection of a variety suitable for a location require at least 3 years of field experimentation. In contrast, a validated model can give precise information on the above aspects in no time. Thus, scientists can significantly reduce the number of agronomic trials and save their time and energy if the simulation models are tested and validated. CERES models of barley, maize, millet, sorghum, wheat and rice, CROPGRO models of dry bean, peanut and soybean, CROPSIM of cassava, SUTOR models of aroids and potato (Tsuji et al., 1994), CROPCAST of maize, soybean and wheat, and precision farming models like PALMS are some of the internationally accepted crop simulation models. In addition to the above, the crop simulation models viz. WTGROWS, INFOCROP, LINTUL and SUCROS are also being tested in India. A crop simulation model divides each component into different files. Each file demands a large quantum of information for accurate prediction of yield and yield-contributing factors of crop production.

14.3.1 CERES Sorghum Model

The phenology of sorghum as described in the CERES Sorghum model is given as an example to illustrate the effect of mean daily temperature and day length on the phenology. For sorghum, base temperature of 8°C is used.

Stage 1: (*Seedlings emergence to end of juvenile stage*). The plants are insensitive to photoperiod and therefore the thermal time required to complete the phase depends only on the mean daily temperature. The thermal time required to complete Stage 1 is referred to as **genetic coefficient** P1.

Stage 2: (*End of juvenile stage to end of panicle initiation*). The rate of development is controlled by the length of the day i.e. the length of the photoperiod. If the day length is more than 12 hours, this phenophase is delayed. Therefore, there is a threshold photoperiod (P20) above which the daily thermal time up to panicle initiation (DTTPI) increases linearly with increase in length of the photoperiod. The increasing thermal time for a 1 hour increase in length of photoperiod over P20 is called the **photoperiod sensitivity coefficient (P2R)**. The values of P20 and P2R are also considered genetic coefficients.

Stage 3: (*Panicle initiation to end of flag leaf growth*). Stage 3 depends on daily thermal time only. The flag leaf appearance and expansion is completed during the stage.

$$P3 = 0.199\ DTTPI + 300$$

where DTTPI = Daily thermal time required up to panicle initiation.

Stage 4: (*End of flag leaf growth to beginning of the grain filling*). During this stage, the panicle develops rapidly and peduncle grows fast, extending the panicle through the leaf sheath. The duration of this stage is 270 degree days and flowering occurs after 150 degree days.

Stage 5: (*Effective grain filling to physiological maturity*). Stage 5 requires 550 degree days to reach physiological maturity.

From the above description, it is clear that:

(i) The thermal time required during Stage 2 depends upon the length of the day.
(ii) The thermal time required for Stage 3 depends upon thermal time requirement of Stage 2.
(iii) The thermal time requirement during Stage 4 and Stage 5 do not vary among the different cultivars of sorghum.

Therefore, crop development has to be monitored as per the descriptions given in the user guide. Crop description varies depending upon the type of models chosen to run.

14.3.2 ORYZA Rice Model

An ecophysiological model for irrigated rice production was developed under Simulation and Systems Analysis for Rice Production (SARP). The institutes involved were: International Rice Research Institute, Manila, Phillippines; Department of Theoretical Production and Agricultural Research; Department of Ecology, Research Institute of Aerobiology and Soil Fertility, Wageningen Agricultural University, the Netherlands.

An important advantage of the ORYZA model is that it can be used to simulate realistic yields and to assess the impact of planting date, weather and latitude on measured leaf N content. This is in contrast to the models developed for potential production, which possess a fixed pattern of leaf photosynthesis in time.

IARI, New Delhi provides regular trainings on ORYZA rice models. It is the pioneer institute in India in crop simulation modelling.

14.3.3 CERES Rice Model

The Crop Environment Resources Synthesis (CERES) rice model forms the basis for the International Benchmark Sites Network for Agrotechnology Transfer (IBSNAT) and the Decision Support System for Agrotechnology

Transfer (DSSAT). This project is under the department of Agronomy and Soil Science, University of Hawaii, Honolulu in collaboration with ICRISAT, Michigan State University, USA (Project Leader: Prof. J.T. Ritchie). The schematic presentation of the input and output files of the CERES rice model is depicted in Fig. 14.4 (Saseendran, et. al., 1998).

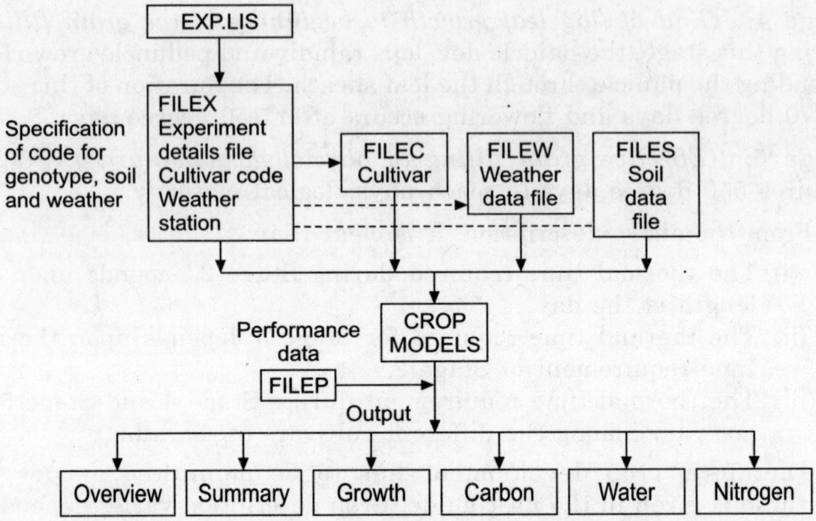

Fig. 14.4 Input and output files of the CERES rice model.

Information on weather, location, soil, crop and management is required to run the CERES rice model. It is used to simulate the rice crop under different environments and to predict potential crop yield on the basis of weather variables viz. daily rainfall, solar radiation, and maximum and minimum temperatures. The model helps to identify the differences between expected possible crop yields and actual yields in a given environment. Once the gap is identified between possible and actual crop yields, optimisation of crop yields can be manipulated through transfer of technology. In the CERES rice model, the entire programme is divided into weather file, soil file, crop file (or genotype coefficient file) and crop management file. The details of different files are given in the following section.

Weather file

This file demands 1-year daily weather data on solar radiation, maximum temperature, minimum temperature and rainfall.

Soil file

This file demands soil data related to the root zone depth of the crop, soil texture, soil structure, field capacity, nutritional status, pH and other related information on soils.

Management file

Data on fertilization of the crop, irrigation and residue left over by the previous crop are needed for this particular file.

Genotype coefficient file

The file contains the cultivar-specific coefficients. Eight genetic coefficients are required for describing the various aspects of performance of a particular genotype (see Table 14.1).

Table 14.1 Genetic coefficients used in CERES rice model

Name	*Description*
Juvenile phase coefficient (P1)	The time period (expressed as growing degree days (GDD)°C over a base temperature of 9°C from seedling emergence, during which rice plant is not responsive to changes in photoperiod. This phase is also referred to as the *basic vegetative phase* of the plant.
Critical photoperiod (P20)	The longest day length (in hours) at which the development occurs at a maximum rate. At values higher than P20, the development rate is slow. Hence there is a delay owing to longer day lengths.
Photoperiodism coefficient (P2R)	The extent to which phasic development leading to panicle initiation is delayed (expressed as GDD in °C) for each hour's increase in photoperiod above P20.
Grain filling duration coefficient (P5)	The time period in GDD (°C) from the beginning of grain filling to physiological maturity with base temperature of 9°C.
Spikelet number coefficient (G1)	Estimated from the number of spikelets per gram of main culm dry weight (less lead blades and sheaths plus spikes) at anthesis. A typical value is 55.
Single grain weight (G2)	The single grain weight (g) under ideal growing conditions.
Tillering coefficient (G3)	The tillering coefficient (scalar value) relative to IR 64.
Temperature tolerance coefficient (G4)	Usually 1 for varieties grown in normal environment.

The genetic coefficients developed by the IRRI, Philippines are incorporated in the model. For a test variety, one has to work out the genetic coefficients. In the absence of actual data on genetic coefficients, they can be derived on the basis of genetic coefficients already included in the model. The genetic coefficients included in the model for some of the varieties are given in Table 14.2.

Table 14.2 Genetic coefficients of IR8, IR20 and IR36

Genetic Coefficients	*IR8*	*IR20*	*IR36*
P1	880.0	500.00	550.00
P2R	52.00	166.00	149.00
P5	550.00	500.00	550.00
P20	12.10	11.20	11.70
G1	75.00	75.00	70.00
G2	0.028	0.028	0.023
G3	1.00	1.00	1.00
G4	1.00	1.00	1.00

The simulation model requires precise information on the following aspects for accurate output of the model:

(a) Genetic coefficients of crop varieties are to be worked out by field experimentation for fitting into the model. They also can be generated by using the genetic coefficient calculator (Gencalc).
(b) Data on soil are to be generated by conducting various laboratory studies. Assumption of data may lead to gap in actual and simulated output.
(c) Precise data on daily weather parameters are needed for operating the model.

If the model is run with the help of necessary information based on experimental data, it will simulate the behaviour of the crop in terms of phasic development, physiology, yield and yield-attributing characters. Some of the variables that can be simulated by using the CERES rice model are panicle initiation, time of flowering, physiological maturity, grain yield, weight per grain, grain number, panicle number, leaf area index, biomass at different phasic development stages, harvest index, and straw yield.

14.4 APPLICATION OF CERES RICE MODEL

A field experiment was laid out at the Regional Agricultural Research Station, Pilicode (12° 12′N, 75°10′E) for testing the crop growth simulation models under the tropical humid climate, where the soils are of shallow sandy loam. Three high-yielding varieties viz. Jaya, Jyothi and Triveni were transplanted on five different dates (8th June, 15th June, 22nd June, 29th June and 6th July) at weekly intervals during *Kharif* 1993–94 and on four different dates (15th October, 22nd October, 29th October and 5th November) at weekly intervals during *Rabi* 1992–93 in a split plot design. The cultural and management practices were followed as per the package of practices recommendations on crops of the KAU. The yield attributes viz. number of panicles per hill, number of grains per earhead and 1000 grain weight, heading and maturity dates, grains/sq.m., leaf area index, grains N%, straw N-uptake, grain N-uptake, total N-uptake, grain and straw yields were recorded in each case. The daily weather parameters viz. maximum temperature, minimum temperature, rainfall, and bright sunshine hours, collected at the Agrometeorological Station (which is very close to the field) were used to run and test the CERES rice model, which provides output support for International Benchmark Sites Network for Agrotechnology Transfer (IBSNAT). The model does not take into account the grain losses that take place due to incidence of pest and diseases. The daily bright sunshine hours were converted into solar radiation (MJ/m/min), which is one of the weather inputs along with the other three variables viz. maximum and minimum temperatures (in °C) and rainfall (in mm). The model also takes into account the variety, age of seedlings, and method of planting, plant population, fertilizer management, soil characteristics and floodwater management, including the effect of bund height. Using the above inputs, the model simulates thirteen parameters, which also include grain and straw yields.

The simulated grain yield through the CERES rice model was in agreement with the observed findings during Kharif 1993 while it was not the case in Kharif 1994 (see Figures 14.5 and 14.6). The low grain yield in 1994 was due to maximum incidence of gall midge (see Table 14.3), brown spot and rice smut, coupled with high rainfall during the flowering phase (Rao et al., 1999).

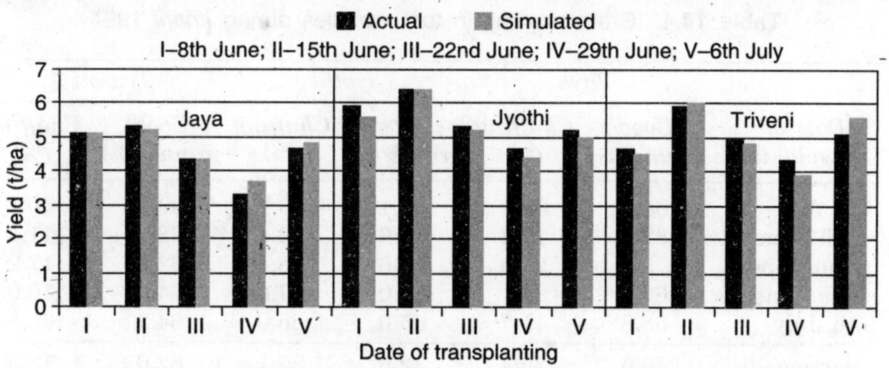

Fig. 14.5 The actual and simulated grain yield of test varieties during kharif 1993 at RARS, Pilicode.

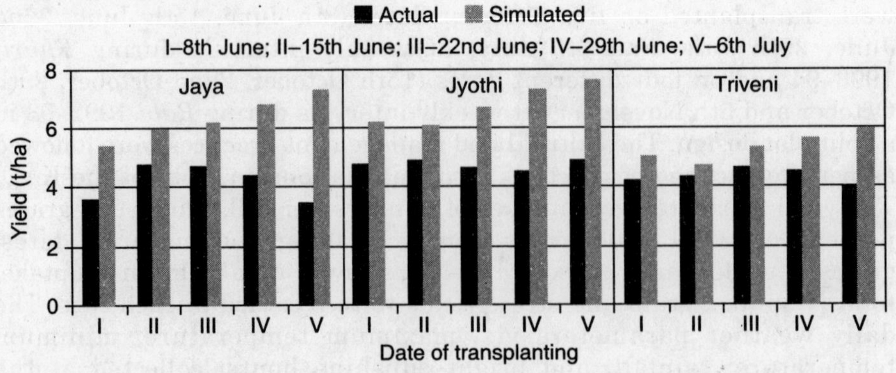

Fig. 14.6 The actual and simulated grain yield of test varieties during kharif 1994 at RARS, Pilicode.

Table 14.3 Incidence of gall midge during *kharif* 1993–1994 at Pilicode

Date of Transplanting	*Percentage of gall midge incidence*					
	Kharif 1993			*Kharif 1994*		
	Jaya	*Jyothi*	*Triveni*	*Jaya*	*Jyothi*	*Triveni*
8th June	0.9	1.5	0.5	46.9	15.9	13.1
15th June	16.6	17.5	2.7	36.8	52.1	32.9
22nd June	25.1	19.0	13.3	47.2	45.3	41.2
29th June	31.8	48.8	12.5	30.3	40.3	41.3
6th July	28.3	30.7	14.4	18.4	14.1	14.7
Average	20.5	23.5	8.7	36.3	33.5	28.6

The above conditions resulted in grain chaffing during *kharif* 1994 (see Table 14.4).

Table 14.4 Grain chaffing in test varieties during *kharif* 1994

Date of Transplanting	*Jaya*		*Jyothi*		*Triveni*	
	Good grains (%)	*Chaffing (%)*	*Good grains (%)*	*Chaffing (%)*	*Good grains (%)*	*Chaffing (%)*
8th June	68.7	31.3	81.1	18.9	90.3	9.7
15th June	82.7	17.3	83.6	16.4	88.7	11.3
22nd June	75.2	24.8	71.6	28.4	82.8	17.2
29th June	67.9	32.1	63.0	37.0	84.0	16.0
6th July	58.3	41.7	65.0	35.0	64.2	35.8
Average	70.6	29.4	72.9	27.1	82.0	18.0

As the model runs under floodwater maintained through irrigation at three different early stages, the grain yield was not comparable between observed and simulated levels during *rabi*. Hence, the CERES rice model needs to be validated for *rabi*, while it could be used for *kharif* under better cultural and management practices in the humid tropics for assessing potential yields based on weather variables viz. daily rainfall, maximum and minimum temperatures, and sunshine hours. The crop growth processes and phenological stages were also simulated for the Jaya variety by using CERES rice version 3.0 model (Saseendran et al., 1998). It was inferred that the model can be used for making various strategic and tactical decisions related to agricultural planning in the State of Kerala.

14.5 STATUS OF CROP SIMULATION MODELS IN INDIA

Several institutes are actively involved in running crop simulation models in the country. Many scientists were trained abroad to acquaint them with different models. The scientists at ICRISAT, IARI and CASAM are also involved in training the technical personnel of sister institutes. The scientists under AICRP on Agrometeorology have been motivated to take up modelling work with multidisciplinary groups of scientists, wherever facilities are available. Indigenous crop simulation models are yet to be developed. At IARI, some work is in progress in the above direction. Hopefully, it will generate a few simulation models in collaboration with other institutes during the ensuing decade. The centres under AICRP on Agrometeorology will also involve themselves in developing operational simulation models at least in a few major crops. The existing crop simulation models need validation for applications under different agroclimatic zones across the country.

REFERENCES/SUGGESTED READINGS

Abrol, Y.P. and Gadgil, S. (Eds.), *Rice: In a Variable Climate*, APC Publications, New Delhi, 1999, p. 243.

CASAM, *Crop Modelling*, Eds. Varshneya, M.C. and Salunke, S.S., Centre of Advanced Studies in Agricultural Meteorology, College of Agriculture, Pune, 1998, p. 251.

Hunt, L.A., Pararajasingham, S.J., Jones, J.W., Hoogenboon, G., Imamura, D.T. and Ogoshi, R.M., 'GENCALC software to facilitate the use of crop models for analysing field experiments', *Agron. J.*, **85**, 1090–1094, 1993.

KAU, *Package of Practices and Recommendations: Crops*, 12th ed., Directorate of Extension, Kerala Agricultural University, Thrissur, 2002, p. 278.

Miller, B.C., Theodore, C.F. and James, E.H., 'CARICE, A rice model for scheduling and evaluationg management actions', *Agron. J.*, **85**, 938–947, 1992.

Penning de Vries, F.W.T., Jansen, D.M., ten Berge, H.F.M. and Bakema, A., *Simulation of Ecophysiological Processes of Growth in Several Annual Crops*, Centre for Agricultural Publishing and Documentation (Pudoc), Wageningen, 1989, p. 271.

Rao, B.V.R., *Operational Agricultural Meteorology (Problems and Priorities)*, Indian Society of Agronomy, Indian Agricultural Research Institute, New Delhi, 1988, p. 59.

Rao, G.S.L.H.V.P. and Subhash, N., 'Use of CERES Rice model to assess potential yield', *IRRN*, **21**, 2–3, 1996.

Rao, G.S.L.H.V.P., Sebastian, S. and Subhash, N., 'Crop growth simulation models of rice under humid climates', *Advanced Technologies in Meteorology*, Eds. Gupta, R.K. and Reddy, J.S., Tata McGraw-Hill, New Delhi, 1999, pp. 436–439.

Saseendran, S.A., Singh, K.K., Rathore, L.S., Rao, G.S.L.H.V.P., Mendirath, N., Narayan, L. and Singh, S.V., 'Evaluation of the CERES rice version 3.0 model for the climate condition of the State of Kerala', *Meteorol. Appl.*, **5**, 1998, 385–392.

Singh, S.V., Rathore, L.S., Saseendran, S.A. and Singh, K.K. (Eds.), *Dynamic Crop Simulation Modelling for Agrometeorological Advisory Services*, National Centre for Medium Range Weather Forecasting, Department of Science and Technology, Government of India, New Delhi, 1999, p. 358.

Tsugi, G.Y., Uhera, G. and Bales, S.S. (Eds.), DSSAT v.3, University of Hawaii, Honolulu, Hawaii, 1994.

QUESTIONS

1. Define the following:
   (a) Dynamic simulation model
   (b) Genetic coefficient
2. Expand the following:
   (a) CERES (b) SARP
   (c) DTT (d) DSSAT
   (e) IBSNAT (f) GDD
   (g) PAR
3. Distinguish between:
   (a) Regression model and simulation model
   (b) CERES rice model and ORYZA model

4. Write short notes on:
   (a) Empirical statistical models
   (b) Crop weather modelling
   (c) Dynamic simulation models
   (d) CERES–rice model
   (e) Model validation
5. Fill in the blanks:
   (a) ——, ——, —— and —— are the weather variables required to run the CERES rice model.
   (b) Crop simulation models can substitute ——— .
   (c) Temperature tolerance coefficient (G4) is usually —— for most of the rice varieties.
   (d) Tillering coefficient (G3) is worked out in relation to —— rice variety.
   (e) P1 is known as ———.
   (f) P2R is known as ———.
   (g) P5 is known as ———.
   (h) G1 is known as ———.
   (i) Genetic coefficients can be generated using ——— software.
   (j) Knowledge in ——— and ——— are required to run simulation models with a given software.
6. Match the following:

(a) CERES	Rice
(b) CROPSIM	Potato
(c) PALMS	Maize
(d) ORYZA	Sorghum
(e) SUTOR	Cassava
(f) CROPCAST	Precision farming
(g) SOYGRO	Groundnut
(h) PNUTGRO	Soyabean

7. Elucidate crop simulation models in India with case studies.

15 INFLUENCE OF WEATHER ON PEST AND DISEASE OUTBREAKS

Considerable crop losses are caused due to pests and diseases in the humid and sub-humid tropics. Many of the restrictions on productivity and geographical distribution of plants and animals are imposed by climatic constraints and biotic pressure exerted by pests and diseases. The geographical distribution of pests is mainly based on climatic factors. The climatic conditions show a gradient from place to place and there is a related gradient in the abundance of a particular pest or disease. The periodic or seasonal nature of incidence and outbreaks of several pests and diseases of many crops can be ascribed to weather conditions as the triggering factors. These epidemics of diseases are principally weather-dependent, either in terms of local weather conditions being favourable for growth and development of the causal organisms, or the prevailing winds helping to disseminate airborne pathogens or spores of pathogens such as mildew, rusts, scabs and blights.

The migration and dispersion of insect pests depend on the wind speed and direction, besides the nature of air currents. Most of the plant pathogenic viruses are transmitted by insects. Weather conditions suitable for the development of these vectors favour the transmission of such diseases. A large number of pests and diseases which affect plants are kept in check by seasonal fluctuations in atmospheric temperature or relative humidity and other factors.

Insect pest outbreaks occur as a result of weather conditions which facilitate their uninterrupted multiplication. The weather and climate greatly influence the quantity and quality of food provided by the host crop(s) to the associated species of pests. The abundance or scarcity of the pestiferous species is thus dependent on climatic conditions indirectly too.

The surface air temperature, relative humidity, dewfall, sunshine, cloud amount, wind, rainfall and their pattern and distribution are the primary weather factors influencing the incidence or outbreaks of pests and diseases of crops. In the humid tropics, the weather variables, namely, air temperature, intermittent rainfall, cloudy weather, and dewfall, may play a crucial role in the outbreaks of pests and diseases, as these are generally very conducive factors. In the arid, semi-arid, moist and sub-humid environments too, temperature, rainfall and relative humidity are the dominant factors for outbreaks of crop pests and diseases. The impact of various weather components on pests and diseases is experienced in a location- and crop-specific manner. Among the major pests associated with crops are insects, mite and nematode species, whose impact is of a serious nature in terms of their abundance and damage potential. If the occurrence of pest or disease in time and space can be predicted in advance with reasonable accuracy on the basis of relevant weather parameters, appropriate and timely control measures can be programmed and adopted. Crop losses can thus be minimised. Appropriate insecticide or fungicide interventions can certainly reduce the pesticide load in the environment and the related pollution and health hazards.

Hence, insect pest and disease forewarning systems play a major role in integrated pest management and the sustenance of agricultural production at desired levels.

15.1 SEASONAL ABUNDANCE

The majority of insect pests and diseases are commonly seen during the rainy season, while aphids, thrips, scales, mealy bugs, whiteflies and mites are relatively more frequent during summer. Blast disease of rice and powdery mildews of many crops are common in winter.

15.1.1 Rice Pests and Diseases

Among the rice pests, gall midge appears to be strongly weather-dependent, while the population of other pests is somewhat season-bound and occur in different crop growth stages. The incidence of gall midge is common during kharif as compared to rabi, due to the high humidity requirement of the pest, particularly in the early growth stages. The freshly hatched maggots of this fly-pest require a minimum RH of 80% for establishing infection in the plants. The pest incidence is correlated positively with rainfall and negatively with maximum temperature (Thomas et al., 1975). The pre-monsoon showers during March, April and May trigger an increase in the gall midge population at the time of nursery and affect the crop in the main field. Continuous rainfall and low maximum surface air temperature during the tillering phase may induce gall midge infestation. Similar weather conditions may not be favourable for leaf blast disease in rice (Premanathan et al., 1999). Rice blast is severe during winter when the minimum temperature goes below 20°C and dew nights of 16–20 hours.

Moderate rainfall, RH of around 80% and a wider range of atmospheric temperature of 20–33°C appear to be highly conducive to BPH (Abraham and Nair, 1975). Such conditions occur during the winter crop (Aug.–Sep. to Dec.–Jan.), and the pest multiplies rapidly during this season. The summer crop (*Puncha*) is prone to serious damage from the heavy populations arising out of the preceding winter crop. Brown spot also appears to be severe in *rabi* compared to that of *kharif*. The number of rainy days and minimum temperature influence disease incidence, and the spread is much more if the dew nights are more during *rabi*.

15.1.2 Coconut Pests and Diseases

Several pests and diseases are noticed in coconut. A few are discussed here on the basis of their seasonal occurrence.

Pests

The cockchafer beetle is commonly known as *root grub* or *white grub*. The population peaks occur in June and July and less in summer (February–May). This pest is more dependent on soil environment rather than atmospheric variables, as the immature stages develop in the soil environment. Sandy or sandy loam soils are much favourable for the multiplication of this pest under favourable microclimatic conditions. Rhinoceros beetle, red palm weevil and black-headed caterpillar are the other major pests. The black-headed caterpillar generally occurs throughout the year, but during the summer months their populations attain peak levels. This is mainly due to the temperature and humidity conditions becoming progressively favourable for the pests and unfavourable for the associated natural enemies. Distinct seasonal peaks are not discernible for the rhinoceros beetle, but their populations are relatively heavier during the rainy season (June–September). The red palm weevil, which is the most destructive pest of coconut in the country, shows maximum infestation levels during the dry weather prevailing from January to May.

Diseases

The common crop diseases are as follows:

Bronze wilt

This wilt disease is common in non-traditional areas, where hot summer with severe moisture stress is noticed. During periods of drought, the leaves become bronze-coloured and finally drop off. It is a severe menace in Tamil Nadu, Andhra Pradesh and Karnataka.

Stem bleeding in coconut

The stem bleeding disease in coconut is common in the northern districts of Kerala and comparatively less or negligible in the southern districts. Continuous heavy wet spell followed by long dry spell is one of the factors for the incidence of stem bleeding in coconut. The fungus enters through cracks on the stem. The disease is characterised by the oozing of a dark red brown viscous liquid through these cracks, and it is severe when the soil moisture depletion takes place in summer.

15.1.3 Cashew Pests

The tea mosquito bug is the most serious pest of cashew in south India, causing inflorescence blight and drying up of tender shoots and nuts. The most favourable period for rapid multiplication and population build-up of the pest is during December–February, synchronizing with the flushing and flowering phases of cashew. The pest population of the tea mosquito bug is low in April and May. It also appears that the pest is totally absent in rainy season (see Figures 15.1(a) and (b)). Incidence of unprecedented inflorescence and twig blight was noticed across the cashew tract of Kasaragod and Kannur districts of Kerala during 1998–99, resulting in most severe crop losses due to large-scale drying up of young shoots and inflorescences. Immature nuts were also dried up and seen hanging with panicles. This phenomenon was found to be due to severe infestation by the tea mosquito bug, closely followed by a moderate fungal pathogen infection.

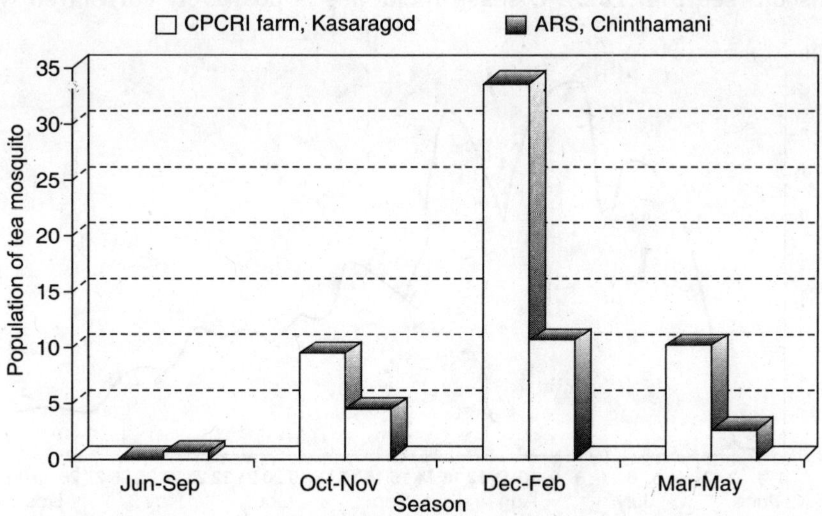

(a)

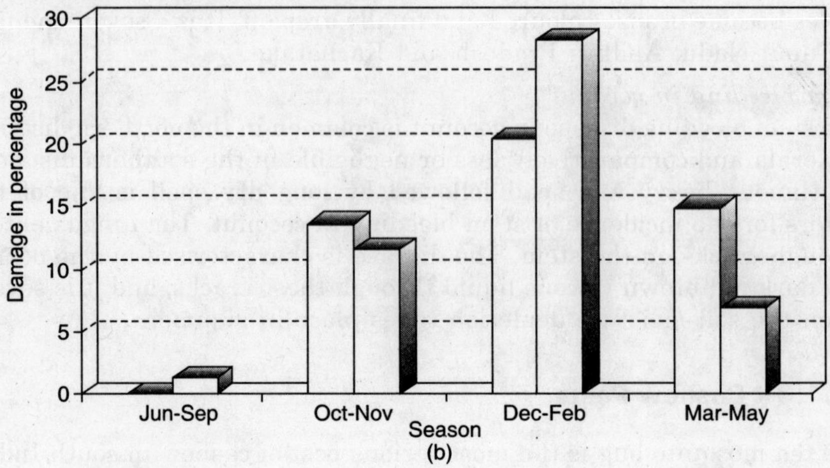

Fig. 15.1 Seasonal variations in tea mosquito population.

15.1.4 Phytophthora Diseases of Crops

The crop diseases which are caused by *Phytophthora* are strongly weather-dependent. The bud rot in coconut, foot rot in black pepper and the abnormal leaf fall of rubber are caused due to *Phytophthora,* resulting in substantial crop losses, particularly in south India. High rainfall and heavy cloudiness are factors that incite these diseases.

Foot rot of black pepper

A drastic increase in disease incidence is seen during the southwest monsoon (see Fig. 15.2). Disease incidence is positively correlated with

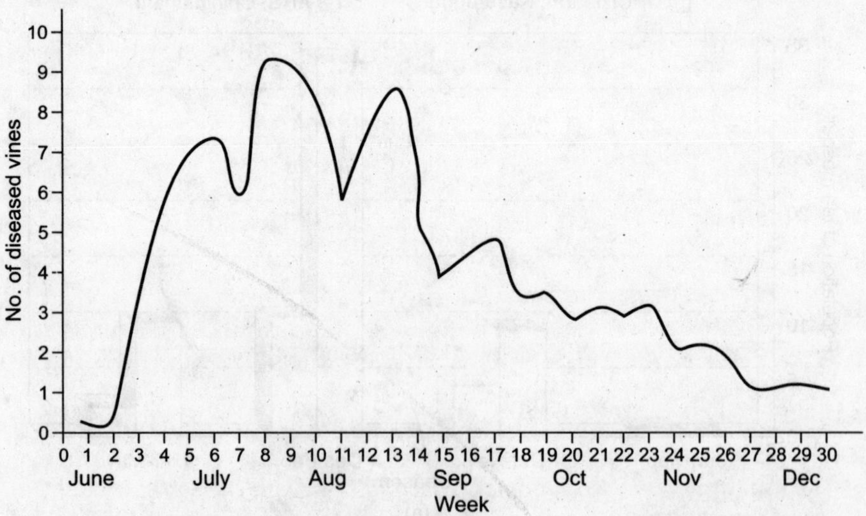

Fig. 15.2 Weekly incidence of foot rot of black pepper from June to September.

relative humidity, rainfall and number of rainy days, while maximum temperature and bright sunshine show a significant negative correlation with disease incidence. In a continuous wet spell, high relative humidity triggers disease incidence.

Abnormal leaf fall in rubber

In Kerala, the disease is characterised by annual recurrence during the southwest monsoon period. The outbreak of disease is seen invariably when a five-day rain spell with an overcast sky exists. The epidemics can be normally expected within 9–15 days after the first overcast day, depending on the onset of southwest monsoon. The initiation of the disease is usually noted by mid-July. However, the disease incidence may deviate according to variations in the southwest monsoon.

Bud rot of coconut

The bud rot disease of coconut occurs sporadically during the rainy season. The disease is common in the humid climates during the monsoon and post-monsoon seasons. No correlation was found between temperature and outbreak of bud rot, but definite correlations were established between rainfall and occurrence of disease. Reports outside India suggested the role of lightning as a causal or pre-disposing factor to the false bud rot.

15.1.5 Leaf Spot in Banana (Sigatoka)

Disease occurrence is seen at all stages of the crop during the southwest monsoon, and it is relatively free during the rest of the year. The incidence and spread are noticed during the periods when the relative humidity is greater than 90% with more than 5 hours of bright sunshine per day after a prolonged wet spell. The disease spreads faster in cloudy weather when the prevalent maximum and minimum temperatures are around 30°C and 20°C respectively.

15.1.6 Desert Locust

The desert locust is the most destructive pest of international importance, extending from Africa in the west to the Indo-Pak region in the east. The breeding, migration and convergence of locusts are regulated by seasonal rainfall pattern and wind direction. In India, the locust outbreaks occur during July–September. The breeding of the pest in India takes place with the onset of the first pre-monsoon showers. With the advent of the monsoon, they lay eggs and become swarms in 4 to 5 weeks and destroy the entire vegetation along the zone of convergence.

There are two distinct breeding regions for the locusts, namely, spring breeding region, which receives rainfall in winter (January to April), and the summer breeding region, which receives rainfall during May to July.

The spring breeding region extends from the northern parts of Africa and Arabia through Iran up to Balochistan. The summer breeding region extends from the south of Sahara, Libya and Egypt in Africa through southern Arabia and Pakistan to western India. The locust migrates back and forth rather regularly between these two regions to make effective use of the rainfall received in these areas. The ascent of the locust swarms from the breeding sites and the descent in the area of convergence are regulated by upward and downward air currents respectively.

15.2 PEST AND DISEASE PREDICTION MODELS

Several regression models have been developed for the prediction of crop pest and disease outbreaks based on weather variables. If it is found that there is a seasonal match in any pest or disease, a relationship could be established between the weather variables and the incidence of such pest and disease through correlation analysis and fitting regression equations. These are more location-specific and often fail to explain the incidence of the same pest or disease in other agroclimatic zones, as the crop environments are totally different. The regression models also fail to predict the incidence of pest and diseases if the initial inoculum or minimum pest population is absent under the field conditions, since the models do not consider biological activity as one of the parameters. Unless the threshold values of weather variables affecting different stages of pathogen or insects in their life cycle are understood properly and considered for developing the models, these may not serve any useful purpose. Hence, a thorough knowledge (see Fig. 15.3) on interaction between host (H) and pest (P) should be gained before attempting any model, as the same physical environment (E) influences both host and pest. It also appears that forecasting pest incidence on the basis of weather variables is more complex, as compared to crop diseases, due to complex interactions with numerous biotic factors in the former case.

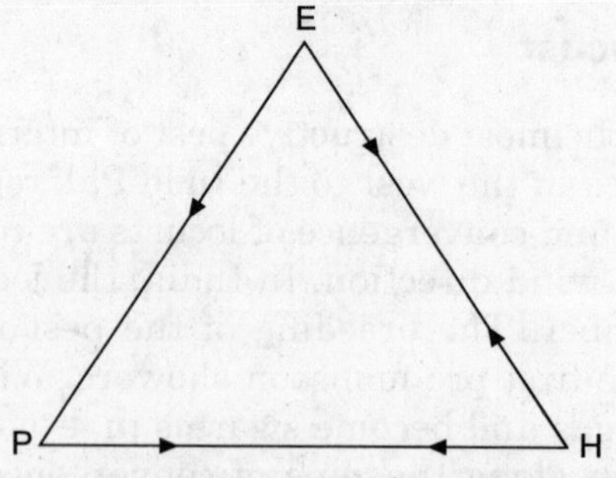

Fig. 15.3 Host, pest and environment interaction.

It is well established that leaf wetness duration is the causal factor for triggering disease, provided initial field inoculum is satisfactory. The leaf wetness duration is dependent on the temperature and the actual vapour pressure in atmosphere just above the crop canopy. Based on the above concept, several simulation models have been developed and farmers are advised on the initial fungicidal spray. Intensive studies have been carried out in India by the ICRISAT in the case of late leaf spot of groundnut. The minimum threshold levels of leaf wetness should be worked out for each disease, and it is a pre-requisite to forewarn various crop diseases. Such studies are lacking for other important crop diseases across the country. In the case of insect pests, the temperature and humidity parameters in relation to precipitation and light intensity are more important considerations for developing models on outbreak intensities.

To predict growth and development of insect pests and diseases, the Degree-Day (DD) Simulation Models are used extensively. The degree-days represent the accumulation of thermal units above a particular minimum (threshold) temperature for the 24-hour period. The threshold temperature (base temperature) represents the limit below which the developmental processes are arrested. To predict the stage of development from DD values, the thermal constant (sum of DD) for an event must be established through experimentation. The Degree-Day (DD) Simulation Model is as follows:

$$DD = [(T_{max} + T_{min})/2 - T_{base}]$$

where

T_{max} is maximum temperature in °C
T_{min} is minimum temperature in °C
T_{base} is base temperature (threshold) of insect pest or disease

15.2.1 Powdery Mildew of Grapes

The disease can be forewarned on the basis of maximum temperature and rainfall (Jahagirdar et al., 2001) using the following model:

$$DI = 86.69 + 6.41\ Max.T(2) + 0.30\ RF\ (4)\quad R^2 = 0.97$$

where

DI is the disease index

Max. $T(2)$ is the maximum temperature two weeks before the incidence of disease

RF(4) is the rainfall four weeks before the incidence of disease.

In the case of late-pruned crop, the above equation holds good for forewarning the incidence of powdery mildew in northern Karnataka two weeks ahead. It can be used for dissemination well ahead through agromet advisory system based on medium-range weather forecast.

15.2.2 Leaf Spot in Banana

The rate of increase of disease at a given time can be predicted using rainfall, relative humidity and percentage of infected leaves. The above model was developed in Taiwan (Doraisamy et al., 2002), and is as follows:

$$Y = 0.124 - 0.00489\, z + 0.000462\, (A) + 0.00515 I$$

where

Y is the rate of increase of disease at a given time
z is the percentage of infected leaves at previous recording
A is the product of accumulated rainfall in four weeks
I is the number of days with RH above 90% in a week.

15.2.3 Stem Bleeding in Coconut

Based on the aridity index during summer, a multiple linear regression was worked out for estimating the number of palms affected due to disease incidence (Radhakrishnan and Rao, 1988). The prediction equation is as follows:

$$Y = 243.0155 + 2.4895\, x_1 - 8.8681\, x_2 + 3.7445\, x_3 + 1.6488\, x_4 + 0.2550\, x_5 \quad (R^2 = 0.80)$$

where
Y is the estimated number of palms due to disease incidence
x_1–x_5 are the aridity indices from January to May respectively.

15.2.4 Phytophthora Disease of Black Pepper

Based on weather variables, the number of black pepper vines affected due to Phytophthora was worked out (Nair et al., 1988) using the past data, and it is as follows:

$$Y = -3.5644 - 0.6545\, x_1 - 1.4342\, x_2 + 0.6946\, x_3 - 0.0069\, x_4 + 0.0085\, x_5 - 0.1742\, x_6 \quad (R^2 = 0.80)$$

where
Y is the weekly estimated number of affected vines due to disease incidence
x_1–x_6 are maximum temperature, minimum temperature, maximum relative humidity, rainfall, number of rainy days and number of bright sunshine hours respectively.

Such equations can be developed if the data are available for a large number of years. They are useful for forewarning the pest disease incidence if lead time is worked out between the weather variables and disease/pest incidence. It will also be a useful tool for quantifying the impact of weather variables on disease incidence.

15.3 SIMULATION MODELS

A simulation model is a mathematical representation of a system which is dependent on several biological and environmental factors. Several simulation models have been developed and they are in use for operational purposes outside the country in case of a few crop diseases. These are listed below (Doraisamy et al., 2002):

BLITECAST for late blight of potato (Krause et al., 1975 and Krause and Hyre, 1975)

FAST for early blight of tomato (Madden et al., 1978 and Fry and Fohner, 1985)

EPIVEN for *Cercospora* leaf spot of peanut (Parvin et al., 1973)

EPIDEM for apple scab (Jonès et al., 1980)

EPIPRE for *Puccinia striiformis* on wheat (Smeets, 1994)

Couger blight for fire blight of apple (Smith, 1993)

BOTMAN for *Botrytis cinerea* on greenhouse vegetables (Shtienberg and Elad, 1997)

BLASTAL for leaf blast of rice (Hashimoto et al., 1984)

PYRICULARIA for leaf blast of rice (Gunther, 1986)

EPIBLAST for leaf blast of rice (Kim and Kim, 1994 and Rao and Krishnan, 1991)

BLASTAM for leaf blast of rice (Koshimizu, 1988)

15.3.1 Forecasting Alternaria Solani on Tomatoes (FAST) Model

It calculates a daily rating of the severity of risk of an early blight outbreak in tomatoes. During the growing season, the FAST model is run daily for each weather station, and spray advisories are issued 1 to 3 times per week as needed. The modified program, called TOMCAST (Tomato Forecaster), could be used not only for early blight prediction, but also to advise growers on the time to apply fungicide for control of anthracnose and Septoria leaf spot. Temperature and leaf wetness duration are the inputs to the model.

15.3.2 Forewarning Late Leaf Spot in Groundnut

The leaf wetness counter is an electronic device, with an artificial leaf as sensor, developed by ICRISAT, Hyderabad. Based on the wetness of the artificial leaf, the calibrated electronic counter gives leaf wetness index and the cumulative count. If the cumulative count for the previous 7 days is 450 or more, it flashes "SPRAY" and farmers are advised to spray against the disease. A maximum of three sprays are recommended if the duration of the groundnut variety is more than 90 days. This system was popularised among the groundnut farmers in Andhra Pradesh. It is of double advantage, because by controlling leaf spot disease the occurrence of stem rot can also be prevented. This is because the latter disease is promoted by leaves infested by fallen leaf spot.

15.4 HOT SPOT AREAS FOR PESTS AND DISEASES

For a particular crop pest or disease, some regions may be highly prone to incidence due to favourable environmental conditions. The demarcation of such regions for a particular pest or disease in relation to the threshold weather variables would be of immense use to the planners to formulate suitable strategies against their outbreaks. In hot spot areas it would be rewarding to introduce crop varieties with adequate levels of resistance against the pest or disease concerned. The work done in this direction on the tea mosquito bug (Rao, 2002) is discussed here as a case study.

The tea mosquito bug is a very severe pest in cashew, and the incidence is seen during the flushing and flowering phases which coincide from September to February, depending upon genotype and geographical co-ordinates. The night minimum surface air temperature plays a major role in the incidence of pest population. The peak pest population follows the night temperature (see Fig. 15.4(a)).

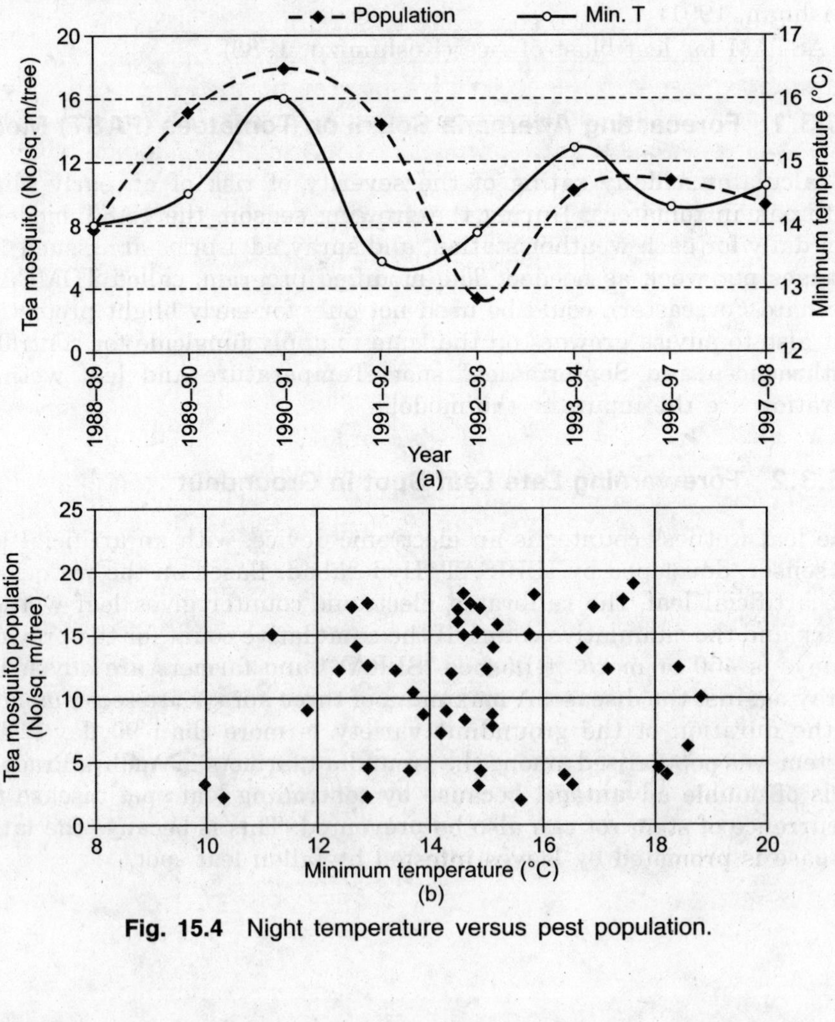

Fig. 15.4 Night temperature versus pest population.

The pest population was low when the minimum temperature was below 12°C and vice versa. Similarly, the occurrence of pest population was low beyond 18°C (see Fig. 15.4(b)). Occasional cloudy weather may be a pre-requisite and one of the important factors for pest population build-up. From the above, it can be inferred that night temperature between 13 and 18°C is the optimum for severity of tea mosquito bug incidence in cashew. Cloudiness and relatively dry weather with afternoon relative humidity between 40 and 60% may also be optimum for triggering pest population. The cold (<12°C of minimum temperature), hot (>35°C of maximum temperature) and wet (continuous heavy rains) weather conditions may not be conducive for pest population outbreaks.

The hot spot areas of tea mosquito bug across the cashew tract of the whole country have been demarcated, taking into account the optimum night temperature during the flushing and flowering stages. The relationship between intensity of pest occurrence and night temperature is given in Table 15.1.

Table 15.1 Relationship between pest occurrence intensity and night temperature.

Night surface air temperature	*Intensity of tea mosquito bug*
14 and 18°C	Moderate to severe
12 and 14°C	Low to moderate
10 and 12°C	Low
<10°C	Nil

The pest population may be moderate to severe across the west coast, from the northern districts of Kerala to the southern districts of Maharashtra, where cashew is predominantly grown. The pest population may be low to moderate across the east coast and high ranges (Ambalavayal in Kerala and Chintamani in Karnataka), and the crop is relatively free from pest population across the cashew tracts of West Bengal and the north-eastern States (see Fig. 15.5), as the pest population may not survive in these areas in cold weather.

15.5 WEATHER AND PESTICIDE APPLICATION

The deposition and retention of pesticidal spray droplets or dust particles on the crop canopy are dependent on the weather conditions prevailing at the time of application. Heavy rains following pesticide application may cause run-off of the spray deposits, and it would be therefore advantageous to avoid such contingencies by following weather forecasts. Pesticide application during noon hours in the summer months is not advisable due to aggravation of their phytotoxic potential. Wind speed and direction are to be considered before applying pesticides. In strong winds, pesticides may get blown away from the target crops. It is more so in case of pesticide sprays using aircrafts. Spraying or dusting should not be attempted in the direction opposite to wind. The best time for applying dust is early morning, when the atmosphere is charged with dew droplets. Pesticide sprays, if applied in the morning hours in dewfall areas during winter, are likely to get diluted, causing poor control and wastage of pesticide inputs.

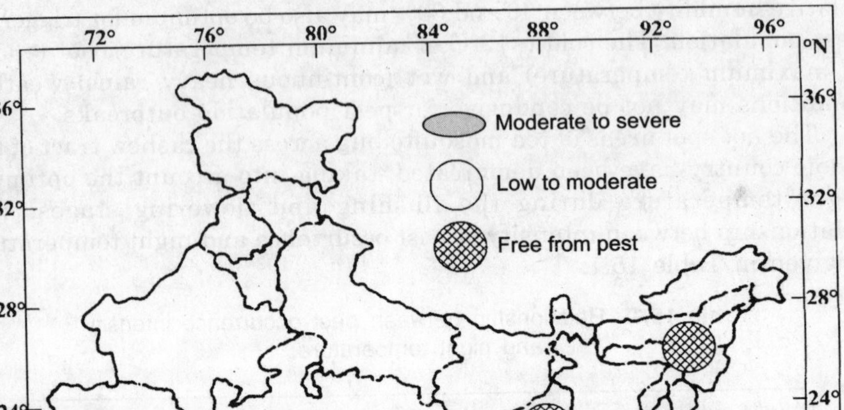

Fig. 15.5 Hot spot areas of tea mosquito bug in cahew across India.

15.6 ARE PEST AND DISEASE PREDICTIONS OPERATIONAL?

It is obvious that weather plays a major role in initiating outbreaks of pests and diseases. In temperate climate, pest population forecasting is relatively more reliable due to lack of complexities in the interactions involving the biotic and abiotic environments in such areas. The endemic areas could be very well identified by systematic collection of data on pest and disease incidence, through surveillance, along with the threshold weather variables. As the pest and disease incidence builds upon host plants under a favourable physical environment, it is difficult to understand the system and to forewarn the further build-up of pest and disease, unless concerted efforts are made to build-up a sound database. For the last three decades, several models, namely, linear and non-linear regression equations, logistic models and computer-based simulation models have been used for predicting or forewarning pest and diseases. Most of the above models are seen in literature from abroad, where these are put in operation. In India the forewarning and early warning systems against major crop pests and diseases are still not operational, despite the availability of some models. Databanks on pest population trends in time and space in relation to meteorological parameters are to be developed on major crop pests. Avenues for predicting the occurrence of pests or diseases with reasonable degree of accuracy have to be explored and appropriately refined models developed for adoption.

REFERENCES/SUGGESTED READINGS

Abraham, C.C. and Nair, M.R.G.K., 'The Brown Planthopper outbreaks in Kerala', The Rice Entomology Newsletter, IRRI **(2)**, 36, 1975.

Abraham, C.C., 'Pests of Coconut and Arecanut', *Advances in Horticulture: Plantation and Spice Crops, Part 2*, Eds. Chada, K.L. and Rethinam, P., Malhotra Publishing, New Delhi, pp. 709–726, 1994.

Boote, K.J., Batchelor, W.D., Jones, J.W., Pinnschmidt, H. and Bourgeois, G., 'Pest damage relations at the field level', *Systems Approach for Agricultural Development*, Eds. Penning de vries, F.W.T. et al., Kluwer Academic Publications, The Netherlands, pp.277–296, 1993.

Doraisamy, S., Rabindran, R., Muthulakshmi, P., Gunasekaran, K. and Salomi, J.H., *Seasonal climate forecasting for pest and disease management*, Short course on capturing the benefits of seasonal climate forecasts in agricultural mangement, Tamil Nadu Agricultural University, Coimbatore, pp. 88–103, 2002.

Jahagirdar, S., Venkatesh, H. and Jamadar, M.M., 'Influence of antecedent weather variables on powdery mildew of grapes in northern Karnataka', *Abstract and souvenir of national seminar on agrometeorological research for sustainable agricultural production*, Association of Agrometeorologists, Gujarat Agricultural University, Anand, p. 112, 2001.

Nair, P.K.U.K., Sasikumaran, S., Pillay, V.S. and Rao, G.S.L.H.V.P., 'Influence of weather on foot rot disease of black pepper (*Piper nigrum*

L.)', *Agrometeorology of Plantation Crops*, Eds. Rao, G.S.L.H.V.P. and Nair, R.R., Kerala Agricultural University, Thrissur, pp. 90–92, 1988.

Premanathan, T., Rao, G.S.L.H.V.P. and Mahana, N.K., 'Forecasting of rice blast disease epidemic in northern region of Kerala', *Oryza*, **36**(4), 364–368, 1999.

Radhakrishnan, T.C. and Rao, G.S.L.H.V.P., 'Impact of drought on stem bleeding disease incidence of coconut', *Agrometeorology of Plantation Crops*, Eds. Rao, G.S.L.H.V.P. and Nair, R.R., Kerala Agricultural University, Thrissur, pp. 90–92, 1988.

Rao, G.S.L.H.V.P., 'Weather and tea mosquito bug', *Climate and Cashew*, Kerala Agricultural University, Thrissur, pp. 53–62, 2002.

Rao, G.S.L.H.V.P. and Nair, R.R. (Eds.), *Agrometeorology of Plantation Crops*, Kerala Agricultural University, Thrissur, 1988, p. 187.

Thomas, B., Abraham, C.C., Karunakaran, K. and Gopalakrishnan, R., 'Effect of planting seasons and the associated weather conditions on the incidence of rice gall midge *Pachydiplosis oryzae* (Wood-Mason)', *Indian J. Met. Hydrol. Geophys.*, **26**(2), 237–240, 1975.

QUESTIONS

1. Write short notes on:
   (a) Seasonal abundance of pests and diseases
   (b) Pest and disease prediction models
   (c) Simulation models
   (d) Hotspot areas of pests and diseases
   (e) Weather and pesticide application
2. Fill in the blanks:
   (a) ——— conditions are the triggering factors for outbreak of crop pests and diseases.
   (b) Migration and dispersion of insect pests depend on —— and ——.
   (c) Geographical distribution of pests is mainly based on —— factors.
   (d) Majority of insect pests and diseases are commonly seen during the ——— season.
   (e) Aphids, thrips, scabs, mealy bugs and mites are common during the ——— season.
   (f) Blast disease of rice and powdery mildews of many crops are common during the ——— season.
   (g) Crop diseases caused by —— are strongly weather-dependent.
   (h) ——— pest is of international importance.
   (i) ——— index is used for forewarning late leaf spot in groundnut.
   (j) FAST model is used to forecast the ——— disease in tomato.

(k) ——— range of night temperature is optimum for severe incidence of tea mosquito.
(l) ——— and ——— meteorological factors are to be considered before applying pesticides.

3. State true or false:
   (a) Tea mosquito incidence is severe in cashew along the west coast.
   (b) Data banks on pest surveillance is required for developing pest forewarning models.
   (c) Pesticide sprays during dewfall mornings are effective in the control of pests and diseases.
   (d) Wet weather conditions are conducive for incidence of tea mosquito.
   (e) Leaf wetness index is a good measure for forewarning pests in groundnut.
   (f) $DD = \frac{T_{max} - T_{min}}{2} - T_b$.
   (g) Desert locust develops in India during the monsoon.
   (h) Bud rot in coconut is prevalent round the year in the humid tropics.
   (i) Incidence of phytophthora of black pepper is severe during the rainy season.
   (j) Pre-monsoon showers are conducive for incidence of gall midge in rice.
4. Match the following:

(a) BOTMAN	*Puccinia striiformis*
(b) BLASTAL	Late blight
(c) FAST	Late leaf spot
(d) BLITECAST	Tea mosquito
(e) EPIPRE	*Alternaria solani*
(f) Leaf wetness	*Botrytis cinerea*
(g) Hot spot areas	Leaf blast

5. Elucidate the following:
   (a) Pest and disease forewarning systems
   (b) Influence of weather on pest and disease outbreaks

16 CLIMATE CHANGE

Climate change and climate variability have become major concerns of humankind all over the world in recent decades. Recurrent drought and desertification seriously threaten the livelihood of over 1.2 billion people who depend on land for most of their needs. Weather-related disasters viz. drought and floods, icestorms, snowstorms, duststorms, hailstorms, landslides, thunder clouds associated with lightning, and forest fires are not uncommon in any region of the world. The year 1998 was one of the recent weather-related disaster years. It caused hurricane havoc in Central America and floods in China, India and Bangladesh. Canada and New England in the U.S. suffered heavily due to icestorm in January, and Turkey, Argentina and Paraguay due to floods in June. Vast fires in Siberia burned over 3 million acres of forests. Human and crop losses are the worst phenomena in such weather disasters, affecting the global economy to a considerable extent. The 1997–1998 El Nino events, the strongest of the last century, are estimated to have affected 110 million people and cost the global economy nearly US $100 billion. Statistics compiled from insurance companies for the period 1950–1999 show that major natural catastrophes, which are mainly weather- and climate-related, caused estimated economic losses of US $960 billion. Most of the losses were recorded in recent decades (Obasi, 2003). Increase in aerosols (atmospheric pollutants) due to emission of greenhouse gases (including black carbon, CFCs, HCFCs, HFCs and PFCs), ozone depletion, UV-B filtered radiation, cold and heat waves, global cooling and warming, and "human hand" in the form of deforestation and loss of wetlands, in the process of imbalanced development for betterment of humankind, may be causal factors for climate variability and climate change.

Indian economy is mostly agrarian, and Indian foodgrains production depends on the onset of monsoon and its further behaviour. The year 2002 was a classic example of how Indian foodgrains production is dependent on the rainfall of July. It was declared as the *all-India drought year*. The rainfall deficiency was 19% against the long period average of the country, and 29% of the area was affected due to drought during the above year. The *kharif* foodgrains production was less by around 16%. Similar was the case during the all-India drought in 1987. It reveals that monsoon failure over India is of great concern, leading to economic loss of a considerable extent.

16.1 GLOBAL WARMING

The rate of increase in global warming was much slow (less than 0.5°C before 1850) before industrialisation in Europe and US, and it is projected to be faster now. The Intergovernmental Panel on Climate Change (IPCC) in its second interim report in 1995 projected the rate of warming for the 21st century to be between 1 and 3.5°C. High global mean average temperatures were recorded consecutively for 16 months during 1997 and 1998. 2005 was the second warmest year after 1998, and 2001 was the third warmest. 1999 was the sixth warmest on record, though it was a La Nina year, supposed to be cooler. The 1990s were the warmest decade of the 20th century and it is likely that the rate and duration of warming of the 20th century were larger than at any other time during the last 1000 years (Obasi, 2003). The above situation may lead to a faster rate of global warming than projected (see Fig. 16.1). Perhaps, the most striking evidence of the effects of global warming are the changes in flora and fauna. In parts of the northern hemisphere, the growing

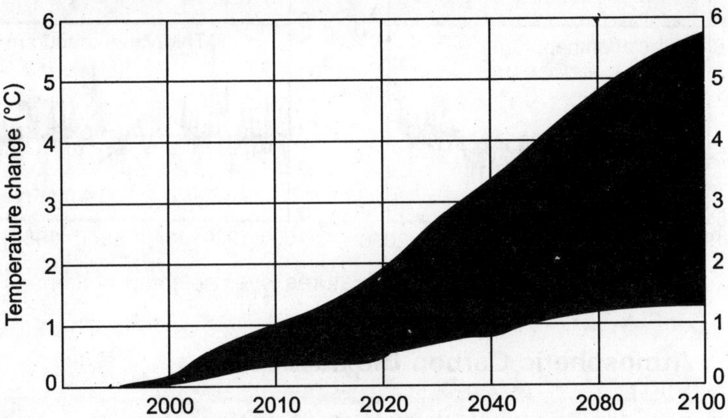

Fig. 16.1 Projected rise in global temperature.

season has lengthened by nearly 11 days since the early 1960s. Some of the changes in the growing season are linked to milder winters that have been part of the overall pattern of global warming since around 1970 (Obasi, 2003). A significant warming of 0.4°C per 100 years was noticed in the last century over India, but the rate of increase slowed down in the last three decades (see Fig. 16.2). This warming was mainly due to an increase in the maximum temperature (Kumar et al., 2002).

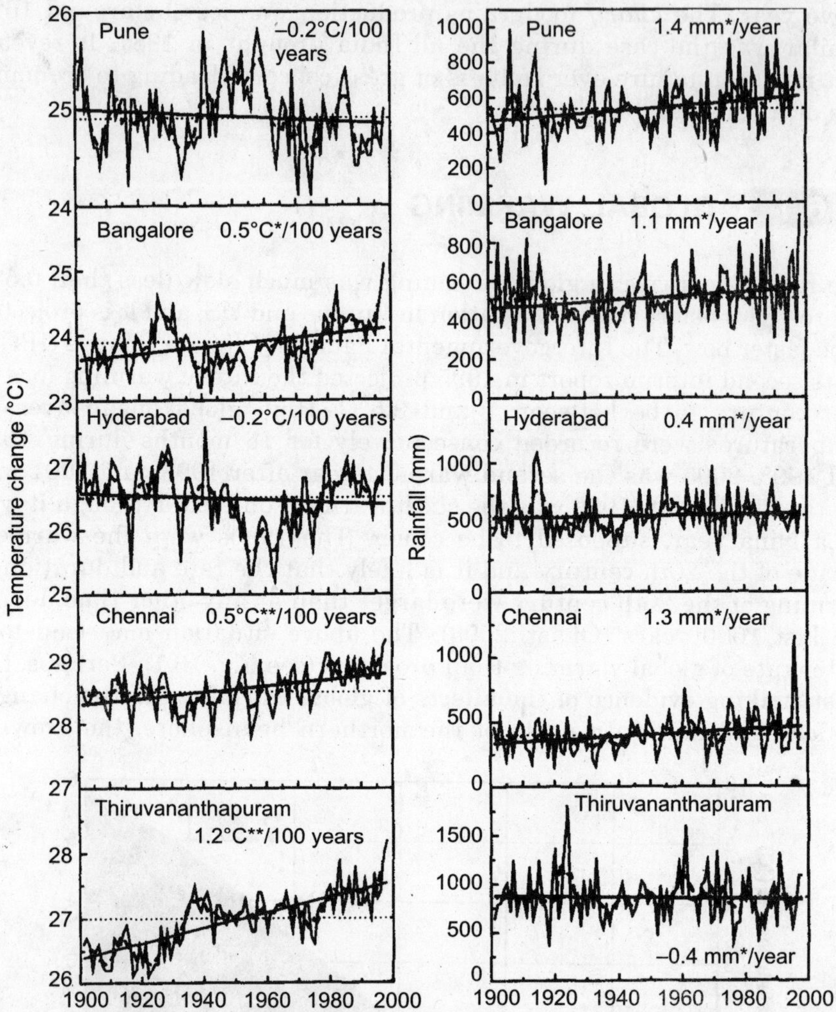

Fig. 16.2 Rainfall and temperatures over selected stations.

16.1.1 Atmospheric Carbon Dioxide

The concentration of carbon dioxide in the atmosphere was in a steady state at 280 ppm till the preindustrial period up to 1950. It is rising since then at the rate of 1.5 to 1.8 ppm per year. The concentration of carbon dioxide in the atmosphere increased from 280 ppm before 1950 to 370 ppm in 2000. It is likely to be doubled by the end of the 21st century. The measurements at IARI, New Delhi showed that the concentration of carbon dioxide increased from 330 ppm in 1984 to more than 370 ppm in 2000 (see Fig. 16.3), indicating a trend similar to that found globally (Uprety, 2003). Over the same period, atmospheric concentrations of methane and nitrous oxide increased by 151% and 17% respectively. This resulted in an increase of global temperature by 0.6°C, causing global warming. As a result of the warming, the global mean sea level has risen by between 10 and 20 cm. The rise has been estimated to be as much as 10 times the average increase in the last 3000 years (Obasi, 2003). An increase of 2.5 mm per year is noticed in the level of the Indian Ocean since the 1950s. The trend appears to be higher on the east coast when compared to that of the west coast (Das and Radhakrishnan, 1991). The climate projections indicated that if the atmospheric CO_2 is stabilized at 400 ppm, 450 ppm and 550 ppm, the rise in the atmospheric temperature range will be 0.8–2.4, 1.0–3.1 and 1.5–4.4°C respectively.

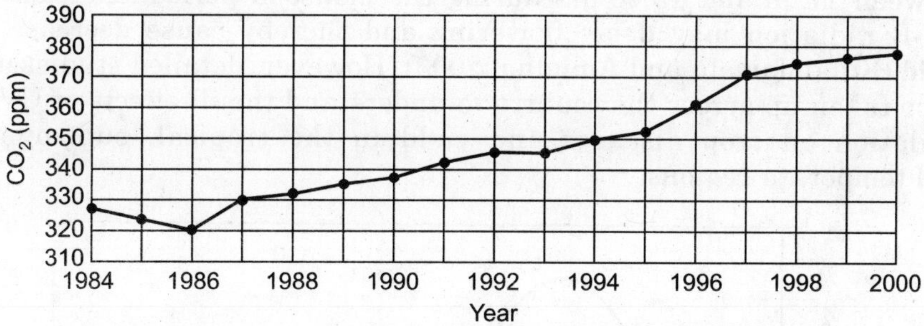

Fig. 16.3 Increase in CO_2 in the atmosphere.

16.1.2 Ozone Depletion

The global average thickness of ozone that is considered necessary for survival is 300 Dibson units, which is currently equivalent to 3 mm. In contrast it is about 100 Dibson units, which alarms us that we are in great danger. The ozone level fell to 90 Dibson units, which is historically the second lowest value, on 30-09-98. It nearly equals the lowest value ever recorded, 88 Dibson units, which was recorded on 28-09-94 over Antarctica. The ozone losses are caused by chlorine and bromine compounds, which are released by chlorofluorocarbons and halons. Year-to-year variations of size and depth of the ozone hole depends on the variations in meteorological conditions. The unusual cold temperature (5–9°F) over the Antarctic zone enables greater activation of reactive chlorine that ultimately causes more ozone loss and lower ozone levels.

16.1.3 UV-B Solar Radiation

Plants use sunlight for photosynthesis, and as a consequence, are exposed to the UV radiation that is present in sunlight. The UV-A, UV-B and UV-C are three regions having different wavelengths under UV radiation (see Fig. 9.1). Due to ozone depletion, the UV-B radiation in the range of 280–320 nm reaches the ground, and if the plants receive the same, it is detrimental to crops. Little is known on the impact of ozone depletion and increasing UV-B radiation on the ontogeny of tropical plants. The average dose of UV-B radiation at Madurai (10°N) indicates that two peaks have been observed, one during summer and the other between August and September (see Fig. 16.4). The UV-B radiation at Vellanikkara (10°31′N) during the noon hours crossed the level of 2 MED for 2 hours between 12:00 and 14:00 hr irrespective of the duration of sunshine more or less. On a cloudy day (01-08-02), the UV-B radiation never crossed 2 MED, which is arbitrarily fixed as the upper limit, beyond which it may affect biological activity. The UV-B radiation is noticed from 8:00–9:00 hr to 18:00 hr (see Figs. 16.5(a), (b) and (c)). The maximum duration of UV-B radiation (>2MED) varied between 11:30 and 15:30 hrs during the monsoon period. Increase in UV-B radiation may delay flowering and thereby cause decrease in yield (Kulandaivelu and Amudha, 2003). However, detailed studies are to be taken up across the country to understand the ill-effects of UV-B radiation on crop diseases and yield in the tropical, subtropical and temperate regions.

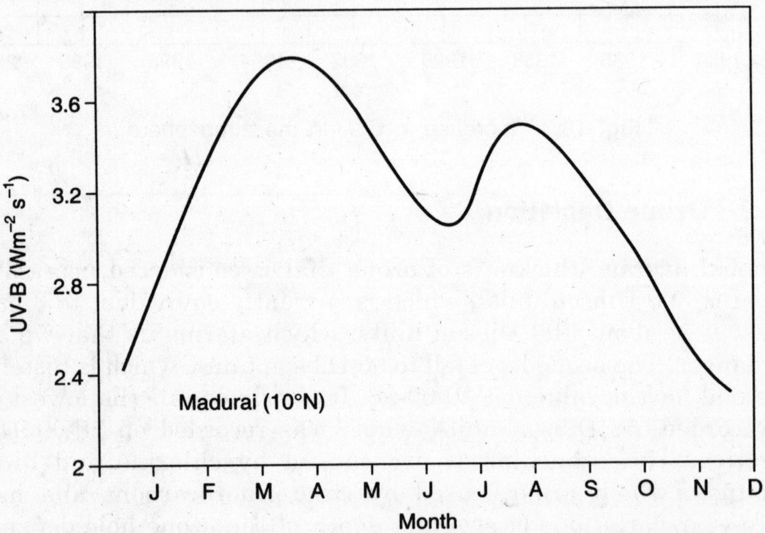

Fig. 16.4 Monthly UV-B radiation at Madurai (Tamil Nadu).

Period	6–7	7–8	8–9	9–10	10–11	11–12	12–13	13–14	14–15	15–16	16–17	17–19	Total
Sunshine	0.0	0.0	0.5	0.2	0.5	0.5	0.8	0.0	0.5	0.7	0.4	0.0	4.1

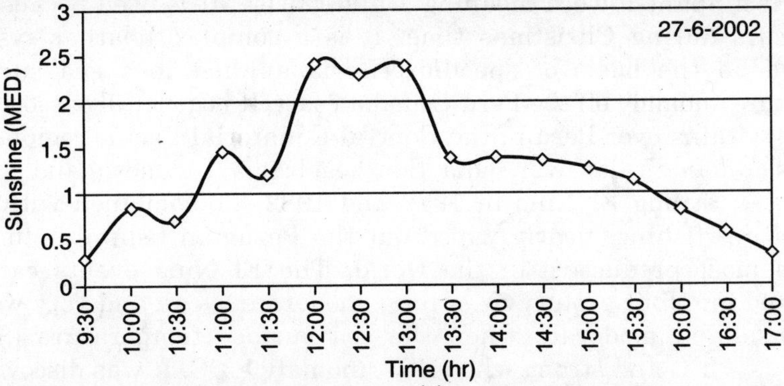

(a)

Period	6–7	7–8	8–9	9–10	10–11	11–12	12–13	13–14	14–15	15–16	16–17	17–19	Total
Sunshine	0.0	0.0	0.6	1.0	1.0	1.0	0.8	0.6	1.0	0.6	0.8	0.3	7.7

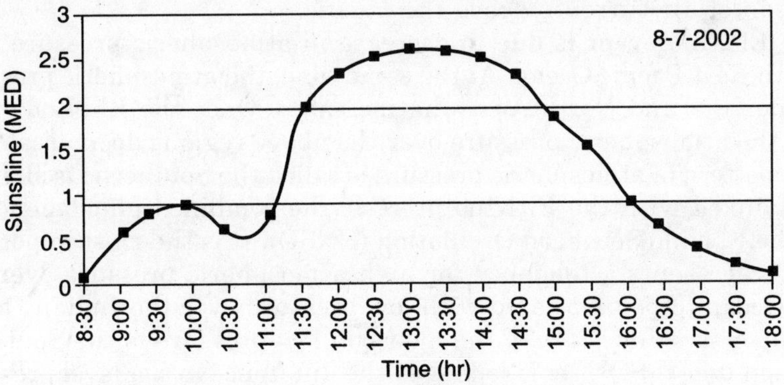

(b)

Period	6–7	7–8	8–9	9–10	10–11	11–12	12–13	13–14	14–15	15–16	16–17	17–19	Total
Sunshine	0.0	0.0	0.1	0.4	0.0	0.1	0.1	0.0	0.0	0.0	0.0	0.0	0.7

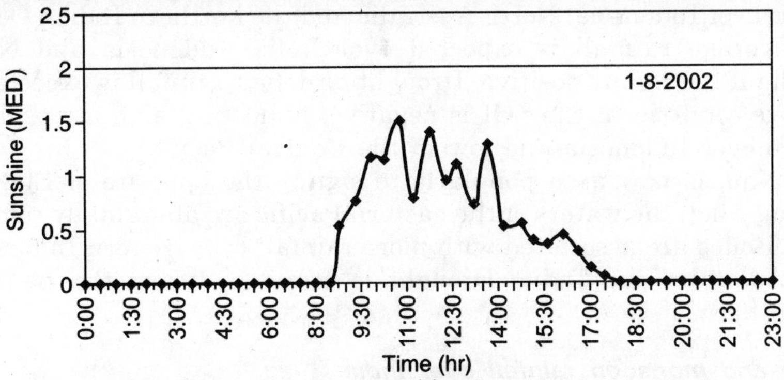

(c)

Fig. 16.5 Hourly UV-B radiation at Vellanikkara (Kerala) during the monsoon.

16.1.4 El-Nino and La Nina

El Nino is a Spanish word meaning 'Child Christ'. It is used because El Nino occurs during Christmas time. It is a complex weather system identified on the basis of specific criteria applied to a sea surface temperature anomaly off the Peru-Ecuador coast. It is generally associated with heavy rains over Peru in the Central Equatorial Pacific region and droughts or deficit rains over India, Southeast Asia, Indonesia and North Australia. A strong El Nino in 1972 and 1982–83, combined with the effects of overfishing, nearly wiped out the Peruvian fishing industry, once the most productive in the world. The El Nino events can be categorised into four groups viz. strong, moderate, weak and very weak. It is considered moderate when the sea surface temperature (SST) anomaly is > 2°C and strong when SST anomaly >3°C. It was discovered that the Pacific Ocean warming usually coincided with the warming that occurred off the coast of Peru, which was connected with large shifts in the rainfall patterns across the tropical Pacific and broad changes in the circulation of air currents above the Pacific.

The El Nino event is due to decrease in atmospheric pressure over the South East Pacific Ocean. At the same time, the atmospheric pressure over Indonesia and North Australia increases. Once the El Nino event is over, the atmospheric pressure over the above regions decreases. This see-saw pattern of atmospheric pressure is called the Southern Oscillation. As it is linked with the El Nino process, the combined phenomenon is called the El Nino Southern Oscillation (ENSO). It is the most important one and represents a tendency for high atmospheric pressure over the Pacific Ocean to be associated with low pressure over the Indian Ocean and vice versa. A measure of the monsoon low pressure is the Southern Oscillation Index (SOI) represented by the difference in sea level pressure over Tahiti, an island in the South Central Pacific, and Darwin in North Australia, which represents the northern part of the Indian Ocean. The positive SOI denotes high pressure over the Central Pacific and low pressure over Indonesia, North Australia and the northern Indian Ocean. Above-average rainfall is expected over India, Indonesia and North Australia if the SOI is positive. Drought or deficit rainfall is expected in the above countries if the SOI is negative, indicating high atmospheric pressure over Indonesia and low in the Central Pacific.

La Nina is now used popularly to signify the opposite of El Nino, occurring when the waters of the eastern Pacific are abnormally cold. La Nina episodes are associated with more rainfall over eastern Indonesia, North Australia and India. Drought is common during the La Nina years in Peru.

El Nino and monsoon rainfall over India

Indian monsoon rainfall has shown stable long-term characteristics since historical times, with extremes being a part of its natural variability.

However, palaeoclimatic evidence indicates that the monsoon had been more active during interglacial conditions (Kumar et al., 2002). However, the El Nino events are used in India for forecasting monsoon rainfall with the help of long-range weather forecasting techniques, as El Nino events influence the Indian monsoon. During the period 1871–2001, 11 out of a total of 22 drought years were El Nino years, whereas none of the 19 excess rainfall years was an El Nino year (see Table 16.1).

Table 16.1 El Nino/La Nina association with all-India summer monsoon rainfall anomalies during 1871–2001 (number of years, categorywise)

Category	*Deficit monsoon*	*Normal monsoon (–ve)*	*Normal monsoon (+ve)*	*Excess monsoon*	*Total*
El Nino	11	11	4	0	26
La Nina	0	1	9	8	18
Others	11	23	42	11	87
Total	22	35	55	19	131

(Source: Kumar et al., 2002).

Note: A year is classified as either deficient, normal (negative), normal (positive) or excess monsoon year, when the all-India summer monsoon rainfall is below –10%, between –10% and zero, between zero and +10%, or above +10%, respectively.

The above relationship is yet to be established, as the interrelationship between El Nino and the Indian monsoon is not always the same, as seen during the monsoon of 2002.

16.1.5 Climate Shifts in the Humid Tropics

Kerala falls under the humid tropics. In the humid tropics, where monsoon rainfall is stable, the climate shifts are not uncommon. Kerala falls under the climatic type of "B_4–humid" based on climatic normals. The State had shifted from B_4 to A (perhumid) climatic type in 36 years, indicating 26.7% of the years on the wetter side. In contrast, the State had shifted from B_4 to B_3, B_4 to B_2 and B_4 to B_1 in 30 (22.2%), 26 (19.3%) and 5 (3.7%) years respectively, indicating the drier side in 45.2% of the years. Only three years fell under perhumid (A) category in the last two-and-a-half decades (see Table 16.2). The moisture index was also in decreasing trend (–27.6%) from 1951 to 2005, indicating that the climate shifted from B_4-humid to B_3, B_2 and B_1-humid in recent decades, though there was no such trend for the period as a whole (see Fig. 16.6(a) and (b)). It revealed that the State of Kerala experienced change from wet to dry within the B category of climate types in recent decades. The climate shift towards the drier side was more in 1981–90 as only 1 year recorded normal climate type (B_4) and 8 years between B_3 and B_1 on the drier side. Such a trend was seen before 1900. However, 1981–90 was a typical decade as only 1 year was seen on the wet side (perhumid). It reveals that 1981–90 was the driest decade. If it is studied on a seasonal basis, the scenario of climate shifts will be different. The majority of years fell under the semi-arid and arid climate types during winter and summer months towards the northern districts of Kerala.

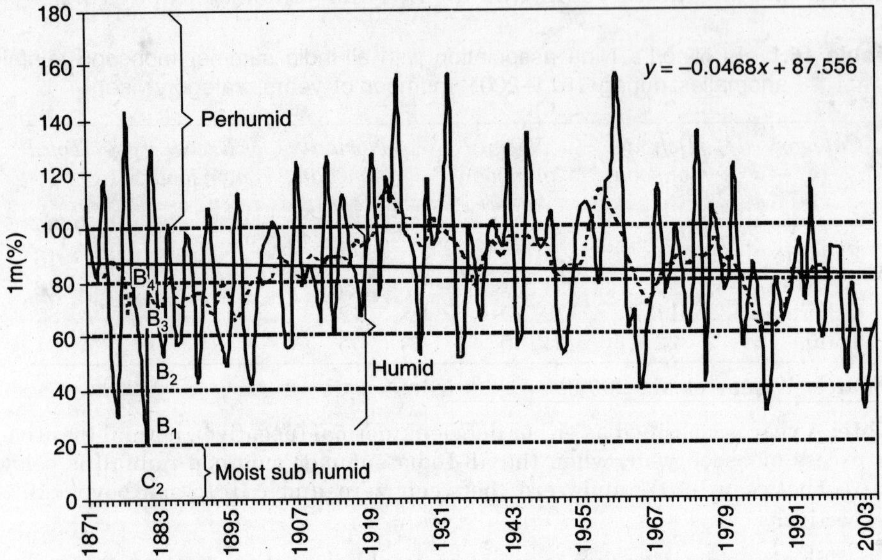

Fig. 16.6(a) Moisture index from 1871 to 2003 over Kerala.

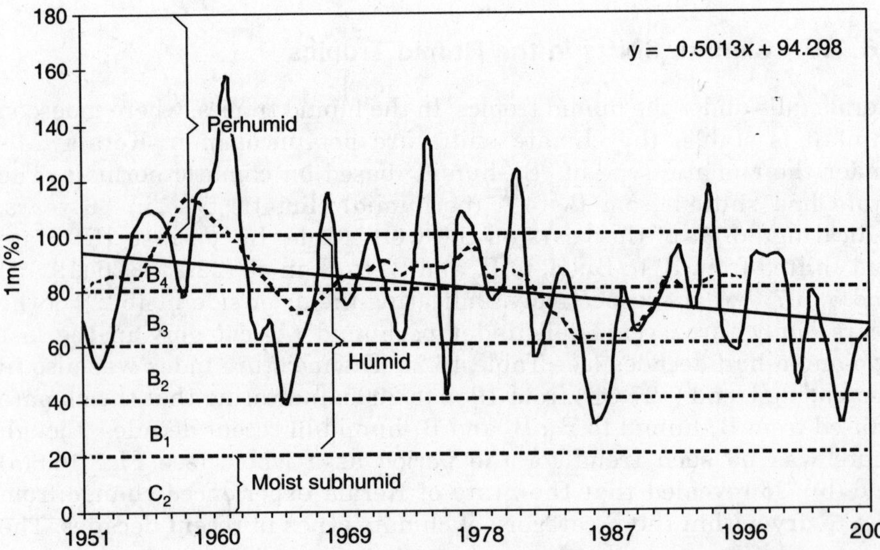

Fig. 16.6(b) Moisture index from 1951 to 2005 over Kerala.

Table 16.2 Decade-wise number of climate shifts over Kerala during 1871–2005.

Decade	*Climatic type*				
	B_1	B_2	B_3	B_4	A
1871–80	1	2	1	3	3
1881–90	0	5	1	2	2
1891–00	0	4	3	1	2
1901–10	0	2	1	5	2
1911–20	0	0	3	4	3
1921–30	0	1	1	4	4
1931–40	0	2	1	4	3
1941–50	0	1	1	4	4
1951–60	0	1	3	1	5
1961–70	1	0	5	1	3
1971–80	0	3	1	3	3
1981–90	2	1	5	1	1
1991–00	0	2	2	5	1
2001–05	1	2	2	0	0
Total	5	26	30	38	36
Climatic shifts (%)	3.7	19.3	22.2	28.1	26.7

16.2 WEATHER ABNORMALITIES

The weather abnormalities which can affect crops adversely in India are as follows:

(i) Droughts and floods
(ii) Heat and cold waves
(iii) Excessive or defective insolation
(iv) Thunderstorms, lightning and hailstorms
(v) Duststorms
(vi) Cyclones and anticyclones
(vii) High winds
(viii) Forest fire outbreaks
(ix) Ice and snowstorms
(x) Sea level size

16.2.1 Droughts

Droughts are dealt with in Chapter 11. The summers of 1983 and 2004 were declared as *disastrous drought years*. During these perods the major part of plantation crop production was adversely affected over Kerala.

16.2.2 Floods

The years in which actual rainfall is "above" the normal by twice the mean deviation or more is defined as *years of floods* or *years of excessive rainfall*. Like droughts, the definition of floods also varies from one situation to another and from one region to other. Some of the flood years based on monsoon rainfall are as follows:

India	Kerala
1878, 1892, 1917, 1933, 1942, 1956, 1959, 1961, 1970, 1975, 1983, 1988 and 2007	1994, 1998, 2005 and 2007

The rainfall deficiency during monsoon 2002 over Kerala was 35%, which was one of the highest rainfall-deficit years. In contrast, several districts in Kerala experienced heavy rain during October 2002 due to severe cyclonic storm over the Karnataka coast. It led to floods and devastated plantations in Kannur and Kasaragod districts. Kannur received 370 mm of rainfall on 14th October 2002, which was the highest single-day rainfall received since 1924.

Only few years are mentioned here as examples. Every year across the country, droughts and floods are noticed in one part or the other. If the economy of the whole country is affected due to floods or droughts, these are recorded only as catastrophes. Floods and their effects on crops can be mitigated through:

(a) Early flood warning
(b) Growing tall trees or plants along with coastal belts
(c) Selection of suitable crops and varieties
(d) Construction of protective dams and other structures.

The impact of draught on foodgrains is felt much more when compared to that of floods.

16.2.3 Heat and Cold Waves

Heat and cold waves can be classified as disasters since many deaths occur due to them, particularly among weaker sections of society who cannot afford to ensure protection against extreme heat or cold. May–June is the period when the heat wave occurs over northern and central India.

As per Indian convention, the departure of the maximum temperature to the order of 6 to 7°C above normal is termed 'moderate heat wave', and, that to 8 and above 8°C is called 'severe heat wave'. Once the monsoon sets in, the centre of high frequency shifts towards northwest India.

Heat wave develops over Rajasthan and Punjab due to their being in the interior. The strong northwesterly winds along with land trajectory over hot regions inhibit the setting in of the sea breeze over the eastern coastal belt. That is why there is high incidence of heat waves and their persistence over coastal Orissa and Andhra Pradesh. Maximum heat wave incidence is in May and June. In July and August, southeast Chennai becomes another centre of high frequency of heat waves. The heat wave that occurred during March 2004 over Himachal Pradesh devasteted several temperatures. Huge crop loss was estimated.

Cold wave is defined with respect to the normal minimum temperature experienced over a given area. Every year a number of deaths are reported during cold wave spells. Ground frost, which adversely affects various crops, is experienced under cold wave conditions. Nearly 75% of cold waves in the country occur during January to February. The year 2002–03 (December 2002 and January 2003) experienced one of the worst cold waves across the northern plains of the country. Uttar Pradesh and Bihar were the worst-affected States due to the cold wave in January 2003. Bihar declared the cold wave as a natural calamity. It was seen for the first time after 12 years. Excluding the extreme south peninsula, the country is susceptible to cold waves. The northwestern parts of the country and areas further north are more favourable regions for the genesis of cold waves, and the incidence of 5 to 20 days of frost in each month during the cold season is common.

The damage was more in low-lying areas where cold air settles and remains for a long time on the ground. Crops like mango, guava, pineapple, boro rice (in Assam) and maize (in Bihar) suffered heavy losses. Various fish species had high mortality rates and arrested growth. However, temperate fruits like apple, peach and cherry prospered due to extended chilling. Similar was the effect on crops in several northern states due to cold wave in winter 2008.

Proper crop selection; use of wind breaks or shelter belts; frequent irrigation; covering young plants with thatches or plastic shelters; and maintenance of maximum water depth in, and aeration of, fish ponds, will help to minimise the impact of cold waves. A proper weather forewarning system is of utmost importance.

Cold wave warnings

It is generally possible to give prior warning of the incidence of cold waves from the daily charts prepared by the IMD. On receipt of the warning messages, a farmer or horticulturist will know that a cold wave is approaching, and protective steps may have to be taken to save his crops from frost damage. Slopes are less liable to frosts as the cold air current collects in valleys. Therefore, slopes are to be preferred for horticultural gardens. Frost albums and frost warnings, windbreaks and heating can save the crops to a greater extent from frost damage. They are being practiced in developed countries. However, the above control practices against cold waves are not seen in India.

16.2.4 Excessive or Defective Insolation

Insolation is a function of the inclination of the solar rays, the length of the day and the transparency of the atmosphere. During clear days in summer, soil temperature in the equatorial and tropical regions reach very high levels, particularly on days when air movement is weak. High soil temperature in the coconut root zone in the absence of soil moisture during summer may restrict root activity. The same is true in almost all crops. The crops grown under sandy soils are the worst-hit, as the soil temperature is higher in sandy and sandy loam soils when compared to that of laterite soils.

Experiments made at Pune show that on such days a thin cover of white substance like chalk can bring down the soil temperature at noon by as much as 18 to 20°C. A black cover is useful for maximum absorption of the insolation, thus heating up the soil layers as much as possible during winter days with weak insolation. Wetting also cools the soil surface temperature by about 15°C. But this cooling effect by evaporation lasts only as long as there is enough moisture in the soil. Experiments at Pune have also shown that a thin layer of chalk or a surface mulch of "basse" helps not only to keep the soil layers cool but also to conserve soil moisture.

In the humid tropics like Kerala, insolation is less during rainy season (June–September) for about 120 to 150 days depending upon locality. It may retard crop growth and yield, while the same weather conditions become favourable for the incidence of crop pests and diseases. Cloudiness is injurious to crops during certain critical phases like flowering. To some extent, the same may be true in the northeastern States. The low amount of insolation in these States during the monsoon season may be an inhibiting factor for crops which need more sunlight.

16.2.5 Thunderstorms, Lightning and Hailstorms

These are comparatively local in character but are often associated with depressions affecting larger areas. The relatively more violent ones usually occur in the pre-monsoon and post-monsoon seasons. There is a similarity between duststorms, thunderstorms and hailstorms in their fundamental mechanism. Only duststorms are noticed in the absence of sufficient moisture in the atmosphere. When enough moisture is present, a thunderstorm results. The hailstorm is a particularly violent thunderstorm. Hailstorms are very destructive to standing crops and even to livestock and human beings. Their frequency increases in summer. Once the monsoon sets in, the phenomenon stops occurring. The protection of crops against violent hailstorms is rather difficult, but if forewarned the farmer can consider harvesting his crops if they are almost ready for harvest. Weather insurance is the only option in such weather abnormalities.

16.2.6 Duststorms

When the conditions over the desert and semi-arid areas are unstable and humidity is low, convective clouds do not build up to greater heights. The downdrafts from these storms raise loose dust and cause duststorms. The mechanism for the formation of duststorms is the same as that of thunderstorms. The vertical growth is arrested due to low humidity aloft. The downdrafts are initiated quickly by the fall of super-cooled drops from levels a little above freezing level. The updrafts carry dust aloft through a large part of the cloud. The falling water drops evaporate quickly due to the temperature and low humidity in the low levels, and hence hardly any precipitation reaches the ground. The duststorms of northern and western India are locally known as 'andhis'. They are common over Delhi, Jammu, Haryana, Punjab, Rajasthan, Uttar Pradesh, Madhya Pradesh and Bihar. It is estimated that the quantity of the raised dust may be of the order of 50 to 500 kg/ha/day.

16.2.7 Cyclones and Anticyclones

The more serious cyclonic storms usually form in the Bay of Bengal and the Arabian Sea in the transition months (April to May and October to November). They enter inland, and cause heavy rainfall and widespread destruction of life and property, due to high winds and occasional tidal waves. Their effects are most severe in the coastal tracts. The east coast of India is affected more frequently when compared to that of the west coast. On an average, one or two severe cyclones may be expected in the pre-monsoon period and two or three in the post-monsoon periods. One of the features of the meteorological phenomena like cyclonic storms and depressions is that their normal direction and speed of movement are generally understood, so that the expected time of their impact on a particular locality and the associated heavy rains and squally weather can reasonably be forecast a day or two ahead. On receipt of the warning, precautionary measures are taken and damage is prevented.

Anticyclones are opposite to cyclonic characteristics and are composed of subsiding air. No condensation and cloud formation are seen in anticyclone areas. The anticyclone that forms in the upper atmosphere over western Rajasthan is one of the reasons for not getting enough rains in Rajasthan during the rainy season (June to September).

16.2.8 High Winds

Wind transports the moisture and heat of the atmosphere from one location to another and therefore has some effect on crop production. The wind constitutes a hazard to agriculture in different ways:

(i) Wind can mechanically damage crops if speed is too high.
(ii) Hot wind will encourage high rate of evapotranspiration.

(iii) Wind may cause soil erosion.
(iv) Wind may speed up the chilling of plants under conditions of low temperature.

Crop damage by winds may be minimised or prevented by the use of windbreaks or shelterbelts. These are natural or artificial barriers to wind flow to shelter animals and crops. Wind has its most important effect on crop production indirectly through transport of moisture and heat in the air. Wind dispersal of pollen and seeds is natural and necessary for native vegetation and may be helpful for certain crops. Windbreaks composed of trees, shrubs, hedges or fences are widely used to protect both crops and animals from the winds. The permanent windbreaks are rows of trees planted perpendicular to the prevailing winds.

16.2.9 Forest Fire Outbreaks

The main causes of forest fires are lightning, self-combustion and human activity. Out of these, lightning is the major cause of forest fires. However, there are instances when fires have occurred during dry summer due to self-combustion.

Fire-inducing weather conditions are lack of rainfall, high temperature, low humidity and high wind speed. The types of trees, their moisture content and the stage of foliage also contribute to forest fires. The humidity and wind speed within the forest, and air temperature immediately outside the canopy, are also contributing factors. Fire-weather forecasts can be given on the basis of climatic indices.

16.3 WEATHER AND CLIMATE MODIFICATION

This is one of the areas in agricultural meteorological research. To maintain optimum environment and to increase crop production, some of the techniques used in weather and climate modification may be quite useful. However, great efforts are required to evolve the techniques in weather and climate modification, as the modified environment may pose several other problems. The following are some of the techniques through which better crop yields can be harvested.

16.3.1 Heat Evasion

During summer or in the absence of rains, the heat load on some of the plants is above the tolerance limits in the tropics and sub-tropics. Shading of plants is a common method of evading solar radiation. A number of shade structures are used and they are opaque. The shade can be of wood or fibre. The material acts as a thermostat and can be applied or removed as required. It keeps the temperature low and retards evapotranspiration.

Liming on lower portion of stems in the case of coconut and rubber during summer months may have beneficial effects. Some of the farmers practice it in several parts of Kerala.

16.3.2 Heat Trapping

The opposite of heat evasion is heat trapping, which is extremely beneficial in the temperate climate. Planting trees on sunny slopes and erecting alternate rows of low stone walls may be useful techniques in heat trapping. The low stone walls reflect the light back towards the lower portions and shaded sides of the trees. Additionally, the thermal capacity of the wall material will increase the local source of heat by night. Heat-sensitive crops can be covered with plastic enclosures during the night. These enclosures retard the loss of heat from the surroundings of the crop. This is an effective measure against cold wave damage. In the case of tomato, the crop is covered with polythene in the afternoon, so as to retain relatively high temperature in the night in the temperate regions, where temperature is below optimum for flowering. This will help in better fruiting.

16.3.3 Mulching and Ploughing

Mulching and ploughing are effective methods to retard heat from the soil and to save crops from successive cooling. Mulch is an effective mechanism in plantations to prolong the availability of soil moisture and to reduce the ill-effects of soil moisture stress on yield if it is done at the right time. Sand acts as a good mulch in black cotton soils to combat drought effectively. Farmers in Karnataka practice it and they have good crops in drought years.

Though mulches have many advantages, there are some limitations as well. The major limitation is that they are unable to save the aerial sections of plants from heat loss through thermal radiation.

16.3.4 Windbreaks and Shelterbelts

Wind influences plants physiologically and mechanically. As high winds increase evapotranspiration loss, erection of windbreaks (any structure that reduces wind speed) and shelterbelts (rows of trees planted for wind protection) can reduce evapotranspiration too. Reports also indicate that crop yields are better if they are protected against wind through windbreaks or shelterbelts. The beneficial effects of windbreaks may be stated as follows:

(i) Checking air movement and thus affording protection to fields and orchards.

(ii) Checking top-soil erosion. This is more effective in sand soils.

(iii) Evapotranspiration is reduced over the crop field by reducing wind speed.
(iv) Providing shade to the farm shed.
(v) Providing timber and fuel, as older trees are replaced by younger ones.

In contrast, it is also possible to harvest the winds, wherever wind speed is high, for driving windmills to pump water and generate power, which will indirectly help in the sustenance of agricultural production. In addition to the above, the *in situ* soil moisture conservation technique helps in increasing the amount of soil moisture stored in the root zone. The efficient non-conventional energy technologies will also go a long way in mitigating the ill-effects of climate change.

16.3.5 Solarization

Solarization is a method of hydrothermal disinfection, which can be categorised under modifications of soil microclimate. Solarization can be practiced during summer months in open fields by covering moist soil with a transparent polythene sheet and exposing it to direct sunlight. Increased growth response is observed in plants cultivated in solarized soil, in addition to effective control of soil-borne fungal pathogens and weeds. Soil should be kept moist during solarization to increase thermal sensitivity, as it improves heat conduction (KAU, 2002).

16.4 CLIMATE EVENTS

1877	All-India drought year.
1891	El Nino year.
1899	All-India drought year.
1925	El Nino year.
1929	Commission for climatology instituted by IMO.
1950	First Convention of the World Meteorological Organization (Formerly known as International Meteorological Organization–IMO). World Meteorological Day celebrated for the first time on 23rd March.
1950s	Radars, satellites and computers contributed to high quality research in atmospheric processes, which in turn led to technological achievements.
Late 1960s and early 1970s	Unprecedented drought in the Sahel. Evidence of a period of prolonged cooling raised concern about future climate.

1972	El Nino year.
1976	First authoritative statement by the WMO on the potential impact of an increasing accumulation of greenhouse gases in the atmosphere on our future climate.
1979	All-India drought year. First World Climate Conference led to the establishment of World Climate Programme (WCP); Collaboration of WMO with United Nations Environment Programme (UNEP) and the International Council for Science (ICSU) in research.
1982–83	Severe El Nino year–caused $13 billion of damage worldwide.
1983	Worst drought year during summer over Kerala. Plantation crops production suffered to a large extent in the same year and subsequent years, depending upon the type of crop.
1987	All-India drought. Indian foodgrains production declined.
1988	Established Intergovernmental Panel on Climate Change (IPCC).
1990	First Assessment Report of IPCC led to the United Nations Conference on Environment and Development (UNCED).
1990–91	Mild El Nino year.
1992	Framework Convention on Climate Change was signed in Rio De Janeiro (Brazil) during UNCED.
1995	Second Assessment Report of IPCC contributed to the negotiation of the Kyoto Protocol to the United Nations Framework Convention on Climate Change (UNFCCC).
1997	El Nino year.
1998	The warmest year ever recorded. Declared as the weather-related disaster year.
1999	The sixth warmest year. Super cyclone hit Orissa coast on 27th October.
2001	Third Assessment Report of IPCC. New and stronger evidence that warming over 50 years is attributable to human activities. The third warmest year.
2001–02	El Nino year.
2002	All-India drought. Indian foodgrains production declined.
2003	Worst heat wave in Europe in the last 130 years.
2004	Worst drought year during summer over Kerala. Tsunami disaster on 26th December in South and Southeast Asian countries.

2005 The second warmest year. Declared *the year of hurricanes*. Mumbai recorded more than 940 mm on 26th July (The highest one-day rainfall event). The costliest and deadliest cyclones viz. Katrina (August 31) and Rita (September 24) hit New Orleans, Texas and Louisiana in the U.S. and caused $10.5 billion and $11.3 billion damage respectively.

2006 Severe cyclone 'Ogni' hit AP coast in October and devastated the rice crop. Typhoon 'Durian' hit the Tabaco Albay province, South Manila, Phillippines on November 30 and caused heavy damage.

2007 Fourth Assessment Report of IPCC. Warmest winter in NHS. One of the warmest years. Bangladesh suffers heavily due to supercyclone 'Sidr' in November.

REFERENCES/SUGGESTED READINGS

Bolin, B., *Climatic Change and Their Effects on the Biosphere*, World Meteorological Organization, Geneva, 1980, p. 49.

Colls, J., *Air Pollution—An Introduction*, E&FN Spon, London, 1997, p. 341.

Das, P.K. and Radhakrishnan, M., 'Analysis of Indian tidegauge records', *Proceedings of the Indian Academy of Sciences*, pp. 177–194, 1991.

Das, S.W. and Thakur, R.S., *Supplement to Changes in Global Climate Due to Natural and Human Activities*, Allied Publishers, New Delhi, 1997, p. 171.

Gribbin, J., *Climatic Change*, Cambridge University Press, New York, 1979, p. 280.

KAU, *Package of Practices Recommendations: Crops*, 12th ed., Kerala Agricultural University, Thrissur, 2002, p. 278.

Kulandaivelu, G. and Amudha, P., 'Solar UV-B Radiation—How effective is it to tropical species?', *Second International Congress of Plant Physiology,* January 8–12, 2003, Eds. Pant, R.C. and Ghildiyal, M.C., Division of Plant Physiology, Indian Agricultural Research Institute, New Delhi, 2003, pp. 94–100.

Kumar, K.R., Kumar, K.K., Ashrit, R.G., Patwardhan, S.K. and Pant, G.B., 'Climate Change in India: Observation and Model Projections', *Climate Change and India—Issues, Concerns and Opportunities*, Eds. Shukla, P.R., Sharma, K.S. and Ramana, P.V., Tata McGraw-Hill, New Delhi, pp. 24–75, 2002.

Lockwood, J.G., *Causes of Climate*, Edward Arnold, London, 1979, p. 260.

Obasi, G.O.P., *Our Future Climate: World Meteorological Day Message*, World Meteorological Organization, Geneva, 2003, pp. 2–4.

Rao, G.S.L.H.V.P., *Climate Variability and Coconut Production in Humid Tropics*, Kerala Agricultural University, Thrissur, 2002, p. 38.

Rao, G.S.L.H.V.P., Subash, N., Gopakumar, C.S. and Sebastian, S., 'Climate variability and coconut production', *National Symposium on Climate Variability,* Indian Institute of Tropical Meteorology, Pune, 1994.

Shukla, P.R., Sharma, S.K. and Ramana, P.V. (Eds.), *Climate Change and India—Issues, Concerns and Opportunities*, Tata McGraw-Hill, New Delhi, 2002, p. 317.

Singh, S.K. (Ed.), *Dictionary of Global Climatic Change*, Common Wealth Publishers, New Delhi, 1998, p. 280.

Uprety, D.C., 'Rising atmospheric carbon dioxide and crops: Indian studies', *Second International Congress of Plant Physiology,* January 8–12, 2003, Eds. Pant, R.C. and Ghildiyal, M.C., Division of Plant Physiology, Indian Agricultural Research Institute, New Delhi, pp. 87–93, 2003.

QUESTIONS

1. Define the following:
   (a) Cold wave (b) Heat wave
   (c) Flood (d) Drought
2. Distinguish between:
   (a) Southern Oscillation and Southern Oscillation Index
   (b) Weather abnormalities and weather hazards
   (c) Windbreaks and shelterbelts
   (d) El Nino and La Nina
   (e) ENSO and SST
3. Write short notes on:
   (a) Global warming
   (b) Ozone depletion
   (c) Climate events
   (d) Sea surface temperatures
   (e) El Nino and monsoon rainfall over India
   (f) Economic losses due to climate variability and climate change
   (g) Atmospheric carbon dioxide
   (h) UV-B radiation
   (i) Climate shifts in the humid tropics
   (j) Windbreaks and shelterbelts
4. Fill in the blanks:
   (a) Monsoon low pressure is measured with the ———— index.
   (b) ——— and ——— countries receive rainfall if SOI is +ve.
   (c) La Nina is opposite to ————.
   (d) ———— occurs over Peru in La Nina years.

(e) There is a similarity between thunderstorms and ———.
(f) ——— and ——— were the recent all-India drought years.
(g) ——— was the year during which IPCC came into force.
(h) ——— was the warmest year.
(i) ——— was the warmest decade during the 20th century.
(j) ——— was the driest decade in the humid tropics (Kerala).
(k) The States of ——— and ——— in India were severely affected due to the cold wave in 2002–03.

5. Match the following:

(a) IMO	1950
(b) WMO	1990
(c) IPCC	1929
(d) UNCED	1979
(e) WCP	1988

6. Expand the following:

(a) WCP
(b) IMO
(c) WMO
(d) IPCC
(e) UNCED
(f) UNFCCC
(g) UNEP
(h) ICSU

7. Elucidate the following:

(a) Weather abnormalities
(b) Weather and climate modification
(c) Climate change and its effects

17 REMOTE SENSING

Remote sensing is a tool to monitor the earth's resources using space technology in addition to ground observations. It can be used in soil mapping, land use pattern, forest mapping, geological and hydrogeological purposes, drought and flood monitoring in addition to mapping of crop coverage and crop output estimates. In essence, the remote sensing techniques can be used for the earth's resources sensing. It has several applications in the field of agrometeorological research. The earth and cloud surface temperatures, radiation, rainfall, soil moisture and crop yield estimates based on spectral indices can be worked out. Attempts can also be made to monitor insect pest and disease surveillance using remote sensing techniques along with field observations. Remote sensing inputs combined with crop growth simulation models are very useful tools in crop yield forecasting. The methods are being evolved constantly for analysing crop area coverage and crop yields using remote sensing data and agroclimatic indices, which will be inputs in crop simulation modelling. A mega research projects is currently ongoing at the national level entitled forecasting agricultural output based on space, atmosphere and land (FASAL) parameters. It is popularly known as 'FASAL' programme.

17.1 DEFINITION AND SPECTRAL REFLECTANCE

Remote sensing is the science and technology of making inferences about material objects from measurements made at a distance without coming into physical contact with the objects under study. The human eye and camera are the most basic examples of remote sensing as per the definition. Remote sensing technology uses the visible, infrared and microwave regions of radiation to collect information about the various objects on the earth's surface. The responses of the objects to different regions of the electromagnetic spectrum are different. These typical responses are used to distinguish objects such as vegetation, water, bare soil, concrete and other similar features. The reflectance percentage is the maximum in infrared band for vegetation and less in the case of water (see Fig. 17.1). The reflectance values for other materials in the infrared band vary in terms of water and vegetation.

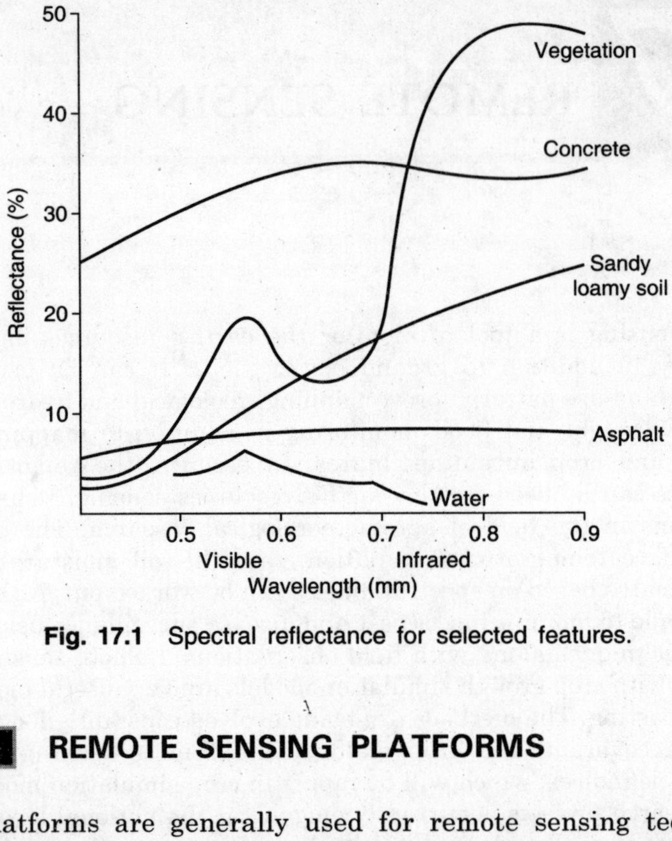

Fig. 17.1 Spectral reflectance for selected features.

17.2 REMOTE SENSING PLATFORMS

Three platforms are generally used for remote sensing techniques: ground-based, air-based and satellite-based. Infrared thermometer, spectral radiometer, pilot balloms and radars are some of the ground-based remote sensing tools while aircrafts are air-based remote sensing tools. Since the ground-based and air-based platforms are very costly and have limited use, space-based satellite technology has become handy for wider application of remote sensing techniques. There are two types of microwave remote sensing techniques, viz. active and passive (see the section under radars and satellites in Chapter 13 on weather forecasting) in addition to visible and infrared remote sensing. Digital image processing, using powerful computers, is the key tool for analysing and interpreting remotely sensed data. The advantages of satellite remote sensing are:

- Synoptic view—wide area can be covered by a single image or photograph (One scene of the Indian Remote Sensing satellite (IRS) series covers about 148 × 178 sq. km area).
- Receptivity—One can get the data of any area repeatedly (IRS series covers the same area every 16–22 days).
- Coverage—Inaccessible areas like mountains, swampy areas and thick forests are easily covered.

Space-based remote sensing is the process of obtaining information about the earth from the instruments mounted on the Earth Observation Satellites. The satellites are subdivided into two classes, and the types of satellites are as follows:

17.2.1 Polar Orbiting Satellites

These satellites operate at an altitude of between 550 and 1,600 km along an inclined circular plane over the poles. These satellites are used for remote sensing purposes. LANDSAT (USA), SPOT (France), and IRS (India) series are some of the remote sensing satellites.

17.2.2 Geostationary Satellites

These have orbits around the equator at an altitude of 36,000 km and move with the same speed as the earth, so as to view the same area on the earth continuously. They are used for telecommunication and weather forecasting purposes. The INSAT series satellites are launched from India for the above purposes. All these satellites have sensors on board, operating in the visible and near infrared regions of the electromagnetic spectrum. INSAT-3A was launched on 10th April, 2003.

17.3 ROLE OF REMOTE SENSING IN AGRICULTURE

Agricultural resources are important renewable dynamic natural resources. In India, the agriculture sector alone sustains the livelihood of around 70% of the population and contributes nearly 35% of the net national product. Increasing agricultural productivity has been the main concern, since the scope for increasing area under agriculture is rather limited. This demands judicious and optimal management of both land and water resources. Hence, comprehensive and reliable information on land use and cover, forest area, soils, geological information, extent of wastelands, agricultural crops, water resources (both surface and underground) and hazards or natural calamities like droughts and floods is required. Season-wise information on crops, their acreage, vigour and production enables the country to adopt suitable measures to meet shortages, if any, and implement proper support and procurement policies. Remote sensing systems, having capability of providing regular, synoptic, multi-temporal and multi-spectral coverage of the country, play an important role in providing such information. A large number of experiments have been carried out in developing techniques for extracting agriculture-related information from ground-borne, air-borne and space-borne data. Some of the broad agricultural application areas are:

(i) Cropped area
(ii) Crop production forecasting
(iii) Mapping of wastelands
(iv) Drought monitoring and its assessment
(v) Flood mapping and damage assessment
(vi) Monitoring of surface water bodies
(vii) Mapping of land use and cover as well as area under forest coverage
(viii) Soil mapping, and
(ix) Groundwater exploration.

Figure 17.2 shows sugarcane crop inventory developed on the basis of IRS-1A data.

Fig. 17.2 Sugarcane crop inventory using IRS-1A data (February 16, 1989, Mysore District, Karnataka).

17.4 INDIAN REMOTE SENSING PROGRAMME

India, with the experience gained from its experimental remote sensing satellite missions BHASKARA I and II, has now established satellite-based operational remote sensing system in the country with the launch of Indian Remote Sensing Satellite IRS-1A in 1988, followed by IRS-1B (1992), IRS-1C (1995) and IRS-1D (1997). Figure 17.3 shows the IRS-1C satellite. The Indian Space Research Organisation (ISRO) Department of Space (DOS) as the nodal agency for establishing operational remote sensing systems in the country, initiated efforts in the early 1970s for assessing the potential of remotely sensed data through several means.

Fig. 17.3 IRS-1C satellite under LISS-3 camera (inset).

In order to meet the user requirement of remote sensing data analysis and interpretation, ISRO/DOS has set up a system to launch remote sensing satellites once in 3 or 4 years to maintain continuity in data collection. This system and the organisations that are part of it are depicted in Fig. 17.4.

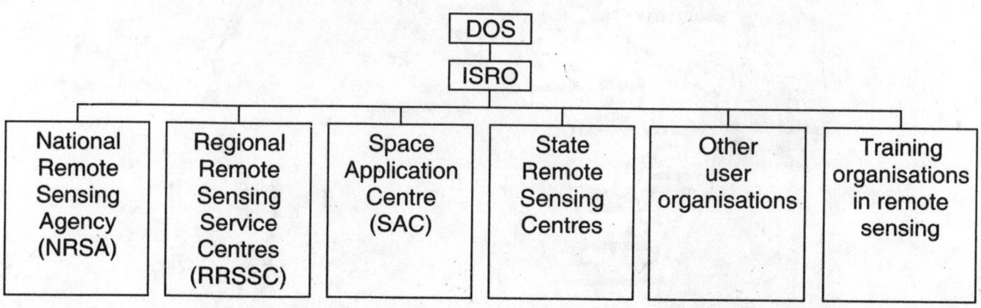

Fig. 17.4 DOS/ISRO remote sensing system and its partner organisations.

17.4.1 National Remote Sensing Agency (NRSA)

For meeting the satellite data requirements of users in the country, the National Remote Sensing Agency (NRSA) at Hyderabad is operating earth stations for receiving data from LANDSAT, IRS, NOAA and ERS satellites. NRSA has also a dedicated aircraft fitted with various modern remote sensors to conduct aerial flights to meet user requirements. NRSA has extensive facilities for remote sensing data analysis and preparation of maps. Several national projects, such as wasteland mapping and land use and cover mapping for delineating agroclimatic zones, are being coordinated by NRSA.

17.4.2 Regional Remote Sensing Service Centres (RRSSC)

Towards providing appropriate facilities for the users in the country for digital analysis and interpretation, 5 RRSSC have been set up by the Department of Space in collaboration with major user departments. They are located at Bangalore, Dehradun, Jodhpur, Nagpur and Kharagpur. The RRSSC have VAX 11/780 computer-based analysis system with image processing capabilities. Multiuser terminals have been provided to enable several users to utilise the facility at a time. The RRSSC at Bangalore and Dehradun have GIS facility with ARC-INFO systems. The RRSSCs also provide training in digital analysis for users in the country (see Fig. 17.5).

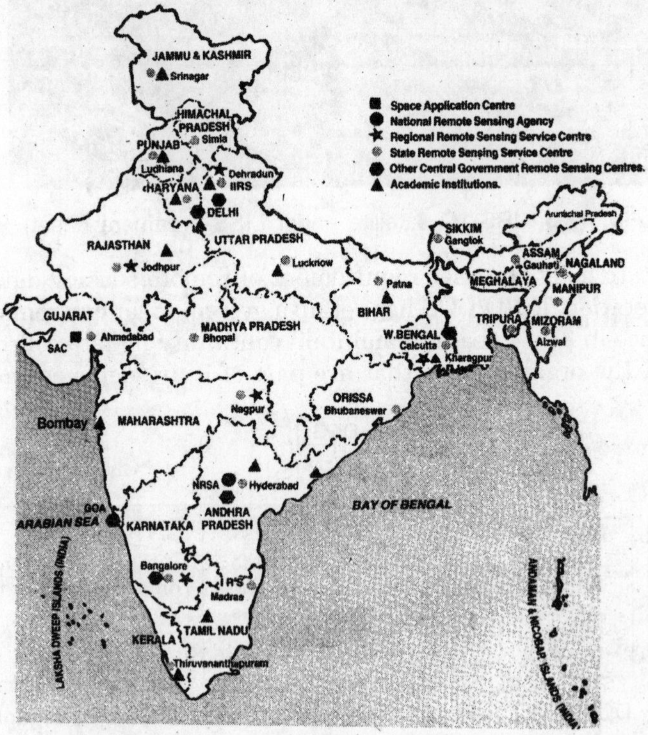

Fig. 17.5 Remote sensing facilities in India.

17.4.3 Space Application Centre (SAC)

The Space Application Centre (SAC), Ahmedabad is equipped with data analysis facilities for taking up R&D projects in the field of remote sensing towards developing suitable operational methodologies. The institute is also engaged in developing appropriate sensors for aircraft and satellite payloads. It is responsible for coordinating national-level projects such as National Drinking Water Mission, Crop Acreage and Production Estimate, and coastal and marine-related projects. The late Professor P.R. Pisharody, known as the father of remote sensing in India, nurtured the institute for a long time. A strong division of atmosphere sciences is attached to this centre and takes up research programmes in Agricultural. Meteorology based as space tools, including research in weather forecasting.

17.4.4 State Remote Sensing Centres

Several state governments have set up state-level remote sensing application centres to cater to the needs of the remote sensing organisations in the state as a part of the overall concept of National Natural Resources Management System (NNRMS). 21 states have established operational application centres. These centres have essentially visual interpretation equipments. States such as Tamil Nadu, Uttar Pradesh, Orissa, Punjab, Haryana and Madhya Pradesh have advanced digital image processing facilities. The State Remote Sensing Application Centres, besides carrying out projects of direct relevance to their states, participate actively in all the national level projects such as mapping of wastelands, exploring ground water potential zones and estimating land use.

17.4.5 Other User Organisations

Major organisations like the Geological Survey of India, Central Ground Water Board, Forest Survey of India, Oil and Natural Gas Commission, Survey of India, National Bureau of Soil Survey and Land Use Planning, All India Soil and Land Use Survey, state departments of Mines and Geology and Forest and Agriculture have set up their own Remote Sensing Cells or Centres for carrying out remote sensing projects. Users can approach the State Remote Sensing Centres or Regional Remote Sensing Service Centres as per their requirements to carry out any remote sensing project.

17.5 TRAINING IN REMOTE SENSING

Availability of trained manpower is identified as most crucial for operationalisation and effective utilisation of remote sensing technology.

The Indian Institute of Remote Sensing (IIRS), formerly known as Indian Photo Interpretation Institute, Dehradun, is a pioneer training institute set up in 1965. Since the early 1970s, major institutions in India have set up training centres in their organisations. Specialised training programmes have also been established at reputed academic institutions. Because of varied requirements, training is imparted at the levels of decision makers, supervisory officers and working scientists. The duration range of training programmes varies from a few days to a few months. Specific training courses are also provided by: National Remote Sensing Agency, Hyderabad; Regional Remote Sensing Service Centres; Space Application Centre, Ahmedabad; Centre of Studies in Resources Engineering, Indian Institute of Technology, Mumbai; Institute of Remote Sensing, Anna University, Chennai; Geological Survey of India (Training Institute); Forest Survey of India, Dehradun; and the Division of Agricultural Physics, IARI. These training centres cater to the specific needs of the users.

17.6 REMOTE SENSING APPLICATION PROJECTS

With remote sensing techniques becoming an essential tool in natural resources survey and monitoring, various central or state government departments, institutes and other agencies have taken up a large number of projects using remote sensing (see Table 17.1) to meet their regular resource surveys or management needs. Remote sensing has achieved operational status as could be seen from the details of user projects taken up so far. Figure 17.6 shows the photograph of a reservoir taken using a remote sensing camera.

Fig. 17.6 Part of LISS-3 scene showing Mayurakshi reservoir in Bihar.

Table 17.1 Projects in remote sensing and their objectives

Project name	*Objective*
National wasteland mapping project	To identify type and extent of wasteland at village level
Land use or land cover mapping	Nationwide mapping for land use or cover for agro-climatic zones on 1:2,50,000 scale
National Drinking Water Mission	Hydrogeomorphological maps (groundwater potential on 1:2,50,000 scale) help in scientific exploration of sources of drinking water
Agricultural drought monitoring	To assess and monitor agricultural drought at district level
Forest cover mapping	Biennial forest cover on 1:2,50,000 scale for entire country
Integrated study to combat drought	Preparation of integrated land and water resources maps at district level to identify long and short term measures to combat drought
Urban studies	To study the urban land use and land cover sprawl and growth of all major cities above 1 million population
Crop acreage and production estimation	To estimate the crop acreage and production
Flood mapping	Near-real-time mapping of flooded areas and estimation of damage
Snow mapping	Delineation of snow cover area and estimation
Environmental impact of mining and thermal power complexes	To develop suitable methodology for monitoring mining and thermal power complexes
Project Vasundhara	Regional mineral targetting by using multi-thematic data for area below 16°N of India
Oceanography	To generate routine information on ocean and coastal parameters To develop methodology for microwave data analysis and parameter retrieval

17.7 REMOTE SENSING IN OPERATIONAL AGROMETEOROLOGY

Meteorological variables such as precipitation, radiation, cloud amount and its temperature, earth's surface temperature and soil moisture at macro-level can be estimated using remotely sensed data. Rao et al. (1989) estimated precipitation and outgoing longwave radiation from INSAT-1B radiance data. It revealed that the accuracy at 100 × 100 sq. km. varied between 40 and 60%. In contrast, the insolation derived from the Heliosat method using satellite data was found to be more close to the actual value obtained using Pyranometer. Rastogi et al. (2000) estimated the leaf area index in wheat using space-borne sensor data and found that the observed and estimated values were in good agreement. Since the remote sensing data is used to estimate several plant parameters, it is feasible to predict the final crop output using the spectral indices. The estimates of cropped area and crop yields using remotely sensed data indicate that the spectral indices in combination with agrometeorological indices could explain the estimates much better. Using the microwave sensors, soil moisture determination over large cropped areas is possible. In addition, land and sea surface temperatures are monitored using digital data and used in seasonal climate forecasts. Natural calamities such as drought and floods, cyclones and landslips, and loss in cropped area due to incidence of pests and diseases, can be effectively monitored using the IRS digital data if large areas are affected, which in turn will enable the government agencies to combat the after-effects of natural calamities effectively. Figure 17.7 shows the satellite imagery of the flood situation in two rivers on a particular day.

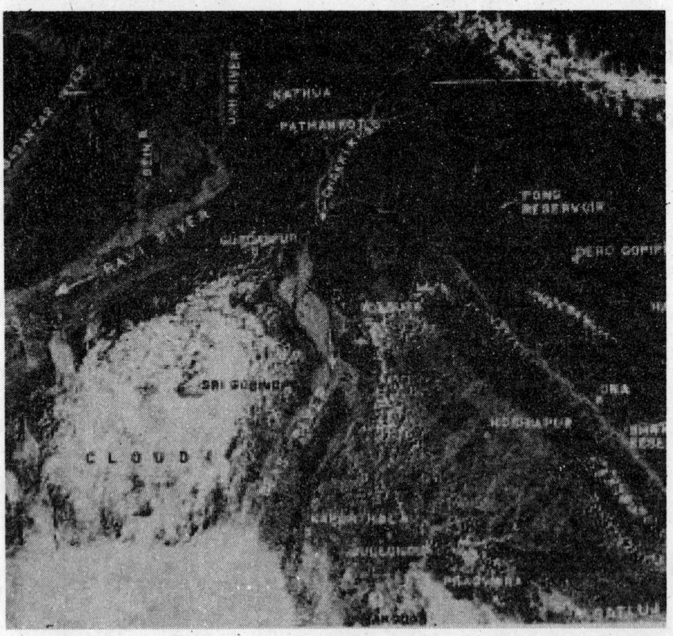

Fig. 17.7 Satellite imagery of 27 Sept. 1988 showing flood situation in Ravi and Beas rivers.

17.8 MILESTONES IN INDIAN SPACE PROGRAMME

The Indian space programme, which has celebrated its silver jubilee, has come a long way from a humble beginning of testing sounding rockets to acquiring the capability of launching giant polar satellite vehicles and putting 1,200 kg-class remote sensing satellite IRS-1D in low-earth orbits. The important milestones in the saga of Indian space programme, that has seen many ups and downs, are shown in Table 17.2.

Table 17.2 Milestones in Indian space programme

1962	Indian National Committee for Space Research formed. Work on Thumba Equatorial Rocket Launching Station (TERLS) began.
1963	First sounding rocket launched from TERLS on November 21.
1965	Space Science and Technology Centre established in Thumba.
1967	Satellite Telecommunication Earth Station set up at Ahmedabad.
1968	TERLS dedicated to the United Nations on February 2.
1969	Indian Space Research Organisation (ISRO) formed on August 15.
1972	Space Commission and Department of Space set up on June 1. Airborne remote sensing experiments commenced. ISRO brought under DoS.
1975	First Indian satellite, Aryabhatta, placed in orbit on April 19. Satellite Instructional Television Experiment (SITE) began. Became a government organization on April 1.
1972–76	Air-borne remote sensing experiments began.
1977	Satellite Telecom Experiments Project (STEP) undertaken.
1979	Experimental satellite for earth observations (Bhaskara-l) placed in orbit on June 7. First experimental launch of SLV-3 attempted on August 10. Satellite could not be placed in orbit.
1980	Second experimental launch of SLV-3 on July 18. (Satellite named Rohini RS-D1) successfully placed in orbit.
1981	First developmental launch of SLV-3. Placed RS-D1 satellite in orbit on May 31. APPLE, an experimental geostationary communication satellite placed in orbit successfully on June 19. Bhaskara-2 placed in orbit on November 20.
1982	INSAT-1A placed in orbit on April 10. Deactivated on September 6.
1983	Second developmental launch of SLV-3 placed in orbit on April 17. INSAT-1B placed in orbit on August 30. Active for the next 10 years.
1984	Indo-Soviet manned space mission in April.
1987	First developmental launch of ASLV on March 24. Satellite could not be placed in orbit.
1988	First operational Indian Remote Sensing Satellite, IRS-1A, launched on March 17. INSAT-1C launched on July 21, abandoned in November 1989. Second developmental launch of ASLV on July 13. Satellite could not be placed in orbit.

(Contd.)...

Table 17.2 Milestones in Indian space programme *(Contd.)*...

1990	INSAT 1-D launched on June 12.
1991	IRS-1B launched on August 29.
1992	Third developmental flight of ASLV launched on May 20. Placed Sross-C satellite in low earth orbit. INSAT-2A (the first indigenously built second generation INSAT satellite) placed on July 10.
1993	INSAT-2B launched on July 23. PSLV's first developmental flight from SHAR on September 20 unsuccessful.
1994	The fourth developmental flight of ASLV placed Sross-C2 satellite in low earth orbit on May 4. PSLV-D2 launched successfully to place IRS-P2 in polar sun-synchronous orbit.
1995	INSAT-2C satellite placed in orbit on December 7. IRS-1C placed in orbit on December 28.
1996	PSLV-D3 launched successfully to place IRS-P3 satellite in intended orbit.
1997	INSAT-2D placed in orbit on June 4. Inoperable on October 4. INSAT series launched (June 4). Becomes inoperable on October 4. IRS-1D placed in orbit on September 29.
1998	INSAT-2DT acquired from ARABSAT in January.
1999	Indian Remote Sensing Satellite, IRS-P4 (OCEANSAT-1), launched by Polar Satellite Launch Vehicle (PSLV-C2) along with Korean KITSAT-3 and German DLR-TUBSAT from Sriharikota on May 26. INSAT-2E launched on April 3.
2000	INSAT-3B, the first satellite in the third generation INSAT-3 series, launched on March 22.
2001	ISRO's Polar Satellite Launch Vehicle, PSLV-C3, successfully launched three satellites, Technology Experiment Satellite (TES) of ISRO, BIRD of Germany and PROBA of Belgium, into their intended orbits on October 22. The first developmental launch of GSLV-D1 with GSAT-1 on board from Sriharikota on April 18.
2002	ISRO's Polar Satellite Launch Vehicle, PSLV-C4, successfully launched METSAT satellite (September 12). It was later renamed Kalpana-1. Successful launch of INSAT-3C on January 24.
2003	INSAT-3A is successfully launched on 10 April. The second developmental launch of GSLV-D2 with GSAT-2 on board from Sriharikota (May 8). Successful launch of INSAT-3E by Ariane from Kourou, French Guyana (September 28). ISRO's Polar Satellite Launch Vehicle, PSLV-C5, successfully launched RESOURCESAT-1 (IRS-P6) satellite from Sriharikota (October 17).
2004	The first operational flight of GSLV (GSLV-F01) successfully launched EDUSAT from SDSC, SHAR, Sriharikota (September 20).
2005	ISRO's Polar Satellite Launch Vehicle, PSLV-C6, successfully launched CARTOSAT-1 and HAMSAT satellites from Sriharikota (May 5). Successful launch of INSAT-4A by Ariane from Kourou, French Guyana (December 22).

(Contd.)...

Table 17.2 Milestones in Indian space programme *(Contd.)*...

2006	INSAT-4C, the heaviest communications satellite made, failed after the rocket carrying it veered off course and exploded.
2007	Polar Satellite Launch Vehicle (PSLV-C7) successfully launched SRE-1 (Space Capsule Recovery Experiment) on January 10. In addition to SRE-1, other satellites viz. India's CARTOSAT-2, Indonesia's LAPAN-TUBSAT and Argentina's PEHUENSAT-1 were also launched from SDSC, SHAR, Sriharikota. On January 22 SRE-1 reentered the earth's atmosphere and descended over the Bay of Bengal about 140 km east of Sriharikota.

REFERENCES/SUGGESTED READINGS

Carter, D.J., *The Remote Sensing Source Book*, McCarta, London, 1986, p. 174.

Curan, P.J., *Principles of Remote Sensing*, ELBS Longman Group, United Kingdom, 1985.

Gopalaswamy, N., *Agricultural Meteorology*, Rawat Publications, Jaipur, 1994, p. 153.

Jain, A.K., *Fundamentals of Digital Image Processing*, Prentice Hall, Saddle River, New Jersey, 1989.

Kamat, D.S. and Sinha, S.K. (Eds.), *Proceedings of seminar on crop growth condition and remote sensing*, June 22–23, ICAR and ISRO, 1984.

Lillesand, T.M. and Kiffer, R.W., *Remote Sensing and Image Interpretation*, John Wiley & Sons, New York, 1979.

Nag, P. and Kudrat, M., *Digital Remote Sensing*, Concept Publishing, New Delhi, 1998, p. 320.

Patel, A.N. and Singh, S., *Principles of Remote Sensing*, 2nd ed., Scientific Publishers (India), Jodhpur, 1999, p. 190.

Rao, A.V.R.K., Kelkar, R.R. and Arklin, P.A., 'Estimation of precipitation and outgoing longwave radiation from INSAT-1A radiance data', *Mausam*, **40** (2), 123–130, 1989.

Rastogi, A., Kalra, N., Agarwal, P.K., Sharma, S.K., Harit, R.C., Navalgoed, R. and Dadhwal, V.K., 'Estimation of leaf area index from satellite data using Price model', *Int. J.Remote Sensing*, **21** (15), 2943–2949, 2000.

Schowengerdt, R.A., *Remote Sensing, Models and Methods for Image Processing*, 2nd ed., Academic Press, London, 1997.

Sharma, V.K., *Remote Sensing for Land Resource Planning*, Concept Publishing, New Delhi, 1991, p. 144.

QUESTIONS

1. Define remote sensing.
2. Differentiate between polar orbiting satellites and geostationary satellites.
3. What are the applications of remote sensing?
4. What do you mean by platforms? Give examples.
5. Fill in the blanks:
   (a) Remote sensing technology uses ——, —— and —— regions of radiation to get information on the earth's surface.
   (b) The reflectance percentage is the maximum in the infrared band for —— and less in the case of ——.
   (c) —— and —— are the two main types of remote sensing techniques.
   (d) —— is the key for analysis and interpretation of satellite remote sensing data.
   (e) Polar orbiting satellites operate at an altitude of between —— and —— km along an inclined circular plane over poles.
   (f) Geostationary satellites operate at an altitude of —— km.
   (g) —— types of satellites are used for weather forecasting purposes.
   (h) The operational remote sensing system in India was started with the launch of the —— satellite.
   (i) —— is the nodal agency for establishing an operational remote sensing system in India.
   (j) NRSA is located at ——.
   (k) RRSSC at Bangalore and Dehradun have —— facility with —— systems.
   (l) Space Application Centre is located at ——.
   (m) Indian Institute of Remote Sensing was formerly known as ——.
   (n) —— is the first Indian satellite placed in orbit.
   (o) ISRO was formed during the year ——.
   (p) —— is the satellite that has been exclusively launched for meteorological purposes.
   (q) —— is the pioneer institute in India for conducting training in remote sensing.
6. State true or false:
   (a) Human eye can be an example of remote sensing.
   (b) Balloon is an example air-based platform.
   (c) Earth resource satellites are geostationary satellites.
   (d) SPA is located in New Delhi.
   (e) Remote sensing techniques cannot be used for resources sensing.

7. Match the following:

(a) ISRO	Dehradun
(b) INSAT series	Platform
(c) IRS series	1969
(d) IIRS	Commnication
(e) Aircraft	Resources

8. Expand the following:

(a) INSAT	(b) IRS
(c) NRSA	(d) SPA
(e) METSAT	(f) ISRO

9. Elucidate the role of remote sensing in the field of agriculture.
10. Write an essay on Indian remote sensing programmes.

SUBJECT INDEX

GEOGRAPHICAL INDEX—INDIA

GEOGRAPHICAL INDEX—OTHER COUNTRIES